# BIOLOGY OF PARASITISM

edited by

**Christian Tschudi**
Department of Internal Medicine
Yale University School of Medicine
New Haven, Connecticut

and

**Edward J. Pearce**
Department of Microbiology and Immunology
Cornell University
Ithaca, New York

**KLUWER ACADEMIC PUBLISHERS**
**Boston / Dordrecht / London**

**Dedicated to TS**

**Distributors for North, Central and South America:**
Kluwer Academic Publishers
101 Philip Drive
Assinippi Park
Norwell, Massachusetts 02061 USA
Telephone (781) 871-6600
Fax (781) 871-6528
E-Mail <kluwer@wkap.com>

**Distributors for all other countries:**
Kluwer Academic Publishers Group
Distribution Centre
Post Office Box 322
3300 AH Dordrecht, THE NETHERLANDS
Telephone 31 78 6392 392
Fax 31 78 6546 474
E-Mail <orderdept@wkap.nl>

Electronic Services <http://www.wkap.nl>

**Library of Congress Cataloging-in-Publication Data**

Biology of parasitism / edited by Christian Tschudi and Edward J. Pearce.
p. cm.
Includes bibliographical references and index.
ISBN 0-7923-7823-7
1. Parasitic diseases. 2. Parasitism. 3. Molecular parasitology. I. Tschudi, Christian, 1954- II. Pearce, Edward J., 1958-

QR201.P27 B55 2000
616.9'6—dc21

00-025365



*Printed on acid-free paper.*

Printed in the United States of America

**Photo credit**: the cover photo shows a common deer tick by Monica Oli, a student of the 1998 Biology of Parasitism course.

# CONTENTS

## CELL BIOLOGY



## IMMUNOLOGY

## Appendix

# CONTRIBUTORS

**Wandy Beatty**
Department of Molecular Microbiology
Washington University
660 South Euclid Avenue
St. Louis, MO 63110-1093

**Deirdre L. Brekken**
University of Texas Southwest
Department of Pharmacology
5323 Harry Hine Blvd.
Dallas, TX 75235-9341

**Barbara A. Burleigh**
Department of Immunology and Infectious Diseases
Harvard School of Public Health
Building I, Rm 713
665 Huntington Avenue
Boston, MA 02115
imbassai2@yahoo.com

**Jaime Dant**
Department of Molecular Microbiology
Washington University
St. Louis, MO 63110-1093

**John A. Darling**
Department of Biology
University of Pennsylvania
415 South University Avenue
Philadelphia, PA 19104-6018

**John R. David**
Harvard School of Public Health
665 Huntington Avenue
Boston, MA 02115

**Stephen J. Davies**
Department of Pathology
UCSF Tropical Disease Research Unit
4150 Clement Street- 113B
San Francisco, CA 94121

**Sabrina D. Dyall**
UCLA Microbiology and
Immunology
1602 Molecular Sciences Bldg.
405 Hilgard Ave.
Los Angeles, CA 90095-1489

**Olivia K. Giddings**
Department of Molecular
Microbiology
Washington University
School of Medicine
660 South Euclid Avenue
St. Louis, MO 63110-1093

**Kristin M. Hager**
Department of Biology
University of Pennsylvania
Goddard Laboratories
415 South University Avenue
Philadelphia, PA 19104-6018

**Stephen L. Hajduk**
Department of Biochemistry and
Molecular Genetics
University of Alabama at
Birmingham
1918 University Blvd.
Birmingham, AL 35294
shajduk@uab.edu

**Laurie R. Hall**
Division of Geographic Medicine
Case Western Reserve University
School of Medicine, W137
2109 Adelbert Rd
Cleveland, OH 44106

**Musa A. Haxhiu**
Division of Geographic Medicine
Case Western Reserve University
School of Medicine, W137
2109 Adelbert Rd
Cleveland, OH 44106

**Ayman S. Hussein**
Imperial College of Science,
Technology and Medicine
Department of Biochemistry
London SW7 2AY, UK

**Patricia J. Johnson**
UCLA Microbiology and
Immunology
1602 Molecular Sciences Bldg.
405 Hilgard Ave.
Los Angeles, CA 90095-1489
johnsonp@ucla.edu

**Lisa N. Kinch**
University of Texas Southwest
Department of Pharmacology
5323 Harry Hine Blvd.
Dallas, TX 75235-9341

**Jessica C. Kissinger**
Department of Biology
University of Pennsylvania
415 South University Avenue
Philadelphia, PA 19104-6018

**Andreas Lingnau**
Department of Molecular
Microbiology
Washington University
School of Medicine
660 South Euclid Avenue
St. Louis, MO 63110-1093

**Maren Lingnau**
Department of Molecular
Microbiology
Washington University
School of Medicine
660 South Euclid Avenue
St. Louis, MO 63110-1093

**Jennie L. Lovett**
Department of Molecular
Microbiology
Washington University
School of Medicine
660 South Euclid Avenue
St. Louis, MO 63110-1093

**James H. McKerrow**
Department of Pathology
UCSF Tropical Disease Research
Unit
4150 Clement Street- 113B
San Francisco, CA 94121
jmck@cgl.ucsf.edu

**Rajeev K. Mehlotra**
Division of Geographic Medicine
Case Western Reserve University
School of Medicine, W137
2109 Adelbert Rd
Cleveland, OH 44106

**Dana G. Mordue**
Department of Molecular
Microbiology
Washington University
School of Medicine
660 South Euclid Avenue
St. Louis, MO 63110-1093

**Edward J. Pearce**
Department of Microbiology and
Immunology
College of Veterinary Medicine
Cornell University
Schuman Hall Room C320
Ithaca, NY 14853-6401
ejp2@cornell.edu

**Eric Pearlman**
Division of Geographic Medicine
Case Western Reserve University
School of Medicine, W137
2109 Adelbert Rd
Cleveland, OH 44106
exp2@po.cwru.edu

**Margaret A. Phillips**
University of Texas Southwest
Department of Pharmacology
5323 Harry Hine Blvd.
Dallas, TX 75235-9341
phillip01@utsw.swmed.edu

**Pradipsinh K. Rathod**
The Catholic University of
America
Department of Biology
620 Michigan Avenue, NE
Washington, DC 20064
rathod@cua.edu

**Mary G. Reynolds**
Department of Biology
University of Pennsylvania
Goddard Laboratories
415 South University Avenue
Philadelphia, PA 19104-6018

**David S. Roos**
Department of Biology
University of Pennsylvania
Goddard Laboratories
415 South University Avenue
Philadelphia, PA 19104-6018
droos@sas.upenn.edu

**David G. Russell**
Department of Molecular Microbiology
Washington University
School of Medicine
660 South Euclid Avenue
St. Louis, MO 63110-1093
russell@borcim.wustl.edu

**Phillip Scott**
University of Pennsylvania
The School of Veterinary Medicine
Department of Pathobiology
3800 Spruce Street
Philadelphia, PA 19104-6008
pscott@vet.upenn.edu

**Murray E. Selkirk**
Imperial College of Science, Technology and Medicine
Department of Biochemistry
Prince Consort Road
London SW7 2AY, UK
m.selkirk@ic.ac.uk

**L. David Sibley**
Department of Molecular Microbiology
Washington University
School of Medicine
660 South Euclid Avenue
St. Louis, MO 63110-1093
sibley@borcim.wustl.edu

**Boris Striepen**
Department of Biology
University of Pennsylvania
415 South University Avenue
Philadelphia, PA 19104-6018

**Christian Tschudi**
Department of Internal Medicine
Yale University, School of Medicine
P.O. Box 208022
New Haven, CT 06520-8022
christian.tschudi@yale.edu

**Elisabetta Ullu**
Department of Internal Medicine
Yale University, School of Medicine
P.O. Box 208022
New Haven, CT 06520-8022
elisabetta.ullu@yale.edu

**Andrew P. Waters**
Leiden University
Medical Centre
Department of Parasitology
Postbus 9600
2300 RC Leiden
The Netherlands
waters@lumc.nl

**Audra E. Yermovsky-Kammerer**
Department of Biochemistry and
Molecular Genetics
University of Alabama at
Birmingham
1918 University Blvd.
Birmingham, AL 35294

# PREFACE

Historically, the Biology of Parasitism Course (BoP) at the Marine Biological Laboratory (MBL) in Woods Hole, Massachusetts, is recognized for its ability to introduce the participants to the newest technologies available for studying the molecular basis of parasite function and host/parasite interactions. This is accomplished by exposing the participants to recognized experts in the field in a setting that is highly conducive to intellectual exchange and experimental discovery.

Currently, the course extends over a period of about nine weeks between early June and mid-August of each year and consists of four two week-long sections covering Immunology, Biochemistry, Cell Biology and Molecular Biology of protozoan and helminthic parasites. In addition, workshops that vary from year to year, include subjects such as Molecular Epidemiology, Protein Structure and Modelling, Genomics and Marine Parasitology. On any given day, students receive a 2 to 3 hour long seminar in the morning from that day's invited lecturer then enter the laboratory, where they perform their experimental work. In addition to instructing the students in ways modern techniques can be applied to answer the most important questions in parasitology, BoP creates the potential for the establishment of long-lived professional relationships within the student group as well as between students and faculty.

This book is intended to present a snapshot of the content and spirit of BoP. By presenting a series of chapters that reflect the formal lectures that students receive on a daily basis, as well as the approaches used during the laboratory section of the course, we hope to share some of the science that occurs there. In addition, Dr. John David, the course's originator, has been good enough to provide a brief history of BoP. What the book cannot do is transmit the intensity of BoP. Each year, the sixteen students are expected to spend the best part of every waking moment for nine weeks thinking about, and experimenting in, parasitology. To help them in this endeavor the course provides the unlikely faculty to student ratio of 3:1 and additional stimulation in the form of numerous teaching assistants recruited from the laboratories of participating faculty.

In addition to working hard, students at BoP traditionally play hard! The course has a "lounge", affectionately named Merv's (after Dr. Mervyn Turner), that serves not only as a center for informal presentations and discussions, but also as the site for many impromptu late night events. It is in this room that the plans for the course's participation in the Woods Hole 4th of July parade are hatched and the necessary costumes are constructed. It would be impossible to conjure the flavor of BoP without also mentioning the innumerable discussions about science, politics, life and religion that occur at the bar in the Captain Kidd.

BoP has received financial aid from numerous sources over the years, but of late The Burroughs Wellcome Fund has been to the forefront of supporting the course. We are extremely grateful for their continuing commitment. Many companies very generously provide us with the free loan of equipment, such as Beckman Instruments, Becton-Dickinson, Bio-Rad, Brinkmann, MJ Research, Amersham Pharmacia Biotech, Savant and Zeiss. We are indebted to our colleagues, many not represented in this book, who work so hard to make the course what it is today, to our predecessors who have provided such a solid foundation for BoP and to the director of the MBL, John Burris, who is so supportive.

Christian Tschudi
Edward J. Pearce

# FOREWORD

It is a privilege to be invited again to write a foreword to "Biology of Parasitism", a book based on the Biology of Parasitism Course and edited by its current directors, Christian Tschudi and Edward Pearce. This course was the brainchild of Joshua Lederberg, Kenneth Warren and Anthony Cerami in the late 1970's. They felt that the field of parasitology would be greatly advanced if current concepts of molecular biology and modern immunology were brought to the field, as had happened decades before for bacteria and yeast at Cold Spring Harbor. The course has proved them correct.

This is the 20th year of the Biology of Parasitism Course, which I was asked to develop and direct for its first five years beginning in 1980. At the time, we hoped to attract students with various backgrounds who might be interested in exploring the exciting world of the biology of parasites (David, JR, 1983. The Biology of Parasitism, in *Parasitology: A global Perspective*. Eds. Kenneth S. Warren and John Z. Bowers. Springer-Verlag New York, pp 236-245). For twenty years 16 students a year (except for 20 for three years) have carried out experiments using the most current techniques and have interacted with many excellent scientists both in the lecture room and in the laboratory.

Of the 332 students that have attended in the past 20 years, 54% have come from outside the United States: 47 from Latin America (5 countries), 72 from Europe (13 countries), 21 from Africa (9 countries), 28 from Asia (9 countries), and 4 from Canada. The directors have been John R. David with Roberta David (1980-1984), Paul T. Englund and Alan Sher (1985-1988),

John E. Donelson and Carole Long (1989-1990), John C. Boothroyd (1991-1993) and with Richard Komuniecki (1992-1993), Stephen L. Hajduk (1994-1997) and Christian Tschudi and Edward J. Pierce (1998-2000).

Because of the Biology of Parasitism Course, the Marine Biological Laboratory (MBL) at Woods Hole became a center in the parasitology field with major annual conferences on Molecular Parasitology and Immunoparasitology. Looking back to 1980 when the course started, it is hard to believe that then it was difficult to gather a dozen experts on the molecular biology of parasites in Woods Hole. How times have changed.

This volume is the third describing various concepts and technologies that have been taught in the course. As can be seen from the table of contents, the course continues to reflect the most up-to-date knowledge about basic molecular biology, immunology, cell biology and biochemistry related to the study of parasites and covers the latest technologies available in these rapidly evolving areas.

The biology of parasites continues to fascinate, as is evident by the rush of young scientists to this field. Many parasite organisms have developed unique interactions with various hosts, living in insects or mammals, vertebrates or invertebrates, evading immune systems and other host defenses. Understanding the basis of this parasitism should not only answer many basic questions in biology, but also help us combat some major world health problems. It is undoubtedly the exciting and complex biology of these organisms, coupled with their importance in global public health that makes this field so attractive for study.

Over the past 20 years, there have been major advances in our understanding of the biology of parasites. Development of methods to transfect genes into protozoa has made possible the exploration of virulence and drug resistance genes and basic gene regulation. The numerous parasite genome projects underway have already facilitated the discovery of new parasite antigens, enzymes and structural components. Cytokines have been shown to play a major role in immunity to parasites and the pathology they cause. Studies involving cell biology and immunology have dissected some of the mechanisms these organisms use to evade the hosts immune response. And the study of arthropod vectors has advanced our understanding of the transmission of these organisms.

Yet in spite of these advances, the morbidity and mortality around the world caused by these organisms has not diminished and, for some, has even increased. This, in part, is due to the spread of drug resistant organisms, ecological changes associated with development, and migration of peoples who are not immune into endemic regions. Further, other than the recent explosive interest by several organizations in the problem of malaria, the number of foundations and agencies supporting research in these diseases has decreased in the past decade, leading to a ominous drop in persons investigating parasites especially helminths and protozoa other than malaria.

Parasites are complex organisms with elaborate life cycles: this is what makes them so fascinating. But because of their complexity, it is not surprising that we have not solved the many riddles they pose in the past two decades. The Biology of Parasitism Course at the MBL and the two annual meetings that gather young people are critical in continuing the enthusiasm necessary to solve these problems, as is the need for new infusions of funds for research. This volume will aid by presenting some concepts and methods for this task.

John R. David
Boston

# 1

# IMPORT OF NUCLEAR ENCODED tRNA INTO MITOCHONDRIA

Audra E. Yermovsky-Kammerer and Stephen L. Hajduk
*Department of Biochemistry and Molecular Genetics, University of Alabama at Birmingham, Schools of Medicine and Dentistry, Birmingham, AL 35294*

## INTRODUCTION

The assembly of functional, biosynthetically active, mitochondria requires RNAs and proteins encoded by the mitochondrial genome. Typically, mitochondrial transcripts include ribosomal RNAs (rRNAs), transfer RNAs (tRNAs) and messenger RNAs (mRNAs). These transcripts are indispensable in the formation of mitochondrial ribosomes, in mitochondrial protein synthesis and the formation of inner mitochondrial membrane complexes involved in electron transport and ATP synthesis.

Only 12 to 15 proteins are encoded by the mitochondrial genome. These proteins provide a minor, though essential, subset of the approximately 600 proteins that are found in these complex organelles. The remaining proteins are encoded by nuclear genes and following translation on cytosolic ribosomes are imported into the mitochondrion. The pathways of protein import into mitochondria have been extensively studied using both in vitro biochemical and in vivo genetics approaches (for a recent review see Neupert, 1997).

In 1967 Suyama proposed that some of the mitochondrial tRNAs of the protozoan ciliate *Tetrahymena* were encoded by nuclear genes and therefore must be imported into the mitochondria (Suyama, 1967). These studies were largely ignored until recently, when it was shown, by a number of laboratories, that a diverse group of organisms import nuclear encoded tRNAs into their mitochondria. The number and the identity of the nuclear encoded mitochondrial tRNAs varies. *Saccharomyces cerevisiae* imports a single $tRNA^{Lys}$ (Tarassov and Entelis, 1992) while *Phaseolus vulgaris* (bean) imports eight tRNAs (Marechal-Drouard et al., 1988), *Salanum tuberosum* (potato) imports eleven tRNAs (Marechal-Drouard et al., 1990), *Tetrahymena* imports 26 tRNAs (Suyama, 1986), and *Triticum aestivum* (wheat) (Joyce and Gray, 1990), *Chlamydomonas* (Boer and Gray, 1988) and *Paramecium* (Pritchard et al., 1990) import an undetermined number of tRNAs into their mitochondria. The two members of the order Kinetoplastidae, *Leishmania tarentolae* (Simpson, et al., 1989) and *Trypanosoma brucei* (Hancock et al., 1990), and *Plasmodium falciparum* (Feagin, 1992) are the only organisms described thus far, that encode all of their mitochondrial tRNAs in the nucleus.

In this article, we will describe some of the features of tRNA import in *T. brucei* and *L. tarentolae*. To put these studies in perspective we also briefly describe the general features of tRNA import in two other systems, *S. cerevisiae* and *Tetrahymena*. The study of RNA trafficking and translocation across membranes is in its infancy. While we search for commonality in the pathways of RNA import it is likely that both the import machinery and the elements within the RNAs will show a high degree of diversity.

## ORGANIZATION AND TRANSCRIPTION OF MITOCHONDRIAL tRNA GENES

Mitochondrial tRNAs are typically encoded by genes within the mitochondrion (Attardi and Schatz, 1988). In mammals the tRNA coding sequences within the mitochondrial genome often border protein coding sequences. Transcription of both DNA strands by the mitochondrial RNA polymerase results in the formation of polycistronic, precursor RNAs. Processing of these precursor transcripts by RNAse P results in the formation of the 5' end of the mature mitochondrial tRNA and the 3' end of the upstream mRNA. Subsequent processing is required for the formation of mature tRNAs and mRNAs. In mammals the mitochondrial genome has been minimized to such an extent that tRNA genes often overlap thus necessitating repair pathways, which complete the formation of functional tRNAs (Attardi and Schatz, 1988).

The organization and transcription of the mitochondrial tRNA genes in trypanosomes is very different from that described above for mammalian mitochondrial tRNAs. First, all trypanosome mitochondrial tRNAs are encoded by nuclear genes. Whether some or all of these genes were transferred from the mitochondrion to the nucleus, as has been proposed for many of the protein coding genes, has not been established. Second, trypanosome mitochondrial tRNA genes are transcribed in the nucleus by RNA polymerase III (Galli et al., 1981). Sequence elements within the trypanosome tRNA genes resemble the consensus Box A and Box B internal promoter elements for polymerase III transcripts. Third, unlike most nuclear tRNA genes the tRNA genes of trypanosomes are often organized in clusters (Campbell, 1989).

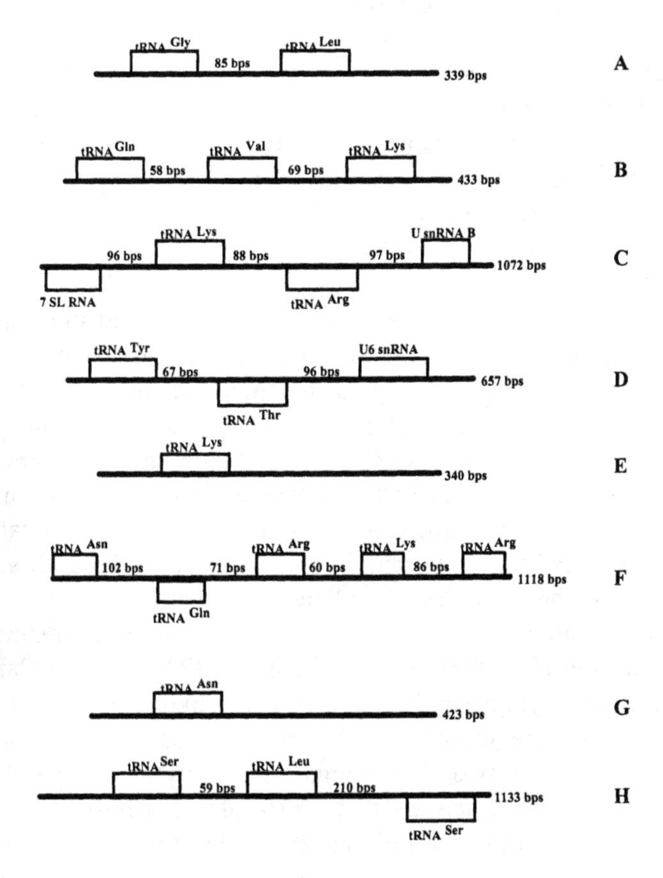


**Figure 1**. Nuclear tRNA genes of *T. brucei*. Schematic representing the genomic organization of the tRNAs genes that have been identified for *T. brucei*. The tRNAs and other small RNAs are labelled along with the length of the intervening regions, the total lengths of the genomic fragments. **A.** Campbell et al., 1989; **B.** Campbell, 1989; **C.** Mottram et al., 1991a; **D.** Mottram et al., 1991a; **E.** Mottram et al., 1991b;

**F.** Mottram et al., 1991b; Hancock and Hajduk, 1992; **G.** Hancock and Hajduk, 1992; **H.** LeBlanc et al., 1999.

Nineteen tRNA genes have been identified in *T. brucei* (Campbell, 1989; Campbell et al., 1989; Hancock and Hajduk; 1992; Mottram et al., 1991; LeBlanc et al., 1999) (Figure 1). The tRNA genes of *T. brucei* are separated by intergenic sequences ranging from 58 to 210 nucleotides. Two of the 19 tRNA genes are not adjacent to other tRNA genes suggesting that not all tRNA genes in *T. brucei* are organized in clusters (Mottram, et al., 1991; Hancock and Hajduk, 1992). Recent studies indicate that the some of the *T. brucei* mitochondrial tRNA genes may be transcribed by polymerase III to produce either monocistronic or dicistronic primary transcripts (Yermovsky-Kammerer and Hajduk, 1999). The possible role of the dicistronic transcripts in tRNA trafficking in *T. brucei* is discussed later in this article. The monocistronic and dicistronic precursors tRNAs are subsequently processed to form the mature 5' and 3' termini of the tRNAs by either nuclear or mitochondrial RNAse P (Hancock et al., 1992; LeBlanc et al., 1999) and the 3' CCA nucleotidyltransferase (Deutscher, 1984).

Functionally, the tRNAs of trypanosomes can be divided into three classes. The first includes those tRNAs that are encoded in the nucleus and are targeted to the cytoplasm but fail to be imported into the mitochondrion. These tRNAs specific to the cytoplasm function exclusively in cytoplasmic protein synthesis and include a $tRNA^{Gln}$ for *L. tarentolae* (Lye et al., 1993). The second class of tRNAs that is imported into the mitochondrion and function exclusively in mitochondrial translation. While a small transient pool of these RNAs is expected in the cytoplasm, these tRNAs do not accumulate in the cytoplasm and do not participate in cytoplasmic translation. A mitochondrial specific $tRNA^{Ile}$ has been identified in *L. tarentolae* (Shi et al., 1994). The third class of tRNAs accumulates in both mitochondria and cytoplasm. These tRNAs are likely to function in protein synthesis in both the mitochondrion and cytoplasm of *T. brucei* and *L. tarentolae*. Based on two dimensional gel analysis of cytoplasmic and mitochondrial tRNAs it appears that >90% of the tRNAs of *T. brucei* are shared by both compartments (LeBlanc et al., 1999).

The mitochondrial tRNA gene organization of kinetoplastids share traits of both nuclear and mitochondrial tRNA genes in other organisms. The clustering of the tRNA genes and the multicistronic nature of transcription is reminiscent of the mitochondrial tRNA genes of mammals and yeast, while the transcription by polymerase III is a singularly nuclear trait.

## HOW ARE tRNAs SELECTED FOR IMPORT?

Mitochondrial tRNAs, encoded within the nucleus, must be selectively sorted within the cytoplasm and recognized by the mitochondrial import machinery. Since only a subset of the nuclear encoded tRNAs are imported sequence elements within the imported RNA may provide recognition signals for either cytosolic carrier proteins or receptors on the mitochondrial membrane.

The sequence requirements for the selective import of tRNAs into *Tetrahymena* have been investigated (Rusconi and Cech, 1996a; 1996b). The import of over 20 nuclear encoded tRNAs has been described for this organism and despite the large number of nuclear encoded tRNAs needed for mitochondrial protein synthesis, the import is selective. The cytoplasm of *Tetrahymena* contains over 45 tRNAs; many of which are not imported. Transfection studies showed that when $tRNA^{Gln}$ (UUG) and $tRNA^{Gln}$ (UUA) were tagged by the addition of two adenine nucleotides in the D-loop, only the $tRNA^{Gln}$ (UUG) was detected in the mitochondrion (Rusconi and Cech, 1996a). The tagged $tRNA^{Gln}$ (UUG) was imported at similar levels as the endogenous $tRNA^{Gln}$ indicating that the sequence tag did not interfere with targeting or import into the mitochondrion.

In order to determine whether mitochondrial-targeting sequences could be identified within the glutamine tRNAs, structural blocks of sequence were exchanged between the $tRNA^{Gln(UUG)}$ and $tRNA^{Gln(UUA)}$. The domain swapping was carefully done to ensure that the tertiary structure of the tRNAs was unaltered (Rusconi and Cech, 1996b). Exchanging the acceptor stem, D stem and anticodon stem did not effect the import of $tRNA^{Gln(UUG)}$ into the mitochondrion. However, mutations in the anticodon abolished import. Additional studies further showed that the third nucleotide (G) of the anticodon was an essential import determinant. While these experiments suggested that the tRNA import machinery was able to discriminate between the glutamine tRNAs based on the anticodon sequence, other sequences may also influence import of the $tRNA^{Gln(UUG)}$. The import of $tRNA^{Gln(UUG)}$ was also lost when the 5' and 3' flanking sequences were deleted (Rusconi and Cech, 1996b). The possible role of sequences flanking imported tRNAs in trypanosomes will be described below.

Plants also import a subset of tRNAs into their mitochondria (Dietrich et al., 1992). Transfection studies using potatoes have revealed another mitochondrial import element. A single point mutation within the aminoacyl stem of the $tRNA^{Ala}$ abolished mitochondrial import. Interestingly, this mutation also inhibited aminoacylation, suggesting that aminoacyl-tRNA synthetases might be involved in import (Dietrich et al. 1996). Studies in yeast also indicate that recognition elements for aminoacyl-tRNA synthetases may be sequence determinants for tRNA import. The mitochondrial genome of *S. cerevisiae* encodes all of the tRNAs necessary for mitochondrial protein synthesis. However, a single nuclear encoded $tRNA^{Lys\ (CUU)}$ is imported (Martin et al., 1979; Tasassov and Entelis, 1992). Since this tRNA is not aminoacylated by the mitochondrial lysyl-tRNA synthetase (Martin et al., 1979), its role in mitochondrial protein synthesis is questionable. Several alternative functions have been suggested, which include a role for this RNA in mitochondrial RNA splicing, reverse transcription or DNA replication. However, as will be described below, $tRNA^{Lys\ (CUU)}$ is imported into the mitochondrion of yeast in an aminoacylated state due to the cytoplasmic aminoacyl synthetase, thus raising the possibility that the imported tRNA may participate in mitochondrial protein synthesis.

In order to investigate the import determinants of the imported $tRNA^{Lys(CUU)}$, structural domains were swapped between a non-imported, mutated $tRNA^{Lys(CUU)}$. Mutant transcripts were tested for import, aminoacylation and binding to the precursor mitochondrial lysyl-tRNA synthetase (Tarassov and Entelis, 1992, Tarassov et al., 1996). All imported transcripts retained their ability to bind the pre-mitochondrial aminoacyl synthetase, while non-imported mutants failed to bind the synthetase. In addition, substitution of the anticodon arm of the $tRNA^{Lys(CUU)}$ with that of the mutated tRNA substrate also abolished import without affecting aminoacylation. A pathway for import of this $tRNA^{Lys(CUU)}$ has been proposed in which the imported $tRNA^{Lys(CUU)}$ is first aminoacylated by the cytoplasmic lysyl-tRNA synthetase and then associates with a precursor form of the mitochondrial lysyl-tRNA synthetase. Both aminoacylation and binding to the mitochondrial precursor protein are essential for import.

Sequence determinants for tRNA import in trypanosomes and Leishmania have also been investigated (Lima and Simpson, 1996; Chen et al., 1994; Yermovsky-Kammerer and Hajduk, 1999). While the majority of the tRNAs in trypanosomes and *Leishmania* are shared between the mitochondrion and cytosol, a few mitochondrial and cytosol specific tRNAs have been described

(Lye et al., 1993; Chen et al., 1994; Shi et al., 1994). *L. tarentolae* were transfected with plasmids encoding either a mitochondria specific $tRNA^{Ile(UAU)}$ or a cytosol specific $tRNA^{Gln(CUG)}$ tagged with altered sequences within the D-loop region of the tRNA. Deletion of or changes to sequences flanking the tRNAs did not influence tRNA localization in these experiments, suggesting that mature tRNA sequences may direct sorting to the mitochondrion. Swapping the D-loop and stem from the cytosolic $tRNA^{Gln(CUG)}$ with the D-loop and stem of the mitochondrially targeted $tRNA^{Ile(UAU)}$ resulted in import of the mutant $tRNA^{Gln(CUG)}$. These results suggest that sequences within the D-loop region may influence selection of a tRNA for mitochondrial import in *Leishmania.* In vitro import studies with isolated mitochondria from *L. tropica* also suggested a structural motif at least partly defined by the D-loop and stem contributed to mitochondrial import (Mahapatra et al., 1998). Other in vivo studies with *Leishmania* showed that a four-nucleotide insertion into the variable loop of $tRNA^{Thr(AGU)}$ abolished import (Chen et al., 1994). These apparently contradictory results may indicate that tertiary interactions outside the D-loop region may be needed for proper cellular localization.

In *T. brucei* we have reported the occurrence of precursor tRNAs within the mitochondrion (Hancock et al., 1992). Sequence analysis indicated that one such precursor was a tandem $tRNA^{Ser/Leu}$ joined by 59 nucleotides of intergenic sequence. The tandem $tRNA^{Ser/Leu}$ was identified in both the cytosol and the mitochondrion and was shown to be a RNA polymerase III transcript that initiated 14 nts upstream of the $tRNA^{Ser}$. We have proposed that the tandem $tRNA^{Ser/Leu}$ was formed by inefficient RNA polymerase III termination at the 3' end of the upstream tRNA resulting in transcription of the intergenic region as well as the adjacent leucine tRNA (LeBlanc et al., 1999). Subcellular fractionation studies supported the localization of the tandem $tRNA^{Ser/Leu}$ within *T. brucei* mitochondria suggesting it was an import substrate. In order to test the role of the tandem $tRNA^{Ser/Leu}$ in mitochondrial import, an in vitro import system was developed (Yermovsky-Kammerer and Hajduk, 1999). The tandem $tRNA^{Ser/Leu}$ are efficiently imported but neither the mature $tRNA^{Ser}$ nor $tRNA^{Leu}$ are imported in this system. Deletions from the 5' and 3' ends of the tandem $tRNA^{Ser/Leu}$ show that import still occurs when one of the tRNAs is deleted at either end. Deletion of the intergenic region at the 3' end by only 8 nts is able to reduce import by 75% (Yermovsky-Kammerer and Hajduk, unpublished). The intergenic region, without associated tRNAs, is also imported though import is reduced 50%.

The sequence element(s) within the intergenic region that directs import has not been established.

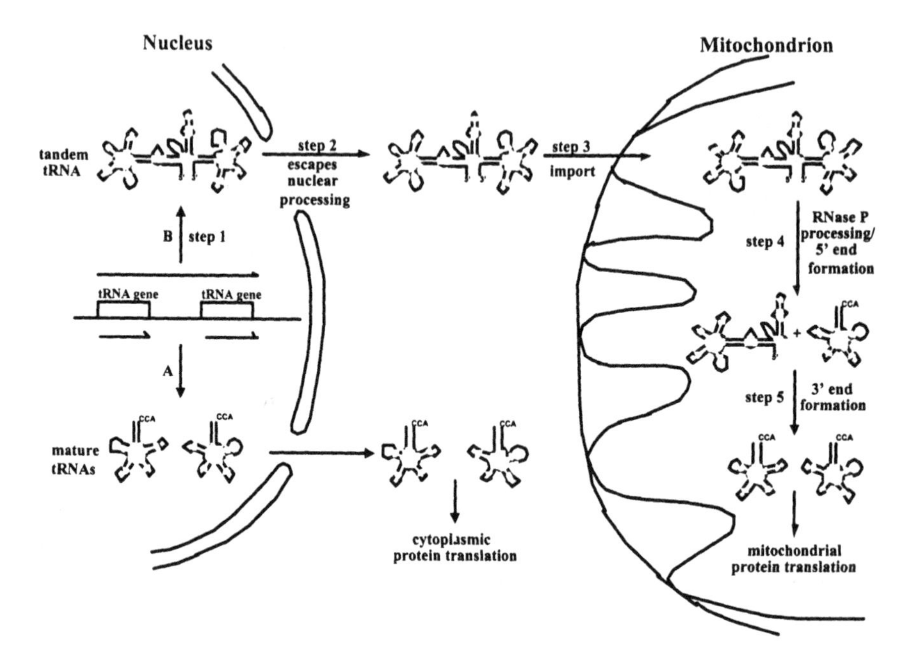


**Figure 2.** A schematic of the predicted pathways for sorting tRNAs to the mitochondrion and cytosol of *T. brucei.* The tRNAs are encoded in the nucleus, arranged as part of a tRNA cluster and are shared between the cytosol and mitochondria. Transcription of these genes by RNA polymerase III results in either a tandem tRNA or mature tRNAs. The tandem tRNA is targeted to the mitochondrion, imported and processed to mature tRNAs within the matrix of the mitochondrion. The monocistronic tRNAs are sorted to the cytoplasm and are not substrates for import into the mitochondrion.

A model for intracellular sorting of the tandem $tRNA^{Ser/Leu}$ is shown in Figure 2 and may encompass most other clustered tRNAs that are shared between the cytosol and the mitochondria. Differential RNA polymerase III termination may directly influence the targeting of the tRNAs to the appropriate cellular compartment. This tRNA gene cluster, contains $tRNA^{Ser}$, a 59 nt intergenic region, and the $tRNA^{Leu}$. Transcription initiates 14 nts upstream of the $tRNA^{Ser}$ and at some frequency fails to terminate immediately downstream of the $tRNA^{Ser}$ gene. Transcription through the intergenic and downstream $tRNA^{Leu}$ terminates at a run of 5 T's at the 3' end of the $tRNA^{Leu}$ gene. Transcription can also initiate immediately upstream of the $tRNA^{Leu}$. Thus, it appears that both a dicistronic tRNA precursor and monocistronic tRNAs can be formed by polymerase III transcription of this locus. In order to escape RNAse P processing in the nucleus, the tandem $tRNA^{Ser/Leu}$ may form an unusual structure that RNAse P fails to recognize. The tandem tRNA is then exported into the cytoplasm and imported into the mitochondrion. Once in the mitochondrion, a mitochondrial RNase P-like activity processes the tandem $tRNA^{Ser/Leu}$ to mature tRNAs that can then participate in mitochondrial protein translation.

The mature $tRNA^{Ser}$ and $tRNA^{Leu}$ may also be formed within the nucleus. Accurate polymerase III transcription initiation and termination would result in tRNAs that could be processed to mature tRNAs in the nucleus by RNAse P and nucleotidyl transferase. We propose that these mature tRNAs are not substrates for mitochondrial import and are retained in the cytosol and participate in cytosolic protein synthesis (Fig. 2).

Currently, there is considerable disagreement concerning the sequence requirements for import of tRNAs into trypanosome and Leishmania mitochondria. Mutations in the tRNA D-loop and stem abolished mitochondrial import (Lima and Simpson, 1996; Hauser and Schneider, 1995; Mahapatra and Adhya, 1996). However, the role of tandem precursor tRNAs in *T. brucei* tRNA import is supported both by the in vitro studies with isolated mitochondria and also by the requirement for 5' extended RNAs for in vivo import (Yermovsky-Kammerer and Hajduk, 1999; Hauser and Schneider, 1995).

## HOW ARE tRNAs IMPORTED INTO MITOCHONDRIA?

Several different models have been proposed for tRNA import. In order to examine the possible mechanism of tRNA import into mitochondria, in vitro systems have been developed for yeast, Leishmania, and trypanosomes. These in vitro systems have provided evidence for cytosolic and mitochondrial proteins in targeting and translocation of tRNAs across the mitochondrial membranes. In addition, cytosolic and mitochondrial matrix

ATP requirements have been evaluated. Finally, the role of the electromotive force across the inner mitochondrial membrane has been studied.

The mechanism of tRNA import into *S. cerevisiae* has been extensively studied. The in vitro import of $tRNA^{Lys(CUU)}$ is ATP dependent, requires an electrochemical membrane potential across the inner mitochondrial membrane, and a cytosolic fraction containing the cytosolic lysyl-tRNA synthetases (Tarassov and Entelis, 1992, Tarassov et al., 1995). To investigate the requirement for an energized inner mitochondrial membrane in the import of $tRNA^{Lys(CUU)}$, isolated mitochondria were treated with carbonyl cyanide n-chloro-phenylhydrazone (CCCP), a protonophore that disrupts the inner membrane potential by allowing protons to freely move across the inner mitochondrial membrane, or valinomycin, a potassium ionophore which also dissipates the inner mitochondrial membrane potential but without affecting the pH gradient across the inner membrane. Both CCCP and valinomycin led to a loss of import capacity for the $tRNA^{Lys(CUU)}$ by the mitochondria. The concentrations of inhibitors needed to block $tRNA^{Lys(CUU)}$ import were similar to those necessary for inhibition of protein import into yeast mitochondria (Tarassov et al., 1995). Finally, the import of the $tRNA^{Lys(CUU)}$ was also inhibited by treatment of the mitochondria with a combination of oligomycin, to block the $F_1F_0$ATPase, and potassium cyanide, a phosphorylation uncoupler. These studies are consistent with the possibility that import of $tRNA^{Lys(CUU)}$ shares a common pathway with protein import. Both in vitro and in vivo import studies have shown that $tRNA^{Lys(CUU)}$ is imported into mitochondria as a mature tRNA (Tarassov and Entelis, 1992). Mutations in the anticodon of $tRNA^{Lys(CUU)}$, particularly the C residue at nucleotide 34 inhibits import suggesting that the anticodon may be important for recognition by the import machinery. Recent studies suggest that $tRNA^{Lys(CUU)}$ is imported as a folded molecule. When strand breaks were introduced into the $tRNA^{Lys(CUU)}$, both halves of the $tRNA^{Lys(CUU)}$ were imported only when the two tRNA moieties were allowed to reassociate (Entelis et al., 1998). Import of the $tRNA^{Lys(CUU)}$ requires an intact protein translocation machinery in the yeast mitochondrial membrane. Mutants carrying either disrupted or deleted alleles were used to determine which protein of the protein translocation machinery might be needed for tRNA import. Binding and translocation across the mitochondrial membranes requires at least two proteins, an outer membrane receptor (Tom19) and an inner membrane receptor (Tim44) (Tarassov et al., 1996).

The in vitro studies with yeast suggest that tRNA$^{Lys(CUU)}$ is aminoacylated by the cytoplasmic lysyl-tRNA synthetase and then associates with a precursor form of the mitochondrial lysyl-tRNA synthetase. The tRNA is co-transported as a protein/tRNA complex. Interestingly, an artificial substrate that mimics the proposed precursor lysyl-tRNA synthetase and tRNA$^{Lys(CUU)}$ complex was previously investigated. Double stranded DNA oligonucleotides were efficiently imported into yeast mitochondria when covalently coupled to a mitochondrial precursor protein (Vestweber and Schatz, 1989) or to a mitochondrial presequence (Seibel et al., 1995). The current evidence overwhelmingly supports a shared protein and tRNA import pathway in yeast mitochondria.

In vitro import systems have also been developed for *L. tropica* and *T. brucei*. The first report of in vitro RNA import for kinetoplastids was in *L. tropica* (Mahapatra et al, 1994). Using isolated mitochondria they reported the efficient import of a 240 nucleotide antisense β-tubulin transcript complementary to the 5'-untranslated region (UTR) and 25 nts of the β-tubulin mRNA. Import was assessed by protection of the RNA substrate from digestion by added ribonuclease. In addition, the imported RNAs were radiolabelled with UTP by the mitochondrial terminal uridylyl transferase. Import was rapid and dependent upon exogeneously added ATP. A high affinity 15 kDa RNA binding protein localized to the outer mitochondrial membrane was also required for import (Mahapatra and Adhya, 1996). While the antisense β-tubulin is clearly not a natural substrate for import, its uptake appears to be specific. Import was saturable at about 1 nM and was competed by several small ribosomal RNAs, tRNAs and the antisense RNA itself. However, yeast tRNAs failed to compete for import. Deletion analysis of the antisense RNA suggested that a hexapurine repeat of GAAA(A/G)G was required for efficient import. Similar purine-rich sequences are found in the D stem-loop regions of kinetoplastid tRNAs and may serve as signals for import.

An in vitro mitochondrial import system has also been developed for *T. brucei* (Yermovsky-Kammerer and Hajduk, 1999). Import is specific since a precursor tandem tRNA$^{Ser/Leu}$ was rapidly and efficiently imported; whereas, mature tRNA$^{Ser}$, tRNA$^{Leu}$, ribosomal rRNAs and non-specific plasmid derived RNAs were not substrates for in vitro import. Half-maximal uptake occurred within 5 minutes and a plateau for import was reached after about 15 min (Yermovsky-Kammerer and Hajduk, 1999). This time frame is similar to that reported for protein import using the same in vitro system

(Priest and Hajduk, 1996). Both the *Leishmania* and *T. brucei* in vitro import systems were shown to have similar ATP requirements and neither required a soluble cytosolic fraction (Mahapatra and Adhya, 1996; Yermovsky-Kammerer and Hajduk 1999). The ability to import RNAs in the absence of a cytosolic fraction differs from the import reported in yeast for the $tRNA^{Lys(CUU)}$, where aminoacyl-tRNA synthetases, provided by a cytosolic fraction, are essential for import. These results argue against the involvement of aminoacylation in tRNA import in trypanosomes. Further evidence that aminoacylation is not coupled to import was obtained in vivo when a mutated, unspliced, $tRNA^{Tyr}$ gene was transfected into *T. brucei.* The unspliced tRNA is not a substrate for its cognate aminoacyl synthetase yet it was imported (Schneider et al., 1994b). Therefore, it appears that aminoacylation of tRNAs is not a prerequisite to import in trypanosomes. It remains possible that binding of a precursor mitochondrial aminoacyl-tRNA synthetase may facilitate import. This is particularly true for studies using crude mitochondrial preparations, which contain cytosolic contamination, and support import of the tandem $tRNA^{Ser/Leu}$ (Yermovsky-Kammerer and Hajduk, 1999).

Import of RNA into kinetoplastid mitochondria requires an external ATP supply as well as an internal ATP pool. Depletion of external ATP by apyrase treatment completely blocked import (Yermovsky-Kammerer and Hajduk, 1999). Import was also abolished by depleting the matrix ATP pool with oligomycin, an inhibitor that prevents the synthesis of ATP by inhibiting the H+ translocating $F_1F_0$-ATPase. Import of the tandem $tRNA^{Ser/Leu}$ in *T. brucei* was also shown to require a membrane potential (Yermovsky-Kammerer and Hajduk, 1999). Import levels were affected by the use of the inhibitor CCCP. Complete inhibition was achieved in the presence of 10 μM CCCP (Yermovsky-Kammerer and Hajduk, 1999). These results indicate that an electrochemical potential across the inner mitochondrial membrane is required for import of the tandem $tRNA^{Ser/Leu}$. Consistent with these results, we have proposed the following model for tRNA import into trypanosome mitochondria (Fig. 3).

We propose that following transcription by RNA polymerase III in the nucleus, the tandem $tRNA^{Ser/Leu}$ associates with a carrier protein. We predict that this protein is itself a mitochondrial precursor protein that contains a mitochondrial targeting sequence. The cytosolic ATP requirement may be associated with either the unfolding of the carrier protein or the tandem $tRNA^{Ser/Leu}$. The mitochondrial presequence of the carrier protein, with the associated $tRNA^{Ser/Leu}$, is able to interact with the outer mitochondrial membrane receptors Tom70-Tom37 and Tom22-Tom20. The carrier protein-$tRNA^{Ser/Leu}$ complex is then translocated into the matrix by the inner mitochondrial membrane channel associated proteins. Once in the mitochondrial matrix, the carrier protein is processed and dissociates from the $tRNA^{Ser/Leu}$ and the $tRNA^{Ser/Leu}$ is processed to mature tRNAs.

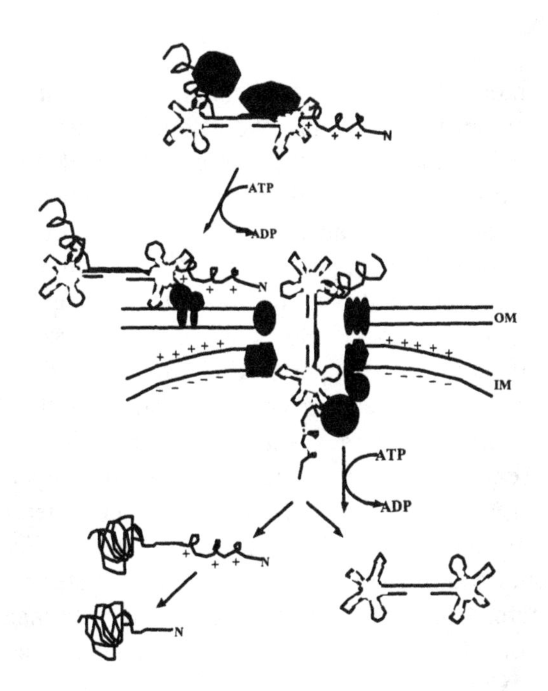


**Figure 3.** Components of a proposed tRNA import pathway. The mitochondrial components of the protein import machinery bind and facilitate translocation of the tandem $tRNA^{Ser/Leu}$ carrier protein complex to the mitochondrial matrix.

What proteins could serve as carriers for the $tRNA^{Ser/Leu}$? There are several interesting candidates. First, the mitochondrial aminoacyl tRNA synthetases. These proteins are nuclear encoded and are present in the mitochondria. They also have a high affinity for the imported tRNAs. While indirect evidence argues against these proteins serving as tRNA carriers the possibility should be carefully evaluated. A second possibility is the protein component of RNAse P. In all of the systems examined thus far, with the exception of yeast, these proteins are encoded by the nucleus and imported. It is possible that these proteins might either directly associate with the $tRNA^{Ser/Leu}$. A number of other RNA binding and modifying proteins that are imported into mitochondria might also serve as carrier proteins. The critical feature of this model is that tRNA import into mitochondria does not require a pathway unique from the protein import pathways.

## CONCLUSIONS

In summary, two models have immerged for import of a highly negatively charged tRNA molecule through the double membranes of the mitochondrion. The first model proposes the co-import of the tRNA, bound with soluble carrier proteins, through the protein import channels of the inner and outer mitochondrial membranes utilizing the import signal on the precursor protein carrier. The import of $tRNA^{Lys(CUU)}$ in yeast complexed with the precursor mitochondrial lysyl-tRNA synthetase supports this model. A slight variation of this model allows for any precursor mitochondrial protein with high affinity for the imported tRNA to serve as a carrier protein. A second model has recently been proposed that suggests that high-affinity, membrane-bound complexes can act as a receptor or carrier for direct import of tRNA in the absence of soluble cytosolic factors (Mukherjee et al., 1999). In *Leishmania*, a 15 kDa protein associated with the outer mitochondrial membrane has been implicated as a tRNA carrier. The binding of the tRNA to the outer mitochondrial membrane is followed by the step-wise transfer of the tRNA first across the outer membrane to the intermembrane space followed by a second translocation step across the inner membrane to the mitochondrial matrix.

Many challenges lie ahead in the study of tRNA import into mitochondria. In trypanosomes and *Leishmania*, where a large number of tRNAs are imported, both proposed pathways may be operative. A detailed understanding of both the sequence recognition elements in the tRNAs and the components of the mitochondrial import machinery are needed. With this information it may be possible to develop transfection vectors that will direct the import of specific RNA into mitochondria. Such a transfection/import system would offer the exciting possibility of gene therapy for diseases caused by mitochondrial defects in mammals and research opportunities in the investigation of the requirements for RNA editing in kinetoplastids.

## ACKNOWLEDGEMENTS

We thank current and past tRNA importers, Kathy Hancock, Allen LeBlanc and Lynn Sherrer for interesting questions and exciting results. In particular, we appreciate the development of the protein import system in *T. brucei* by Jeff Priest, which made much of the work in the Hajduk lab on RNA import possible. Studies on RNA import in the Hajduk lab are supported by National Institutes of Health grant AI21401.

## REFERENCES

Adhya, S., Ghosh, T., Das, A., Kanti, S. and Mahapatra, S. (1997) Role of an RNA-binding protein in import of tRNA into *Leishmania* mitochondria. J. Biol. Chem. 272, 21396-21402.

Attardi, G., and Schatz, G. (1988). Biogenesis of mitochondria. Annu. Rev. Cell. Biol. 4, 289-333.

Boer, P. H., and Gray, M. W. (1988). Transfer RNA genes and the genetic code in Chlamydomonas reinhardtii mitochondria. Curr Genet 14, 583-90.

Campbell, D.A. (1989). Tandemly linked $tRNA^{Gln}$, $tRNA^{Val}$, $tRNA^{Lys}$ genes in *Trypanosoma brucei*. Nucleic Acids Res. 17, 9479.

Campbell, D.A., Suyama, Y., and Simpson, L. (1989). Genomic organization of nuclear $tRNA^{Gly}$ and $tRNA^{Leu}$ genes in *Trypanosoma brucei*. Mol. Biochem. Parasitol. 37, 257-262.

Chen, D.H.T., Shi, X., Suyama, Y. (1994) In vivo expression and mitochondrial import of normal and mutated $tRNA^{Thr}$ in *Leishmania*. Mol. Biochem. Parasitol. 64, 121-133.

Deutscher, M.P. (1984). Processing of tRNA in prokaryotes and eukaryotes. CRC Crit. Rev. Biochem. 17, 45-71.

Dietrich, A., Weil, J. H., and Marechal-Drouard, L. (1992). Nuclear-encoded transfer RNAs in plant mitochondria. Annu Rev Cell Biol 8, 115-31.

Dietrich, A., Marechal-Drouard, L., Carneiro, V., Cosset, A., and Small, I. (1996). A single base change prevents import of cytosolic tRNA(Ala) into mitochondria in transgenic plants. Plant J 10, 913-8.

Entelis, N.S., Krasheninnikov, I.A., Martin, R.P., and Tarassov, I.A. (1996). Mitochondrial import of a yeast cytoplasmic tRNA (Lys): possible roles of aminoacylation and modified nucleosides in subcellular partitioning. FEBS Lett. 384, 38-42.

Entelis, N.S., Kieffer, S., Kolesnikova, O.A., Martin, R.P., and Tarassov, I.A. (1998). Structural requirements of $tRNA^{Lys}$ for its import into yeast mitochondria. Proc. Natl. Acad. Sci. U S A 95, 2838-2843.

Feagin, J.E., (1992) The 6-kb element of *Plasmodium falciparum* encodes mitochondrial cytochrome genes. Mol. Biochem. Parasitol. 52, 145-148.

Galli, G., Hofstetter, H., and Birnstiel, M.L. (1981). Two conserved sequence blocks within eukaryotic tRNA genes are major promoter elements. Nature 294, 626-631.

Hancock, K. and Hajduk, S.L. (1990) Mitochondrial tRNAs are nuclear encoded in *Trypanosoma brucei*. J. Biol. Chem. 265, 19208-19215.

Hancock, K. and Hajduk, S.L. (1992) Sequence of *Trypanosoma brucei* tRNA genes encoding cytosolic tRNAs. Nucl. Acids Res. 20, 2602.

Hancock, K., LeBlanc, A., Donze, D. and Hajduk, S.L. (1992) Identification of precursor tRNAs within the mitochondrion of *Trypanosoma brucei*. J. Biol. Chem. 267, 23963-23971.

Hauser, R. and Schneider, A. (1995) tRNAs are imported into mitochondria of *Trypanosoma brucei* independently of their genomic context and genetic origin. EMBO J. 14, 4212-4220.

LeBlanc, A.J., Yermovsky-Kammerer, A.E., and Hajduk, S.L. (1999). A nuclear encoded and mitochondrial imported dicistronic tRNA precursor in *Trypanosoma brucei*. J. Biol. Chem. 274, 21071-21077.

Lima, B.D. and Simpson, L. (1996) Sequence-dependent in vivo importation of tRNAs into the mitochondiron of *Leishmania tarentolae*. RNA 2, 429-440.

Lye, L.F., Chen, D.H., and Suyama., Y. (1993). Selective import of nuclear-encoded tRNAs into mitochondria of the protozoan *Leishmania tarentolae*. Mol.Biochem. Parasitol. 58:, 33-45.

Mahapatra, S., Ghosh, T. and Adhya, S., (1994) Import of small RNAs into *Leishmania* mitochondria in vitro. Nucl. Acids Res. 22, 3381-3386.

Mahapatra, S., and Adhya, S., (1996) Import of RNA into *Leishmania* mitochondria occurs through direct interaction with membrane-bound receptors. J. Biol. Chem. 271, 20432-20437.

Mahapatra, S., Ghosh, S., Bera, S. K., Ghosh, T., Das, A., and Adhya, S. (1998). The D arm of tRNATyr is necessary and sufficient for import into Leishmania mitochondria in vitro. Nucleic Acids Res 26, 2037-41.

Maréchal-Drouard, L., Weil, J.H., and Guillemaut, P. (1988). Import of several tRNAs from the cytoplasm into the mitochondria in bean *Phaseolus vulgaris*. Nucleic Acids Res. 16, 4777-4788.

Maréchal-Drouard, L., Guillemaut, P., Cosset, A., Arbogast, M., Weber, F., Weil, J.H., and Dietrich, A. (1990). Transfer RNAs of potato (*Solanum tuberosum*) mitochondria have different genetic origins. Nucleic Acids Res. 18, 3689-3696.

Martin, R., Schneller, J., Stahl, A. and Dirheimer, G. (1979) Import of nuclear deoxyribonucleic acid coded lysine-accepting transfer ribonucleic acid (anticodon CUU) in yeast mitochondria. Biochemistry 18, 4600-4605.

Mottram, J.C., Bell, S.D., Nelson, R.G. and Berry, J.D. (1991a) tRNAs of *Trypanosoma brucei*. Unusual gene organization and mitochondrial importation. J. Biol. Chem. 266, 18313-18317.

Mottram, J.C., Shafi, Y., and Barry, J.D. (1991b). Sequence of a tRNA gene cluster in *Trypanosoma brucei*. Nucleic Acids Res. 19, 3995.

Mukherjee, S., Bhattacharyya, S.N. and Adhya, S. (1999) Stepwise transfer of tRNA through the double membrane of *Leishmania* mitochondria. J. Biol. Chem. 274, 31249-31255.

Neupert, W. (1997). Protein import into mitochondria. Annu. Rev. Biochem. 66, 863-917.

Priest, J.W. and Hajduk, S.L. (1996) In vitro import of the Rieske iron -sulfur protein by trypanosome mitochondria. J. Biol. Chem. 271, 20060-20069.

Pritchard, A.E., Seilhamer, J.J., Mahalingam, R., Sable, C.L., Venuti, S.E.,and Cummings, D.J. (1990). Nucleotide sequence of the mitochondrial genome of *Paramecium*. Nucleic Acids Res. 18, 173-180.

Rusconi, C.P. and Cech T.R. (1996a) The anticodon is the signal sequence for mitochondrial import of glutamine tRNA in *Tetrahymena*. Genes Devel. 10, 2870-2880.

Rusconi, C.P. and Cech T.R. (1996b) Mitochondrial import of only one of three nuclear-encoded glutamine tRNAs in *Tetrahymena thermophila*. EMBO J. 15, 3286-3295.

Schneider, A., McNally, K.P. and Agabian, N. (1994a) Nuclear-encoded mitochondrial tRNAs of *Trypanosoma brucei* have a modified cytidine in the anticodon loop. Nucl. Acids Res. 22, 3699-3705.

Schneider, A., Martin, J. and Agabian, N. (1994b) A nuclear encoded tRNA of *Trypanosoma brucei* is imported into mitochondria. Mol. Cell. Biol. 14, 2317-2322.

Seibel, P., Trappe, J., Villani, G., Klopstock, T., Papa, S., and Reichmann, H.(1995). Transfection of mitochondria: strategy towards a gene therapy of mitochondrial DNA diseases. Nucleic Acids Res. 23, 10-17.

Shi, X., Chen, D.H.T. and Suyama, Y. (1994) A nuclear tRNA gene cluster in the protozoan *Leishmania tarentolae* and differential distribution of nuclear encoded tRNAs between the cytosol and mitochondria. Mol. Biochem. Parasitol. 57, 23-37.

Simpson, A.M., Suyama, Y., Dewes, D., Campbell, D., and Simpson, L. (1989) Kinetoplastid mitochondria contain functional tRNAs which are encoded in nuclear DNA and also small minicircle and maxicircle transcripts of unknown function. Nucl. Acids Res. 17, 5427-5445.

Suyama, Y. (1967). The origins of mitochondrial ribonucleic acids in *Tetrahymena pyriformis*. Biochemistry 6, 2829-2839.

Suyama, Y. (1986). Two dimensional polyacrylamide gel electrophoresis analysis of *Tetrahymena* mitochondrial tRNA. Curr .Genet. 10, 411-420.

Tarassov I.A. and Entelis, N.S. (1992) Mitochondrially-imported cytoplasmic $tRNA^{Lys(CUU)}$ of *Saccharomyces cerevisiae*: in vivo and in vitro targeting systems. Nucl. Acids Res. 20, 1277-1281.

Tarassov I.A., Entelis, N.S. and Martin, R.P. (1995) Mitochondrial import of a cytoplasmic lysine tRNA in yeast is mediated by cooperation of cytoplasmic and mitochondrial lysyl-tRNA synthetases. EMBO J. 14, 3461-3471.

Tarassov I.A., Entelis, N.S., and Martin R.P. (1996) An intact protein translocation machinery is required for mitochondrial import of a yeast cytoplasmic tRNA. J. Mol. Biol. 245, 315-323.

Vestweber, D., and Schatz, G. (1989). DNA-protein conjugates can enter mitochondria via the protein import pathway. Nature 338, 170-172.

Yermovsky-Kammerer, A.E. and Hajduk, S.L. (1999) In vitro import of a nuclearly encoded tRNA into the mitochondrion of *Trypanosoma brucei*. Molec. Cell. Biol. 6253-6259.

# 2

# RNA METABOLISM IN TRYPANOSOMES: APPROACHES TO UNRAVEL PRE-mRNA SIGNALS REQUIRED FOR INTRON REMOVAL

Christian Tschudi and Elisabetta Ullu
*Department of Internal Medicine, Yale University School of Medicine, New Haven, CT 06520-8022*

## OVERVIEW AND BACKGROUND INFORMATION

The study of RNA metabolism in trypanosomes has provided the scientific community with a number of unexpected and enticing discoveries, including polycistronic transcription (Ullu et al., 1996), trans-splicing (Ullu et al., 1996), cap 4 modification (Bangs et al., 1992), mitochondrial RNA editing (Hajduk et al., 1997; Simpson and Emeson, 1996; Stuart et al., 1997), coupling of trans-splicing and polyadenylation (Matthews et al., 1994), and more recently RNA interference or RNAi (Ngo et al., 1998). Furthermore, over the past few years we have learned that there is very little, if any, regulation of gene expression at the transcriptional level and that most likely modulation of gene expression is achieved primarily at the post-transcriptional level. In this scenario, the pre-mRNA cis-acting signals for RNA processing, by virtue of their interactions with processing machineries, would be major determinants for regulating gene expression.

### Biogenesis of mRNA in trypanosomes

In protozoa of the family *Trypanosomatidae*, which include African and South American trypanosomes and *Leishmania*, the production of mature mRNA differs in several aspects from the biogenesis of mature mRNA in most eukaryotes. Transcription of the majority of protein coding genes generates a primary transcript that is polycistronic in structure and is colinear with the corresponding chromosomal gene arrangement. Intergenic regions are usually a few hundred nucleotides long and can be as short as 90 to 100 nucleotides. Monocistronic mature mRNAs are generated by cleaving away the intergenic regions and at the same time modifying both ends of the mRNA coding region. This is achieved via two RNA processing reactions, namely trans-splicing of the spliced leader (SL) sequence at the 5' end and cleavage/polyadenylation at the 3' end of the mRNA coding region (Ullu et al., 1996). As a consequence, the 5' ends of all mature mRNAs are formed post-transcriptionally, rather than by transcription initiation as in most eukaryotic organisms.

Until very recently, the only splicing known to occur on pre-mRNA in trypanosomes was trans-splicing. Although first described in *Trypanosoma brucei* about fourteen years ago (Boothroyd and Cross, 1982), trans-splicing was later found to be present in nematodes (Krause and Hirsh, 1987), euglenoids (Tessier et al., 1991), and certain trematodes (Rajkovic et al., 1990). Whereas it was clearly established that these latter organisms in addition have intervening sequences, which are removed by cis-splicing, the conviction developed over the years that trypanosomes lack intervening sequences and therefore the machinery to carry out cis-splicing. We have recently discovered that these tenets are no longer true. Poly(A) polymerase (PAP) genes in *T. brucei* and *T. cruzi* are split by intervening sequences, thus establishing that trypanosomatid protozoa carry out both cis- and trans-splicing (Mair et al., 2000). This chapter will concentrate on this newest development in RNA metabolism in trypanosomes and describe our approaches to unravel pre-mRNA signals required for removal of intervening sequences in trypanosomes.

**Nuclear pre-mRNA splicing**

The precise removal of intervening sequences from pre-mRNAs by cis-splicing is catalysed within complicated ribonucleoprotein complexes known as spliceosomes (Ares and Weiser, 1995; Nilsen, 1994). These large particles are built stepwise on the mRNA precursor from small ribonucleoprotein particles (snRNPs) and proteins. Each of the snRNPs contains an RNA component, a small nuclear RNA (snRNA), a set of common proteins, and several snRNP-specific proteins. To date the view is that the spliceosome contains a large number of distinct proteins (between 50 and 100), as well as five spliceosomal snRNAs (the U1, U2, U4, U5 and U6 snRNAs).

In higher eukaryotes spliceosome assembly is initiated by the binding of U1 snRNP (via base-pairing) to consensus intronic nucleotides at the 5' splice site and recognition of the polypyrimidine tract by the U2 snRNP auxiliary factor U2AF (Fig. 1; Reed, 1996). Subsequently, U2 snRNP engages the pre-mRNA by base-pairing to the branch point sequences. Next, U4, U5 and U6 snRNPs join the spliceosome as a preformed tri-snRNP. Although spliceosome formation is mediated by a number of cis-acting sequences in the pre-mRNA, it is important to realise that it is almost impossible to distinguish exons from introns on the basis of these splicing signals alone. Thus, additional mechanisms must exist to co-ordinate the initial recognition of authentic splice sites. New studies put forward a model in which this recognition is achieved by a large collection of factors, whose interaction with weakly conserved sequence elements in and around the exon all contribute to the assembly of a highly stable complex (Hertel et al., 1997). In particular, a class of proteins called SR proteins, which contain RS domains (regions rich in arginine-serine dipeptides) together with RNA-binding domains, have critical roles in intron-exon definition (Zahler et al., 1992).

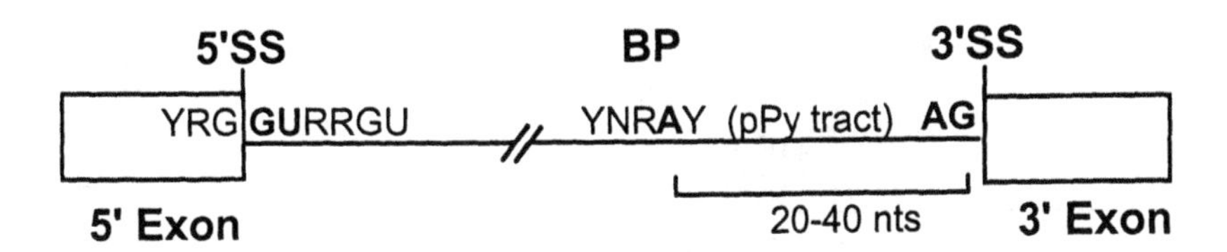


**Figure 1.** Sequence elements required for cis-splicing. The structure of a portion of a typical metazoa pre-mRNA is shown. Conserved sequences at the 5' splice site (5'SS) are indicated. The branch point sequence (BP) has a highly conserved branch site adenosine (A). A 10 to 20 nucleotide-long pyrimidine-rich tract (pPy tract) is located between the branch site and the 3' splice site (3'SS).

## EXPERIMENTAL APPROACH: GENERATION AND TESTING OF MUTATIONS IN THE INTRON OF THE *T. BRUCEI* POLY(A) POLYMERASE GENE

We have recently established that the poly(A) polymerase (PAP) genes in *T. brucei* and *T. cruzi* are split by intervening sequences of 653 and 302 nt, respectively (Mair et al., 2000). The intervening sequences occur at the same position in both organisms and obey to the GT/AG rule of cis-splicing introns (Fig. 2). Interestingly, eleven nucleotide positions downstream of the 5' splice site are conserved between the *T. brucei* and *T. cruzi* intervening sequences. Furthermore, a polypyrimidine tract is found upstream of the 3' splice site. By analogy with other eukaryotes (see above), we would predict that these sequences, in addition to the AG dinucleotide of the 3' splice site and exon sequences, will play a role in removal of the intervening sequence. Here, we describe the steps and experimental protocols to test the importance of these sequences. Mutations are introduced using a two-step PCR method and the effect of the mutations is tested in vivo by transient transfection of procyclic *T. brucei* cells and subsequent analysis of the generated RNA by 5' end RACE (rapid amplification of cDNA ends).

*Comment:*
*In the following experimental procedures, we recommend certain commercial suppliers of equipment and supplies. The preference for a specific brand is subjective and is generally not based on extensive comparison. Thus, readers are welcome to substitute the recommended brands with their own favourite ones.*

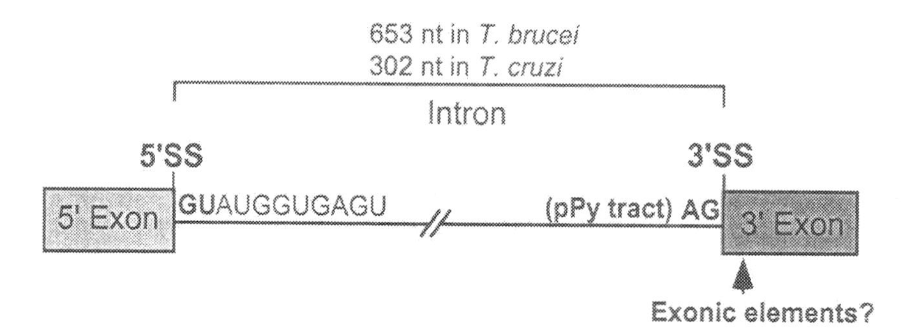


**Figure 2.** Structure of the intron in the poly(A) polymerase gene. The conserved sequences are based on two sequences only, namely the *T. brucei* and *T. cruzi* PAP introns, and therefore may turn out not to be completely accurate.

**The *T. brucei* poly(A) polymerase intron is spliced in vivo**

Although we have performed several control experiments to exclude that the isolated PAP gene represents a cloning artefact or a pseudogene, it was crucial to demonstrate that trypanosome cells actually have the machinery to carry out cis-splicing. To do this, we inserted the *T. brucei* PAP intron plus 101 and 93 bp of 5' and 3' exon sequences, respectively, into the expression vector pLew79 (Fig. 3; Wirtz et al., 1999). In this construct, pPAP.CAT, expression is driven by the promoter of the procyclic acidic repetitive protein (PARP) gene and the PAP sequences are placed between signals for RNA processing at the 5' end by trans-splicing (PARP) and at the 3' end for polyadenylation (3'ALD). In addition, the chloramphenicol acetyltransferase gene (CAT) is fused to the 3' exon of PAP to serve as a reporter gene.

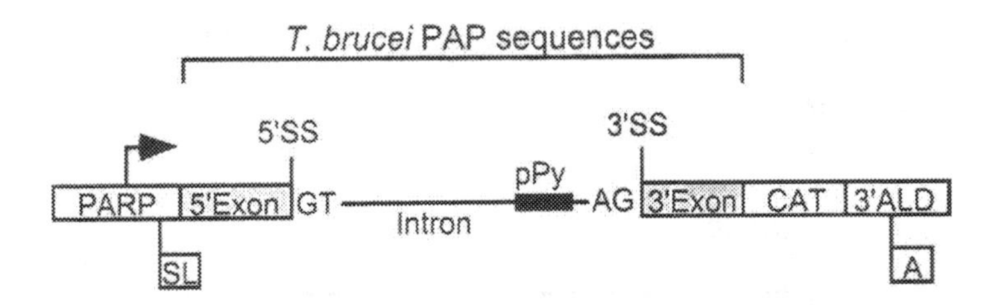


**Figure 3.** Partial structure of the expression vector pPAP.CAT. The PARP promoter indicated by an arrow directs synthesis of pre-mRNA. Trans-splicing will add the SL sequence at the 5' end (SL flag) and 3' end formation/polyadenylation will occur in the aldolase sequences (A flag).

The construct shown in Figure 3 was transfected into procyclic *T. brucei* cells by electroporation and RNA was prepared three hours post-transfection. 5' end RACE with oligonucleotides specific for the CAT and SL sequence, clearly demonstrated that the intron was accurately spliced, thus providing an experimental system to dissect sequences required for this process (see Fig. 7).

**Generation of mutations by the polymerase chain reaction**

There are mainly two different methods available to specifically alter a particular DNA sequence: the use of uracil-containing single-stranded templates (Kunkel, 1985) and directed mutagenesis using the polymerase chain reaction (PCR). We routinely use the PCR-based method, since it does not require the preparation of special templates, and provides a quick and efficient method for generating mutations. As outlined in Figure 4, in two separate reactions, fragments upstream and downstream of the mutation site are PCR-amplified using flanking primers and oligonucleotides containing the mutation to be introduced. The amplified fragments are then placed in the same tube, annealed and amplified in a second PCR step using the flanking primers. The generated full-length fragments are digested with appropriate restriction enzymes, ligated into an appropriate vector and cloned into *E. coli*.

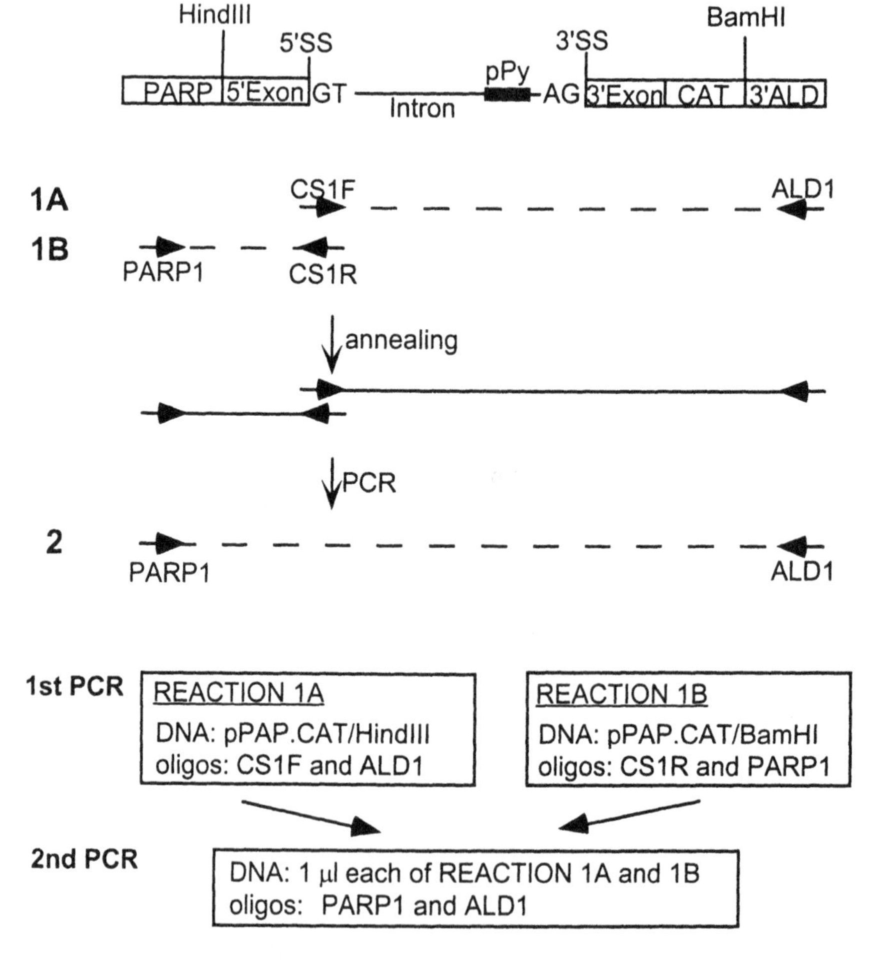


**Figure 4.** Schematic diagram of the PCR mutagenesis protocol. The structure of pPAP.CAT is shown at the top. The principle of the method is illustrated with creating a mutation at the 5' splice site.
CS1F (5'-GATCTCGATGTacATGGTGAGTTC-3') and CS1R (5'-GAACTCACCATgtACATCGAGATC-3) are complementary to each other and the mutation to be introduced, which changes the GT of the 5' splice site to AC, is

shown in small letters. Note that the mutation introduces a restriction site for the enzyme BsrG 1 (5'-TGTACA-3').

In order to guarantee the highest possible yield of mutants, there are two alternative procedures. In the first approach, a circular plasmid is used as a template for the first two PCR reactions. Then the amplified fragments are gel-purified and combined for the second step. Alternatively, we have adopted a procedure, where we initially linearize the plasmid template in two separate reactions with unique restriction enzymes (Bam HI and Hind III) located just inside the flanking primers (Fig. 4). Thus, to amplify the fragment upstream of the site of mutation, a template linearized at the downstream site (Bam HI) is used. Conversely, to amplify the downstream fragment, we use a template linearized at the upstream site (Hind III). This eliminates the necessity to gel purify the intermediate products, but still yields 75% to 100% mutants.

Procedure:

Prepare two aliquots of template plasmid DNA by digestion with two different appropriate restriction enzymes (Hind III and Bam HI in our example; see Fig. 4).

Assemble two separate PCR reactions each containing 1 μl (20 ng) of pre-digested plasmid DNA, 5 μl 10x PCR buffer (Roche Molecular Biochemicals), 1 μl 2 mM dNTPs, 1 μl flanking oligo (100 ng/μl), 1 μl mutagenic oligo (100 ng/μl), 0.5 μl Taq polymerase (5 U/μl; Roche Molecular Biochemicals), and $H_2O$ to 50 μl.

*The mutagenic primers are complementary to each other and consist of the mutation to be introduced flanked on either side by at least 10 nucleotides that are homologous to the template DNA. In designing primers, avoid having a T or A residue at the 3' end of the primer and it may be necessary to increase the length of homology in case of AT-rich sequences.*

Carry out PCR in an automated thermal cycler using the following conditions: 20 to 25 cycles at $94^0$C for 30 sec, $50^0$C for 1 min and $72^0$C for 1 min.

*Note that the PCR conditions, in particular the annealing temperature, might have to be adjusted for each individual application.*

Analyse an aliquot of the reaction by agarose gel electrophoresis to verify that the amplification has generated the predicted products.

Combine 1 μl each of the first two PCR reactions and add 5 μl 10x PCR buffer, 1 μl 2 mM dNTPs, 0.5 μl Taq polymerase and $H_2O$ to 48 μl. Perform 5 cycles at $94^0$C for 30 sec, $65^0$C for 1 min and $72^0$C for 1 min.

*It is our experience that first running 5 cycles without the flanking oligos will promote generation of the full-length fragment.*

Add 1 μl each of the flanking primers (100 ng/μl each; PARP1 and ALD1 primer) and continue the PCR for 25 cycles at $94^0$C for 30 sec, $50^0$C for 1 min and $72^0$C for 1 min.

Analyse the reaction products by agarose gel electrophoresis.

*Comments:*

*The PCR procedure described here will efficiently and reproducibly introduce any desired change into a DNA fragment, including point mutations, deletions and insertions. The main disadvantage is related to the error rate of Taq polymerase, which can introduce unwanted mutations. This means that the entire amplified fragment needs to be sequenced, which is anyway always an excellent idea. Nevertheless, using fragments in the size range of 500 to 1000 bp and low dNTP concentrations (40 μM), we have rarely seen secondary mutations. If possible, we like to generate a restriction enzyme site within the mutated sequence which allows to rapidly score for the introduced mutation (see legend to Fig. 4).*

**Cloning of PCR products**

To remove buffers, primers and dNTPs, which would interfere with subsequent manipulations, PCR products are next purified over commercially available spin-columns from Qiagen (QIAquick) following the manufacturers protocol. The columns contain silica to which the DNA binds in the presence of high salt, whereas contaminants pass through the column and are removed with a quick wash. The DNA is eventually recovered from the membrane by elution with water. Optimal binding of DNA to silica membrane occurs at pH ≤7.5, whereas maximum elution efficiency is obtained with a pH between 7.0

and 8.5. With this procedure it is possible to purify single- and double-stranded PCR products ranging in size from 100 bp to 10 kb.

The next steps are to clone the PCR fragments and to check for the presence of the mutation. Thus, PCR fragments are digested with the appropriate restriction enzymes (Hind III and Bam HI in our example) and purified on an agarose gel using the GENECLEAN II Kit from BIO 101. The principle for this procedure is similar to the one described above for the purification of PCR fragments, in that DNA is bound to a silica matrix in the presence of high salt (>3 M) and is eluted with water. This means the DNA can be used without further purification for subsequent manipulations, i.e. ligation into an appropriate vector.

**Rapid DNA ligation**

To verify that PCR products have been properly purified and to estimate the amount of material recovered, an aliquot (1 μl) of the sample is analysed by agarose gel electrophoresis alongside known amounts of DNA standards. We use the rapid DNA ligation kit from Roche Molecular Biochemicals, which offers the advantage that ligations of sticky- or blunt-end DNA can be performed at room temperature in 5 min, thus avoiding lengthy incubations.

Procedure:

For a 10 μl reaction, mix at room temperature in a 1.5 ml eppendorf tube the following components in the following order:

1 μl vector DNA *[At a concentration of 50 ng/μl for a vector in the size range of 3 to 5 kb or at a concentration of 100 ng/μl for a larger vector].*

Up to 3 μl of insert DNA at a 5-fold or higher molar excess *[For a 0.3 to 0.5 kb fragment to be cloned in a 3 to 5 kb vector, use at least 25 to 100 ng/ligation, increase the amount according to the fragment size. For ligations to double-cut vectors even higher amounts may be needed].*

1 μl of buffer 2 (5x dilution buffer, vortex before using), 5 μl of buffer 1 (2x buffer, vortex before using) and 0.5 μl T4 DNA ligase.

Incubate at room temperature for 5 min and transform an aliquot of the ligation mixture into competent bacterial cells.

*Comments:*

*DNA to be ligated must be in water. For restriction digests or DNA samples which have been treated with Klenow to polish the ends, ethanol precipitate before proceeding. Glycogen can be added to improve recovery, especially* *after polishing the ends. It is a useful practice to always set up one control ligation with the vector DNA by itself and for each insert DNA to set up two ligations with 1x and 5x the amount of fragment DNA. Leftover ligations should be stored at -20$^0$C. The crucial component in this kit is buffer 1 (composition ????). Buffer 2 can be replaced with TE and the enzyme can be replaced with another commercially available T4 DNA ligase.*

**Introduction of plasmid DNAs into *E. coli* cells**

Transformation-competent *E. coli* cells can be purchased from many manufacturers. However, the preparation of competent cells is quite simple and provides good to excellent transformation efficiencies.

**Preparation of competent *E. coli* cells**

Keep a frozen stock of bacterial cells, like DH5α, at -70$^0$C.

Streak bacteria from frozen stock on LB plate (no antibiotic!). Store plate at room temperature and do not use bacteria older than one week.

Inoculate one single colony in 3 ml LB overnight at 37$^0$C. The next day, inoculate 250 ml of warm (37$^0$C) LB with 2.5 ml of the overnight culture.

Grow at 37$^0$C for about 3 hr or until $OD_{550}$ is between 0.4 and 0.6.

Place flask with bacteria in an ice bucket for 10 to 60 min to cool down. Pellet bacteria at 3,000 rpm for 15 min at 4$^0$C.

Pour off supernatant, drain briefly and resuspend cells in 1/3 volume of FB and leave on ice for 10 to 60 min.

Spin at 3,000 rpm for 15 min at 4$^0$C. Discard supernatant and gently resuspend cells in 1/12.5 volume of FB.

Aliquot cells into ice-cold eppendorf tubes, freeze in dry ice and store at -70$^0$C.

FB solution: For 500 ml use 3.73 g KCl, 3.68 g $CaCl_2.2H_2O$, 50 g (about 50 ml) glycerol, and 0.49 g K-acetate. Adjust pH to 6.4 with HCl and filter sterilise. Final concentrations: 100 mM KCl, 50 mM $CaCl_2.2H_2O$, 10% (w/v) glycerol and 10 mM K-acetate.

*Comments:*

*Assess the competency of the cells by transforming 100 μl of cells with 10 ng of an appropriate plasmid vector, i.e. pBluescript, according to the protocol given below. As a negative control, plate cells directly without plasmid transformation.*

**DNA transformation of competent *E. coli* cells**

Label 15 ml conical Falcon tubes and precool on ice for a few min. Thaw competent cells in between your fingers and set on ice. Distribute 100 μl cells per tube.

Add 3 μl of the ligation reaction and mix gently with the tip of the pipette, incubate on ice for 20 min.

Set 2 bacterial plates, containing the appropriate antibiotic, per ligation mix to dry with their top ajar in a $37^0$C incubator.

After the cells have incubated on ice for 20 min, heat shock them at $42^0$C for 45 sec and return cells to ice for 2 min.

Add 0.9 ml LB (NO antibiotic!!!) to cells and shake at $37^0$C at 225 rpm for 1 hr.

Plate 0.1 ml and 0.9 ml of the transformed cells onto two different plates. Do not use force when plating transformed bacteria, be gentle. Keep the plates on the bench till the liquid has been absorbed. Incubate overnight at $37^0$C.

*Comments:*
*Competent cells should be used immediately, once thawed. Leftover cells can be refrozen. However, transformation efficiency will drop by several logs. Transformed cells can be stored on ice, if not plated immediately, or in the fridge overnight or even for a few days.*

**Analysis of recombinant plasmids**

Once bacterial transformants are obtained, the next step will be to verify the predicted structure of the recombinant plasmids by PCR. If the introduced mutations generated a new restriction enzyme site, its presence can be monitored by digesting PCR-amplified fragments with the appropriate restriction enzyme. Alternatively, PCR-amplified DNA can be sequenced directly or plasmid DNA can be prepared from a small overnight culture for subsequent analysis.

**PCR screening of bacterial colonies**

Prepare a 50 μg/ml solution of proteinase K in TE (10 mM Tris-HCl, pH 7.5, 0.1 mM EDTA) and aliquot 10 μl in each PCR tube.

Pick an individual colony from a plate using a plastic tip (do not use toothpicks, because they release some inhibitory substance) and lightly touch the surface of a new Ampicillin plate. Disperse whatever is left attached onto the tip in the 10 μl at the bottom of the PCR tubes. Repeat with nine more colonies. Set the plate containing the array of picked colonies in a bacterial incubator.

Place the tubes in a PCR machine. Incubate at 55$^0$C for 15 min and then at 80$^0$C for 15 min. Remove and briefly spin down the content.

While the tubes are incubating prepare the mix for PCR and aliquot 9 µl into 10 clean PCR tubes (or use a microtiter dish).

*For 10 tubes combine 11 µl 10x PCR buffer, 2 µl each of oligo 1 (100 ng/µl) and oligo 2 (100 ng/µl), 2.5 µl 2 mM dNTPs, 82 µl $H_2O$, and 1 µl Taq DNA polymerase.*

Add 1µl from each colony lysate to the prepared tubes containing the PCR mix, and perform 30 PCR cycles using the following parameters: 94$^0$C for 30 sec, 50$^0$C for 1 min and 72$^0$C for 1 min.

At the end of the PCR, add 2 µl agarose gel loading dye and separate samples on an agarose gel.

**Boiling miniprep DNA**

Pellet one eppendorf tube full of a bacterial overnight culture (30 sec top speed), discard supernatant and resuspend pellet in 360 µl STET (you can vortex).

Add 40 µl lysozyme (10 mg/ml in distilled water), vortex for 1 sec, and boil immediately for 1 min (caps open) in a water bath!!!

Spin 15 min at room temperature in eppendorf centrifuge at top speed, and remove pellet with a wooden toothpick and discard.

Add 400 µl isopropanol to supernatant, mix by vortexing, spin 10 min at room temperature at top speed and discard supernatant.

Wash pellet with 70% ethanol (add 200 µl 70% ethanol, spin briefly and remove supernatant), dry in speed vac, and resuspend in 50 µl $H_2O$ containing 20 µg/ml RNase A.

STET solution: for 100 ml add 8 g sucrose, 5 g Triton X-100 (weight), 10 ml 0.5 M EDTA (pH 8), 5 ml 1 M Tris-HCl (pH 8) and filter sterilise.

*Comments:*

*The boiling method is extremely simple and fast and yields enough DNA for several restriction digests, or a couple of DNA sequencing reactions. Boiling in a water bath, as compared to incubation in a heating block, is important for instantaneous lysis. Wooden toothpicks are essential, since the pellet of bacterial debris and chromosomal DNA does not attach efficiently to plastic tips.*

**Isolation of high quality plasmid DNA**

We have tested a number of commercially available plasmid isolation kits, but so far have not been able to reproducibly isolate DNA of a quality acceptable for transfection of *T. brucei* cells. Thus, we isolate plasmid DNA by CsCl/ethidium bromide equilibrium centrifugation. The following procedure was optimised for 200 ml of bacterial culture grown to saturation overnight. *[This will yield between 0.5 and 2 mg of plasmid DNA.]* For processing of different size cultures, it is necessary to adjust the volumes accordingly.

Procedure:

Spin down cells at 5,000 rpm for 10 min at $4^0$C. Resuspend cells in 10 ml 50 mM Tris-HCl (pH 8.0), 20% sucrose and put on ice, add 1.2 ml lysozyme (10 mg/ml in $H_20$ prepared fresh the same day) and 0.6 ml 0.5 M EDTA (pH 8.0) and keep on ice for 10 min.

Add 12 ml of a solution containing 0.2% Triton X-100, 50 mM Tris-HCl (pH 8.0), 25 mM EDTA (pH 8.0) and RNAse A to 10 μg/ml final concentration. Mix gently, but thoroughly (solution will become viscous) and leave at room temperature for about 15 min. Transfer to Oak ridge tubes or comparable centrifuge tubes and spin at 18,000 rpm for 30 min ($4^0$C).

Decant supernatant to a disposable 50 ml polypropylene tube, adjust volume with $H_2O$ to 24 ml, add 0.6 ml of 20% SDS (to 0.5% final concentration) and incubate at $65^0$C for 30 min in a water bath.

*In the meantime discard pellet and rinse centrifuge tubes with distilled water, since they are needed in the next step.*

Add 4 g K-acetate to each rinsed tube and pour the SDS mixture into them. Mix well and keep on ice for 15 min (a white precipitate will form).

*In the meantime rinse the 50 ml polypropylene tubes with distilled water for reuse.*

Spin at 18,000 rpm for 15 min ($4^0$C) and transfer supernatant to rinsed tubes, adjust volume to 25 ml with distilled water and add 25 ml isopropanol. Mix well and keep on ice for 10 min.

Spin at 5,500 rpm for 10 min, remove supernatant, drain, dry briefly in Speed Vac. Resuspend pellet in 4.3 ml TE and add 4.3 g cesium chloride. Once the salt is dissolved, add 0.1 ml of 10 mg/ml ethidium bromide.

Transfer to Quickseal tubes (Beckman), seal the tubes and spin in ultracentrifuge in a VTi65.2 rotor for 6 to 20 hr at 45,000 rpm at 20°C.

Remove plasmid band, the lower of the two bands, from gradient with syringe and needle and transfer to a 15 ml conical tube. Add 2 volumes $H_2O$ and then 2.5 volumes ethanol, mix and spin 10 min at 5,500 rpm. Remove supernatant, wash pellet with 70% ethanol (do not dry pellet).

*Most often two bands are visible with the lower plasmid band being more intense. RNA will be found at the bottom of the tube or as a pellet alongside the tube. Large amounts of plasmid DNA will be visible in regular light, whereas smaller amounts might have to be visualised by side illumination with short-wave UV.*

Resuspend pellet in 0.9 ml TE and add 0.1 ml 20x SET. Extract with 0.5 ml phenol, vortex, add 0.5 ml chloroform, vortex, spin 5 min at 5,500 rpm.

Remove and discard organic bottom phase and extract aqueous phase with 1 ml of chloroform, vortex, spin and transfer top aqueous phase to a new conical tube and precipitate with 2.5 volumes of ethanol.

Spin 10 min at 5,500 rpm, wash pellet with 70% ethanol, dry and resuspend pellet in 1 ml $H_2O$. Measure $OD_{260}$ of a 1:100 dilution in $H_2O$ to determine the DNA concentration.

**Transient DNA transfection of procyclic *T. brucei* cells**

The effect of the introduced mutations are tested in vivo by transient DNA transfection of procyclic *T. brucei rhodesiense* cells (strain YTat 1.1). The plasmid DNA are introduced into trypanosome cells by electroporation and 3 hr after transfection, the cells are collected and total RNA is prepared. Analysing a time course of RNA accumulation, we found that RNA originating from transfected DNA can be detected as early as 2 hr after transfection and the amount does not increase substantially after 4 hr. In order to control for transfection efficiency, which in our hands varies between 10 and 20%, we routinely co-transfect each construct with a plasmid containing the gene coding for the U6 small nuclear RNA gene. This U6 snRNA gene is marked by the insertion of a short linker sequence, which allows us to monitor expression by primer extension analysis.

Procedure:

The day before the experiment, dilute cells in such a way that they will be just about reaching saturation at the time you decide to transfect them. This is usually 1:5 or 1:6, depending on the strain and media used and how dense the starting culture is. You will need $10^8$ cells per transfection.

Aliquot 10 ml of complete medium (i.e. Cunningham's medium with 20% FCS) in the appropriate number of tissue culture flasks. Aliquot DNA solution (20 to 100 μg) to be tested in individual eppendorf tubes and adjust sample volume to 100 μl with distilled water. Prepare and label the required number of 0.4 cm gap cuvettes for electroporation.

Spin down the cells ($10^8$/transfection) at 3,000 rpm for 5 min. Discard the supernatant and resuspend cells in 1/2 the original volume of cytomix.

After another centrifugation, resuspend cells in 1/20 the original cell volume of cytomix and aliquot 0.5 ml of cells/ DNA sample in the prepared eppendorf tubes.

Transfer cells and DNA to a 0.4 cm gap cuvette with a sterile pipette, cap loosely, and zap twice at 25 μF, 1500 V, waiting 10 seconds in-between zaps.

*These conditions are for a BioRad gene pulser and need to be adjusted when using a different brand of electroporator.*

Transfer cells immediately to medium, rinse the cuvette with some medium from the same flask and incubate at $28^{o}C$.

Cytomix solution: In about 450 ml mix 4.47 g KCl, 0.5 ml of 150 mM $CaCl_2$, 0.87 g $K_2HPO_4$, 2.97 g HEPES, 2 ml of 0.5 M EDTA (pH 8), and 2.5 ml of 1 M $MgCl_2$. Adjust pH to 7.6 with KOH and volume to 500 ml. Filter sterilise.

Final concentration: 120 mM KCl, 0.15 mM $CaCl_2$, 10 mM $KH_2PO_4$, 25 mM HEPES, 2 mM EDTA, 5 mM $MgCl_2$.

**RNA isolation from transfected procyclic trypanosome cells**

*[Each $2x10^6$ trypanosomes yield approximately 1-2 micrograms of total RNA.]*

Spin down cells (10 ml) in 15 ml sterile conical Falcon tube for 5 min at 3,000 rpm at room temperature. Resuspend cells in 1 ml washing buffer (100 mM NaCl, 20 mM Tris-HCl pH 7.5 and 3 mM $MgCl_2$), transfer to 1.5 ml microfuge tube and spin at 4,000 rpm for 1 min. Repeat wash (serum contains DNAse and RNAse). Remove supernatant.

Lyse cells by adding 1 ml TRIZOL Reagent (GibcoBRL), vortex 2x 1 min to brake the DNA and thus reduce the viscosity of the mixture (this also improves recovery of RNA). Leave at room temperature for 10 min. Add 200 µl of chloroform and vortex again for about 10 sec. Keep on ice for 5 min and then separate phases by spinning for 10 min in eppendorf centrifuge (12,000 x g) at 4°C.

After centrifugation, the mixture separates into a lower red, phenol-chloroform phase, an interphase and a colourless upper aqueous phase. RNA remains exclusively in the aqueous phase. Carefully transfer aqueous phase to a new eppendorf tube. Make sure to leave interphase behind! Precipitate RNA by adding 0.5 ml isopropanol, mix several times by inversion and spin in eppendorf centrifuge for 10 min at 4°C (12,000 x g).

*Do not let samples precipitate for longer than 10 min at room temperature, because guanidinium and proteins will start precipitating.*

Remove supernatant, add 1 ml 70% ethanol. Spin for 5 min at 12,000 x g, remove supernatant completely (last few microliters can be removed with a micropipette). Dry pellets in Speed Vac.

*The following steps are essential to completely remove endogenously added plasmid DNAs that otherwise would interfere with subsequent procedures, like RACE and primer extension analysis.*

Resuspend pellet in 180 µl of sterile water by heating at 65°C for 5 min, vortex. If the pellet does not go in solution readily, try to break it by aspirating the solution up and down through a yellow tip attached to a pipetman. Once pellet is in solution add 20 µl of NEB Buffer 2, and 4 µl of DNAse I (RNase free; Roche Molecular Biochemicals). Incubate at 37°C for 1 hr.

Add 20 µl of 20x SET and 220 µl of buffered phenol, vortex, spin for 1 min in eppendorf centrifuge, remove phenol from the bottom and discard. Add an equal volume of chloroform, vortex and spin as above. Remove aqueous phase to a new tube and add 2.5 volumes ethanol. Mix by inversion and store at -20°C, if not to be used immediately, otherwise proceed with the next step.

When ready to use the RNA, spin desired amount of ethanol precipitate in eppendorf centrifuge for 10 min, rinse pellet with 70% ethanol, dry in

Speed Vac and resuspend in appropriate volume of sterile water (make sure RNA is resuspended, see above).

*Comments:*

*Note that RNA is best stored in ethanol. RNA in water should only be stored for short periods of time. From a transfection as described above ($10^8$ cells), you will end up with 770 μl of ethanol precipitate. 100 μl of this ethanol precipitate corresponds to about 10 μg of total RNA. We use 10 μg of total RNA for primer extension analysis and 5 μg each for preparing cDNA for 3' and 5' end RACE. For Northern blot we use 10 μg of total RNA per lane.*

**RNA analysis by 5' end RACE**

To analyse the RNAs generated we use Rapid Amplification of cDNA Ends (RACE), which is a procedure for amplification of nucleic acid sequences from a messenger RNA template between a defined internal site and either the 3' or the 5' end of the mRNA. In 5' end RACE we take advantage of the fact that all mRNAs in *T. brucei* contain the same 39 nt sequence at the very 5' end, thus providing a convenient anchor. In this procedure, mRNAs are converted into cDNA using reverse transcriptase and a primer specific for the CAT sequences (CAT-In). cDNAs are then amplified by PCR using the CAT-5 primer and a primer that targets the SL sequence (Fig. 6).

Procedures:

**First strand cDNA synthesis**

Combine 5 μg of total RNA and sterile water to a final volume of 11 μl in a 0.5 ml microcentrifuge tube and add 1 μl of CAT-in (50 ng/μl).

Heat at 70°C for 10 min and chill on ice for at least a couple of min.

Add the following components: 2 μl 10x PCR buffer (200 mM Tris-HCl [pH 8.4], 500 mM KCl), 2 μl 25 mM $MgCl_2$, 1 μl 10 mM dNTPs and 2 μl 0.1 M DTT.

Incubate the mixture at 42°C for 5 min and add 1 μl Superscript II reverse transcriptase (200 units/μl; GibcoBRL). Continue incubation at 42°C for 50 min.

Terminate reaction by incubation at 70°C for 15 min, chill on ice and add 1 μl *E. coli* RNase H (2 units/μl; GibcoBRL). Incubate at 37°C for 20 min.

*Digestion with RNase H removes the RNA template from the cDNA:RNA hybrid molecule to increase the sensitivity of PCR. At* *this stage the reaction can be stored at -20$^0$C, before proceeding with the PCR reaction.*

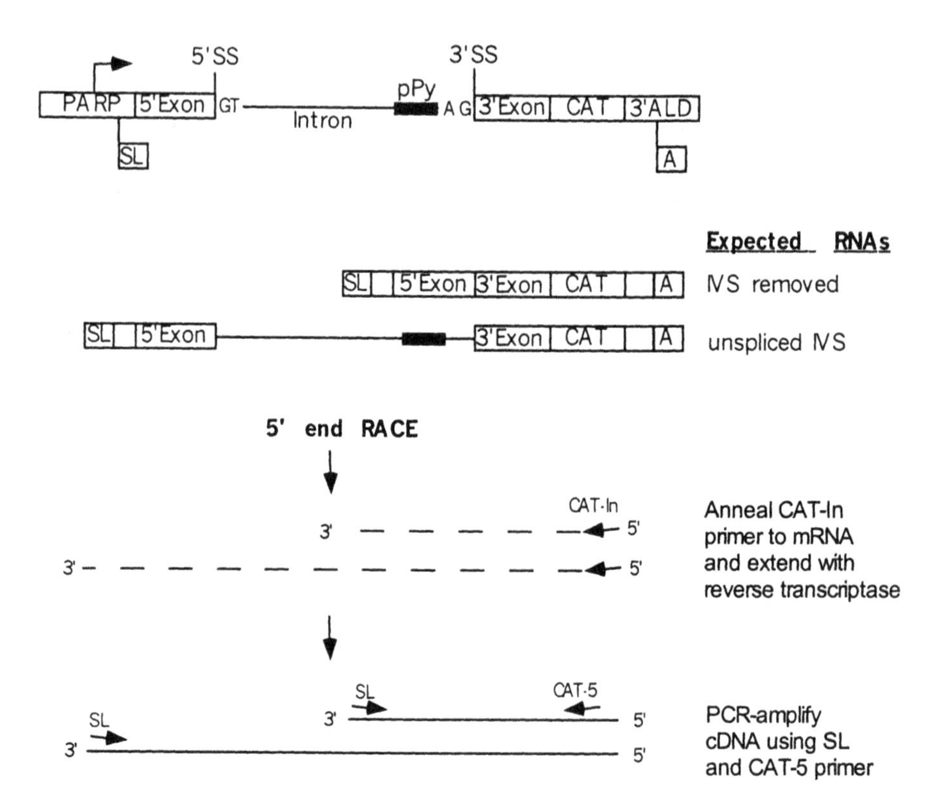


**Figure 6.** 5' end RACE analysis. The pPAP.CAT construct is shown at the top and the structure of the expected RNAs is depicted below.

## PCR amplification of the target cDNA

As we any PCR reaction, optimal conditions will have to be determined empirically and will depend on the target sequence and the particular primers used. Adjustment of the magnesium ion, dNTP, or cDNA concentration may be required. In addition, alterations of the cycling protocol, in particular changing the annealing temperature, might improve product yields.

In a 0.5 ml centrifuge tube combine 5 µl 10x PCR buffer, 3 µl 25 mM $MgCl_2$, 37 µl $H_2O$, 1 µl CAT-5 oligo (100 ng/µl), 1 µl Eco-SL oligo (100 ng/µl), 1 µl 10 mM dNTP, and 1 µl template cDNA.

*We generally set-up two separate reactions: one with 1 µl directly from the cDNA synthesis reaction and the other with 1 µl of a 1:10 dilution of the cDNA.*

Using a thermal cycler, incubate the reaction at 94°C for 3 min, pause and add 0.5 µl Taq DNA polymerase (5 units/µl).

Perform 35 cycles of PCR with 30 seconds at 94°C, 1 min at 60°C and 1 min at 72°C.

*For this particular amplification, increasing the annealing temperature from 55°C to 60°C drastically increased the purity of the PCR products.*

Analyse products on an agarose gel using appropriate molecular weight size markers.

**Primer extension analysis**

To monitor transfection efficiency, we cotransfect the experimental samples with a plasmid encoding a marked U6 snRNA gene. *[Generally, 50 to 100 µg of the test construct is cotransfected with 20 µg of the U6 plasmid.]* The expressed U6 snRNA is then assayed by primer extension analysis, where a complementary $^{32}$P-labelled oligonucleotide is annealed to the RNA of interest and then the RNA is copied into cDNA with reverse transcriptase. Thus, the cDNA extends from the oligonucleotide priming site to the very 5' end of the RNA, and the amount of the extension product reflects the abundance of the RNA.

Procedure:

In a 1.5 ml eppendorf tube assemble 10 µl RNA (corresponding to 10 µg total RNA), 4 µl 5x first strand buffer (250 mM Tris-HCl [pH 8.3], 375 mM KCl, 15 mM $MgCl_2$) and 1 µl $^{32}$P-labelled oligonucleotide.

Incubate at 70°C for 10 min and then place on ice for 10 min. Add 2 µl 0.1 M DTT, 1 µl 10 mM dNTP and 0.5 µl Superscript II reverse transcriptase (GibcoBRL) and incubate at 45°C for 45 min.

To stop the reaction, add 20 µl 0.6 N NaCl/20 mM EDTA and precipitate with 120 µl of ethanol. Wash with 70% EtOH, dry briefly and resuspend in urea dye.

Separate primer extension products on a 6% acrylamide-7M urea gel.

*Comments:*
*One major problem with primer extension analysis is the pausing or stopping of the reverse transcriptase due to secondary structures within the RNA. These artefacts can be reduced by choosing a primer close to the 5' end (within 100 nucleotides) and by performing the reaction at a high temperature (45°C). Methylated cap structures at the very 5' end of the RNA often are the cause of several products varying in size by one nucleotide.*

## RESULTS AND DISCUSSION

During the Molecular Biology Section of the 1999 Biology of Parasitism Course, Hernan O. Aviles, Joseph R. Bishop, Franco H. Falcone, Cristina Gavrilescu, Jacqui L. Montgomery, M. Isabel Santori, Leah S. Stern, and Zefeng Wang introduced and tested eight mutations in the *T. brucei* PAP intron, following the protocols outlined in this chapter. The results of these experiments are displayed in Figure 7. 5' end RACE analysis of RNA isolated from cells transfected with the wild-type construct (pPAP.CAT; lane 11) resulted in four bands, which by direct sequence analysis had following the structure. As predicted from its size, the 307-bp fragment was diagnostic of RNA molecules, where the intron was removed and the SL sequence was added at the 5' end by trans-splicing. The 960-bp fragment represented RNA molecules that were trans-spliced at the 5' end, but the intron was still present, whereas the 543-bp fragment originated from RNA that was cis-spliced at a cryptic site in the intervening sequence. The shortest PCR fragment (178 bp) originated from RNA trans-spliced at the 3' splice site of the intron, an event that also occurs in the processing of the endogenous PAP pre-mRNA (Mair et al., 2000).

Four of the mutations (M1 to M4) were designed based on a sequence comparison between the 5' splice site of the *T. brucei* and *T. cruzi* PAP intron, which revealed that eleven nucleotides, including the GT motif, are conserved (Fig. 7). Furthermore, by analogy with cis-splicing in other eukaryotes, we would predict that these sequences interact through base-pairing with the 5' end of the U1 snRNA (Ares and Weiser, 1995). Indeed, the mutational analysis highlighted the significance of these sequences, since alterations dramatically reduced cis-splicing (Fig. 7, lanes 1 to 4). A mutation of the AG dinucleotide at the 3' splice site resulted in no detectable cis-splicing, as well as no trans-splicing at the 3' splice site (lane 7). The replacement of a candidate polypyrimidine tract did not significantly affect splicing, but led to the activation of a cryptic 3' splice site in the intron (lane 5). On the other hand, mutation of a putative branch point sequence (lane 6) and of the first 20 nucleotides of the PAP 3' exon (lane 8) had no detectable effect.

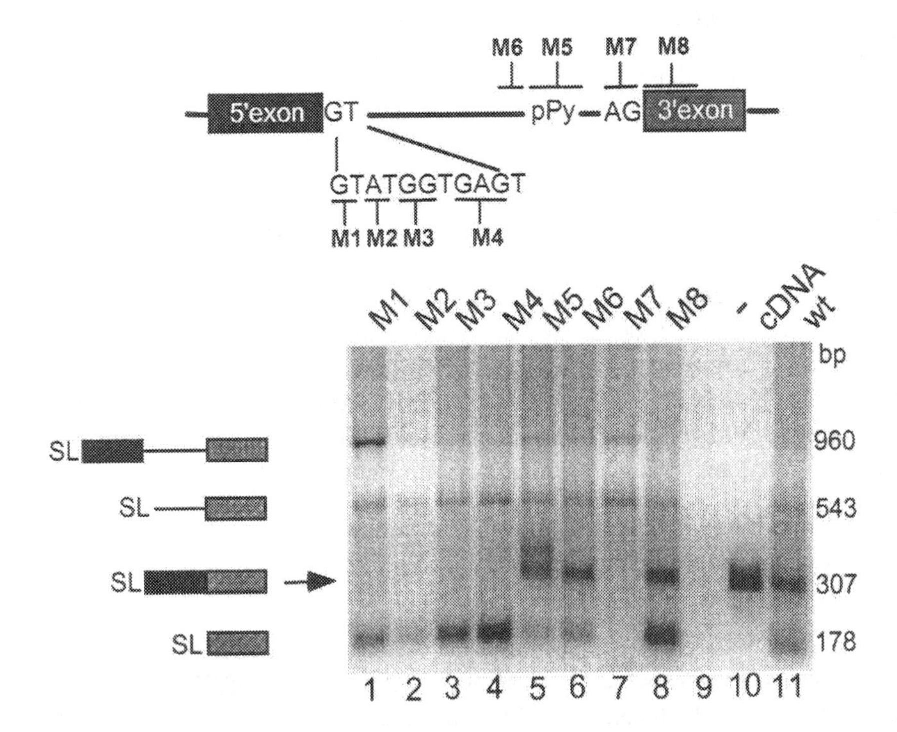


**Figure 7.** Sequences required for removal of the poly(A) polymerase intron in *T. brucei*. A schematic diagram of the mutations is shown at the top. Procyclic *T. brucei* cells were transfected with the DNA constructs as indicated above each lane and 3 hr post-transfection total RNA was isolated and analysed by 5' end RACE as outlined in Fig. 6. PCR products were separated on a 3% agarose gel. As a control, RNA was isolated from cells transfected with vector DNA and a construct lacking the intervening sequence in lanes 9 and 10, respectively. The various PCR products were gel purified and sequence analysis revealed the structure depicted on the left.

Taken together, accurate removal of the PAP intervening sequence requires at least two signals: a conserved sequence motif at the 5' splice site and the AG dinucleotide at the 3' splice site. Interestingly, the 5' splice site sequences have the potential to form an extensive base-pair interaction with the recently identified candidate U1 snRNA in *Crithidia fasciculata* and *Leishmania tarentolae* (Schnare and Gray, 1999). To provide experimental evidence for this interaction, it will be necessary to isolate the *T. brucei* U1 snRNA and to test whether compensatory mutations in the respective positions of the U1 snRNA 5' end can suppress the effect of the 5' splice site mutations.

## CONCLUSION

The approaches described here can be applied directly to a variety of experimental systems. Indeed, we have studied the promoter structure of the U2 and U6 snRNA genes (Fantoni et al., 1994; Nakaar et al., 1994) and have performed an extensive analysis of the pre-mRNA signals involved in trans-splicing in *T. brucei* (Lopez-Estrano et al., 1998; Matthews et al., 1994) using mutational analysis in combination with transient DNA transfection.

## REFERENCES

Ares, M., Jr., and Weiser, B. (1995). Rearrangement of snRNA structure during assembly and function of the spliceosome. Prog. Nucleic Acid. Res. Mol. Biol. *50*, 131-159.

Bangs, J. D., Crain, P. F., Hashizume, T., McCloskey, J. A., and Boothroyd, J. C. (1992). Mass spectrometry of mRNA cap 4 from trypanosomatids reveals two novel nucleosides. J. Biol. Chem. *267*, 9805-9815.

Boothroyd, J. C., and Cross, G. A. (1982). Transcripts coding for variant surface glycoproteins of *Trypanosoma brucei* have a short, identical exon at their 5' end. Gene *20*, 281-289.

Fantoni, A., Dare, A. O., and Tschudi, C. (1994). RNA polymerase III-mediated transcription of the trypanosome U2 small nuclear RNA gene is controlled by both intragenic and extragenic regulatory elements. Mol. Cell. Biol. *14*, 2021-2028.

Hajduk, S. L., Adler, B., Madison-Antenucci, S., McManus, M., and Sabatini, R. (1997). Insertional and deletional RNA editing in trypanosome mitochondria. Nucleic Acids Symp. Ser., 15-18.

Hertel, K. J., Lynch, K. W., and Maniatis, T. (1997). Common themes in the function of transcription and splicing enhancers. Curr. Opin. Cell. Biol. *9*, 350-357.

Krause, M., and Hirsh, D. (1987). A trans-spliced leader sequence on actin mRNA in *C. elegans*. Cell *49*, 753-761.

Kunkel, T. A. (1985). Rapid and efficient site-specific mutagenesis without phenotypic selection. Proc. Natl. Acad. Sci. U S A *82*, 488-492.

Lopez-Estrano, C., Tschudi, C., and Ullu, E. (1998). Exonic sequences in the 5' untranslated region of alpha-tubulin mRNA modulate trans splicing in *Trypanosoma brucei*. Mol. Cell. Biol. *18*, 4620-4628.

Mair, G., Shi, H., Li, H., Djikeng, A., Aviles, H.O., Bishop, J.R., Falcone, F.H., Gavrilescu, C., Montgomery, J.L., Santori, M.S., Stern, L.S., Wang., Z., Ullu, E. and Tschudi, C. (2000). A new twist in trypanosome RNA metabolism: cis-splicing of pre-mRNA. RNA, in press.

Matthews, K. R., Tschudi, C., and Ullu, E. (1994). A common pyrimidine-rich motif governs trans-splicing and polyadenylation of tubulin polycistronic pre-mRNA in trypanosomes. Genes Dev. *8*, 491-501.

Nakaar, V., Dare, A. O., Hong, D., Ullu, E., and Tschudi, C. (1994). Upstream tRNA genes are essential for expression of small nuclear and cytoplasmic RNA genes in trypanosomes. Mol. Cell. Biol. *14*, 6736-6742.

Ngo, H., Tschudi, C., Gull, K., and Ullu, E. (1998). Double-stranded RNA induces mRNA degradation in *Trypanosoma brucei*. Proc. Natl. Acad. Sci. U S A *95*, 14687-14692.

Nilsen, T. W. (1994). RNA-RNA interactions in the spliceosome: unraveling the ties that bind. Cell *78*, 1-4.

Rajkovic, A., Davis, R. E., Simonsen, J. N., and Rottman, F. M. (1990). A spliced leader is present on a subset of mRNAs from the human parasite *Schistosoma mansoni*. Proc. Natl. Acad. Sci. U S A *87*, 8879-8883.

Reed, R. (1996). Initial splice-site recognition and pairing during pre-mRNA splicing. Curr. Opin. Genet. Dev. *6*, 215-220.

Schnare, M. N., and Gray, M. W. (1999). A candidate U1 small nuclear RNA for trypanosomatid protozoa. J. Biol. Chem. *274*, 23691-23694.

Simpson, L., and Emeson, R. B. (1996). RNA editing. Annu Rev Neurosci *19*, 27-52.

Stuart, K., Allen, T. E., Heidmann, S., and Seiwert, S. D. (1997). RNA editing in kinetoplastid protozoa. Microbiol. Mol. Biol. Rev. *61*, 105-120.

Tessier, L. H., Keller, M., Chan, R. L., Fournier, R., Weil, J. H., and Imbault, P. (1991). Short leader sequences may be transferred from small RNAs to pre-mature mRNAs by trans-splicing in *Euglena*. EMBO J. *10*, 2621-2625.

Ullu, E., Tschudi, C., and Gunzl, A. (1996). Trans-splicing in trypanosomatid protozoa. In *Molecular biology of parasitic protozoa*, D. F. Smith and M. Parsons, eds. (Oxford, United Kingdom: Oxford University Press), pp. 115-133.

Wirtz, E., Leal, S., Ochatt, C., and Cross, G. A. (1999). A tightly regulated inducible expression system for conditional gene knock-outs and dominant-negative genetics in *Trypanosoma brucei*. Mol. Biochem. Parasitol. *99*, 89-101.

Zahler, A. M., Lane, W. S., Stolk, J. A., and Roth, M. B. (1992). SR proteins: a conserved family of pre-mRNA splicing factors. Genes Dev. *6*, 837-847.

# 3

# STABLE TRANSFECTION OF *PLASMODIUM BERGHEI*: A CRASH COURSE

Andrew P. Waters
*Department of Parasitology, Leiden University Medical Centre, The Netherlands*

## OVERVIEW AND BACKGROUND INFORMATION

Genetic engineering is an experimental approach that is applied to a wide number of model organisms with the aim of uncovering detailed information about the molecular processes that underpin life at the unicellular and multicellular levels. Most organisms of medical interest are far from model organisms and therefore investigators may have to apply a large amount of ingenuity in order to be able to bring genetic engineering to bear on their organism of choice. Genetic engineering is initiated through transfection, which is defined as the experimental ability to introduce foreign DNA into a cell and following the specific utilisation and/or interaction of that DNA with the genetic apparatus of the cell.

The disease malaria is caused by unicellular apicomplexan protozoal parasites, genus *Plasmodium*, and causes literally incalculable morbidity and mortality. Two thirds of the world's population is at risk of malaria and best guesses at the mortality lie in excess of 2 million per annum. Thus, the aim of genetic engineering of malaria parasites lies in gaining a deeper understanding of the basic processes and strategies that are employed by the parasite to successfully complete its life cycle. These include numerous colonisations of different cell types and an ability to avoid or neutralise the immune systems of both host and *Anopheline* mosquito vector. All features of the parasite might be studied by this approach and the information gained might eventually prove central to the design of effective drugs, immune therapies or perhaps even disabled parasites that might serve as attenuated vaccines.

The history of malaria parasite transfection is brief, beginning seven years ago. It was initially demonstrated that it was possible to introduce by electroporation exogenous DNA encoding firefly luciferase into the extracellular post-fertilisation stages of the avian malaria parasite *Plasmodium gallinaceum*, and within 24 hours observe luciferase activity (Goonewardene et al., 1993). The enzyme was produced as a fusion protein with the ookinete-specific surface protein Pgs28 under control of the homologous promoter and demonstrated for the first time the ability to introduce exogenous genetic material into a *Plasmodium* parasite with subsequent expression.

Work towards the transfection of intracellular parasite forms proceeded on two fronts with those seeking to transiently transform *Plasmodium* with reporter genes and those seeking to obtain drug selectable transformation. Both approaches were coincidentally successful. Early ring stages and trophozoites of *P. falciparum* were successfully transformed with transient expression constructs that expressed chloramphenicol acetyl transferase (Wu et al, 1995). A separate study transformed merozoites and schizonts of the rodent malaria *P. berghei* with a gene encoding a drug resistant form of the enzyme involved in DNA metabolism, dihydrofolate reductase/thymidylate synthase (DHFR-TS). Transformed parasites were successfully selected on the basis of their acquired resistance to pyrimethamine, a specific antimalarial and inhibitor of DHFR-TS (van Dijk et al., 1995). Both studies used electroporation with high field strengths and showed that the additional membranes need not pose an insurmountable problem. All studies used homologous promoters of transcription, untranslated structural mRNA regions and terminators of transcription to generate expression of the introduced transgenes (a gene expressed in a genetically transformed organism) and to date it has proved unreliable to use promoters derived from species other than *Plasmodium*. Subsequently transformed *P. falciparum* was also drug-selected (Wu et al, 1997) and integration demonstrated in both *P. berghei* (van Dijk et al., 1996) and *P. falciparum* (Wu et al., 1996; Crabb and Cowman, 1997). Both *P. berghei* and *P. falciparum* do not readily utilise promoters from other species (Crabb et al., 1997; Tomas et a,. 1998). The simian parasite *P. knowlesi*, the last parasite to be transfected, appears to accept all promoters from any species of *Plasmodium* (van der Wel et al.,

1997). This chapter attempts to demonstrate the necessary steps and experimental protocols that will allow one to successfully transform the *P. berghei* malaria parasite. What follows is a description of the various components.

## CONSTRUCTION AND USE OF VECTORS

### Selectable marker

The essential element for stable transfection is the drug selectable marker: a genetic element that ensures the appropriate production of a protein that confers a selective advantage under environmental conditions that can be experimentally controlled. The first (and until recently only) marker is the DHFR-TS enzyme activity that in the correct configuration (often the result of a single point mutation) can confer resistance to the antimalarial folate analogue, pyrimethamine. Current vectors use the resistant form of the enzyme contained in the genome of *Toxoplasma gondii,* but under control of homologous *Plasmodium* promoters and downstream elements. Thus, we have constructed a "selection cassette" consisting of the open reading frame (ORF) encoding the selectable marker and appropriate upstream and downstream sequences (taken from the *P. berghei* DHFR-TS to ensure that expression is entirely apposite) all contained in pUC/pBSKS backgrounds. This initial "transfection vector" has been engineered to create a range of vectors so that it is possible to clone additional parasite DNA into the same plasmid that can then be used either for targeted integration into the parasite genome or for the expression of transgenes.

### Transgene expression vectors

These vectors contain the selectable marker and an ORF expression cassette that contains promoters, untranslated regions (UTRs) and terminators of transcription as well as a unique site (or a small polylinker) that allows the appropriate introduction of the ORF of choice (Tomas et al., 1998, Crabb et al., 1997). This can be used to express transgenes for biological investigations, vaccine development and the discovery of drug resistance markers (or mechanisms of drug resistance, e.g. sulfadoxine; Triglia et al., 1998).

### Integration constructs

There are two classes of integration vector that exhibit distinct modes of entry into the *P. berghei* genome: i) insertion, via single crossover and ii) replacement via double crossover (Fig. 2). All recombination yet observed is homologous recombination. Random non-homologous recombination is not observed.

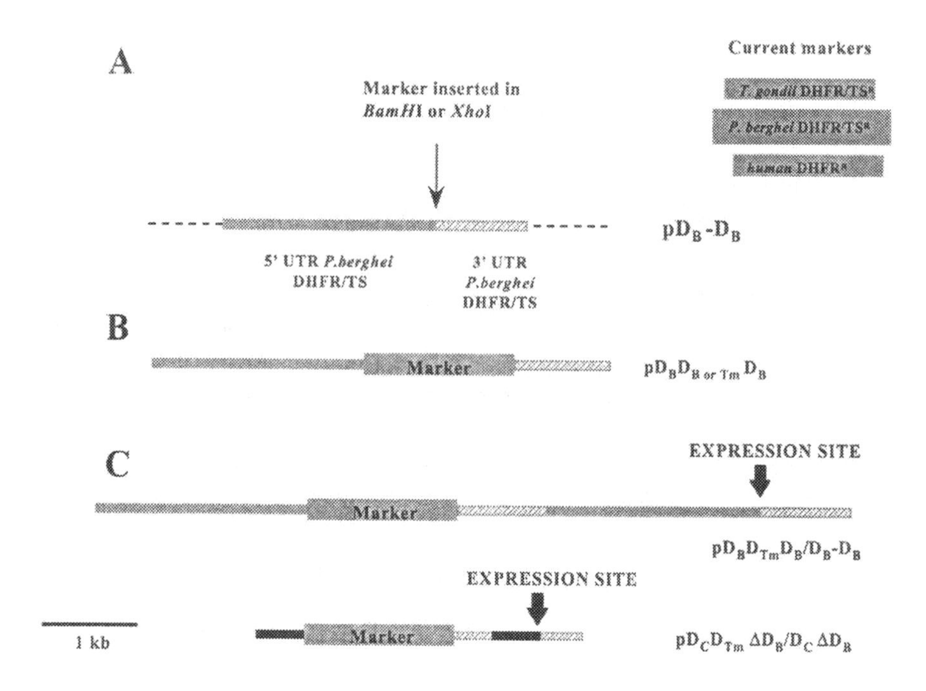


**Figure 1**. Vectors of the expression of transgenes in *P. berghei*. (A) The basic promoter/ terminator-containing vector into which an ORF encoding a selectable marker might be placed. Note all terminators and promoter regions also contain the relevant 5' or 3' untranslated regions, transcription start and termination sites as appropriate. (B) A constructed selectable marker expression cassette. (C) A selection cassette with an empty expression cassette alongside for placement and expression of a second ORF as a transgene. Different promoter combinations are always possible (compare the two examples in C). All vectors contain either a pBSKS or pUC backbone and therefore can be constructed and grown in *E. coli* selecting for resistance to ampicillin. Each element in the vector is identified by a standardised designation that indicates both the gene (D = *dhfr/ts*; A=*ama1*) and a subscript that indicates the species of origin of the element (B = *P. berghei*; Tm = *T. gondii* modified).

Each mechanism offers different experimental possibilities as a consequence of the different mechanisms of integration. i) Single cross over insertion integrates at a locus determined by the target site that is contained on the vector. The vector must be linearised within the target sequence and integration results in target site duplication and introduces the entire vector into the genome including the pUC plasmid vector backbone. Thus this vector class can be used to disrupt or modify genes. It should be noted that because of the duplication of the target sequence upon entry, the mechanism is potentially reversible. The control of such reversibility is a current goal of our research. ii) Vectors that enter the parasite genome by a double crossover event are more complicated to construct. They require two regions of homology with the parasite genome that flank the selectable marker and a greater repertoire of unique restriction sites than insertion both for cloning and to liberate the integrative fragment. These constructs are generally used to irreversibly delete a gene from the genome (especially true of a haploid genome) but can be used to replace genes as well. The limits of the extent of crossover (and therefore deletion) are being explored, but to date regions in excess of 5 kb have been deleted.

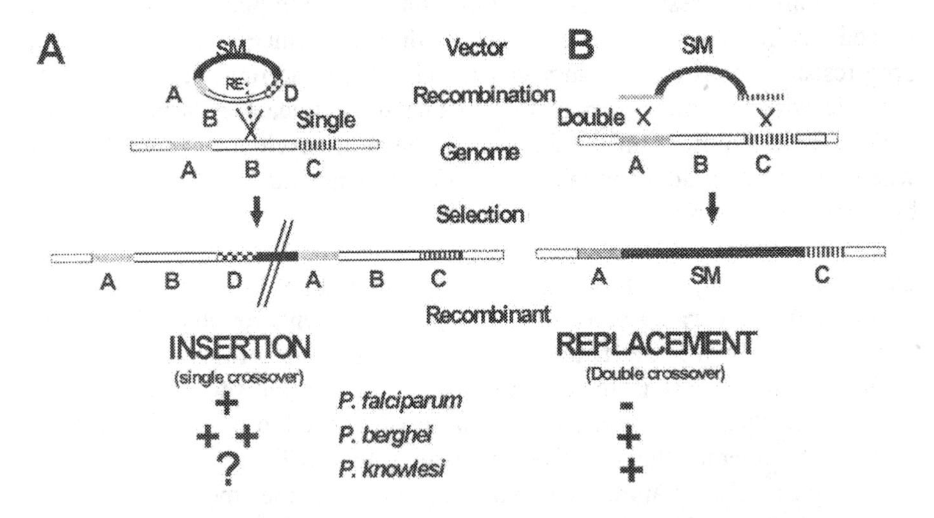


**Figure 2**. The two modes of DNA integration of foreign DNA into the *Plasmodium* genome, insertion and replacement. The modes supported by the different species of Plasmodium that can be transfected are illustrated at the bottom of the figure. SM = selectable marker; RE = unique restriction enzyme site. The remaining letters refer to the segments of an hypothetical gene and illustrate how modification of the gene might be achieved with insertion and disruption might be achieved with replacement.

Additional strategies are also possible for example, that fragment chromosomes and these have recently successfully been applied. The replacement mechanism of integration has not yet been demonstrated for *P. falciparum* and this, to some extent, limits the range of experimental possibilities.

## TRANSFECTION OF THE *P. BERGHEI* PARASITE

It is important to appreciate that transfection of the asexual bloodstage forms of the malaria parasite was always the most plausible option available to experimentalists. This is not so much due to the fact that these forms are responsible for the majority of the pathology associated with infection (and therefore of most clinical relevance) but rather due to the fact that they undergo a cycle (within a cycle) of invasion - development - multiplication - release - invasion of erythrocytes and that their numbers increase exponentially as a result. This fact allows one to design phenotypic selection procedures based on phenomena such as drug resistance. The transformed, drug resistant parasites will survive drug challenge and multiply. This is not possible with any other part of the (in one sense linear) malaria life cycle which would require propagation through the entire cycle to the bloodstages, where selection could again take place (this does not exclude other strategies but does make them less easy to achieve).

Unlike the human parasite *P. falciparum*, a continuous culture system is not available for *P. berghei* bloodstages. A partially asynchronous infection, *P. berghei* can however be matured in vitro to the point of schizont rupture, where it pauses. This is thought to be because the parasite actually relies upon the mechanical shearing forces that it is exposed to in the bloodstream in order to liberate the invasive merozoites that will colonise uninfected erythrocytes. The mature schizont that contains the merozoites is an erythrocyte reduced to virtually a porous sac and it was theorised that this would provide a minimal barrier to the introduction of foreign DNA, especially when compared to the less mature forms in the erythrocytic cycle. Thus, the pausing at rupture in in vitro culture is an ideal feature, which is exploited to maximise the parasite forms, which we target for transformation. Invasion of erythrocytes in culture is an extremely inefficient process therefore, the model relies upon shuttling between an animal host and in vitro culture. The choice is made between mice and rats according to the needs of the experimental procedure. We transfect rat erythrocytic schizonts and reintroduce these by intravenous injection into phenylhydrazine-treated rats. This maximises both the number of parasites that can be transfected (and therefore the selection pool) and their potential to multiply within the recipient host, since the *P. berghei* parasite has a predeliction for reticulocytes.

Transfection is achieved by electroporation: the transient application of a strong electric field that temporarily opens holes in the membrane of the cell that quickly reseal, but at the same time allow the entry of molecules from the external medium (such as foreign DNA). There appears to be little deleterious effect to the cells that survive, but the majority does not survive the procedure. With *P. berghei* we have adjusted the conditions such that there is between 10 to 20% parasite survival. This ensures that those that survive experienced a strong electrical field in the hope that the number of surviving cells that received DNA is maximised. However, transfection is in itself a very inefficient process. We calculated that only 100 to 1000 cells are transformed ($5x10^9$ to $10^{10}$ cells are electroporated, giving an efficiency of about $10^{-7}$). Similar efficiencies are found with *P. falciparum*. Two of the advantages that the *P. berghei* model offers are more rapid growth (24 hr cycle versus 48 hr for *P. falciparum*) and selection in vivo (this latter feature is double-edged). The latter allows the selection of fully functional, robust, transformed parasites but must be a method compatible with host survival (however more toxic selection procedures could be conceived using culture that would be incompatible with host survival). We notice no difference in rate of recovery of recombinant parasites when selecting for episomally transformed parasites and those transformed with integration constructs. These characteristics mean that within 10 days after electroporation the selection procedure is complete and a rat is positively infected: almost without exception with the desired recombinant. Selection in vivo is achieved by single-shot, IP administration of pyrimethamine to the rat on a daily basis for 4 days subsequent to the introduction of the transfected parasites back into the host. This is followed by a period (4 to 5 days) of daily examination of Giemsa-stained blood smears of the infected animals until (drug resistant) parasites are observable.

Procedure:

Time schedule

| | |
|---|---|
| Day -5 | Infect donor rat with *P. berghei* |
| Day -4 | Treat recipient rat with phenylhydrazine |
| Day 0 | Schizont isolation, electroporation and injection in recipient rat |

Day 1-4 Pyrimethamine treatment
Day 7-13 Analysis for developing parasitaemia and transfer to mice

**Culture and preparation of *P. berghei* for transfection**

1. Infection of donor rat: Five days before the electroporation Wistar rats (about 180 to 250 g, not treated with phenylhydrazine) are infected with *P. berghei*. For this, 5 to 6 drops from the tail of *P. berghei* infected mice (5 to 10% parasitaemia) are diluted into 1 ml PBS and injected i.p. into one rat. You need one donor rat per transfection.

2. Preparation of recipient rat: Four days before electroporation rats to be the recipient of the transfected parasites (one rat per transfection experiment) must be phenylhydrazine treated: inject 0.3 to 0.5 ml phenylhydrazine stock solution (25 mg/ml) into each rat (for 200 g and 350 g rats, respectively).

3. Blood culture: Once the parasitaemia of rats infected is minimally 1% (typically 4 days post-infection) in vitro culture of schizonts can be set up according to the following procedure.

In the early afternoon for each rat, thaw 150 ml of RPMI 1640 medium (pH 7.2) with neomycin (50 U/ml) at $37^0$C. Just before use, add FCS (heat inactivated) to 20%. Place 10 ml of RPMI/FCS in a 50 ml Falcon tube and add 0.25 ml of the heparin stock. All further steps are carried out at RT.

Ether anaesthetise the rat, place it on top of a wooden tray and proceed with the dissection of the animal by opening its thorax and exposing the heart. With a 10 ml syringe and 23 gauge needle perform a heart puncture and collect as much blood as possible (about 5 to 8 ml). Carry out this procedure while the heart is still beating. Transfer collected blood to the Falcon tube.

Centrifuge the tubes (no brake) for 8 min at 1500 rpm at room temperature in order to wash the heparin from the red blood cells. Discard the supernatant carefully with a pump. Stop aspirating as soon as the red blood cell layer is reached. This procedure will eliminate the majority of the white blood cells. Care should be taken not to remove the schizonts that are just below the leucocytes.

Mix the red blood cells with about 10 ml of the RPMI/FCS medium and add this mixture to a 500 ml (or 1 liter, see below) Erlenmeyer containing 100 ml of medium. Wash Falcon tube with the remaining 30 ml medium.

Place the Erlenmeyer containing the red blood cells in a shaker in a 37$^0$C room. Attach the tube from the gassing apparatus making sure the shaking is on zero. Open the gas bottle (containing 5% $CO_2$, 10% $O_2$, 85% $N_2$) and ensure steady flow. Leave for 5 min, then switch on the shaker to give mild but not vigorous rotation. Incubate culture for 20 hr in 37$^0$C room (temperature should not be >38.5$^0$C; optimum is 38$^0$C).

If no constant gassing is supplied, use a 1-liter Erlenmeyer flask and gas it extensively before shaking with the above mixture. The culture needs to be saturated with this gas mixture in order to prevent an increase in pH due to low $CO_2$ pressure. Submerge the gas supply (pipette) in the culture to speed up gas exchange in the culture. Keep flask sealed during the incubation on. Shake to keep cells in suspension.

4. Schizont isolation: In order to examine whether the schizonts have developed, a Giemsa-stained slide of the in vitro culture must be prepared the next morning.

Place a 0.5 ml aliquot of the culture into an eppendorf tube and centrifuge it for 5 sec. Discard the supernatant (which should be yellowish, if gassing was adequate) taking care not to remove the schizonts that will be in the upper layer of the red cell pellet. Mix the pellet well but gently not to burst the schizonts. Make the slide and stain it for 10 min (4 drops of Giemsa/1 ml water). Only schizonts ready for rupture and gametocytes should be observed. Proceed with the isolation of these forms if 3 to 9 mature schizonts/field are observed; otherwise the cultures must be left a bit longer.

Schizont separation from non-infected red blood cells is achieved using a 55% Nycodenz gradient. To obtain such a gradient, prepare enough Nycodenz by mixing 55% Nycodenz with 45% PBS. For instance, 400 ml blood will require 10 tubes and 100 ml Nycodenz.

*If there are several cultures (from several rats), they can be mixed at any time to ensure that the same schizont material is used in all electroporations.*

Place 35 ml of the schizont culture into each Falcon tube and add the Nycodenz/PBS preparation by introducing a 10 ml pipette with a electronic pipettman through the schizont culture and, slowly, adding the gradient solution (10 ml) from beneath. Balance the tubes with the remaining of the schizont culture. Using a stable centrifuge (no brake) spin the tubes at 1,000 rpm (200 g) for 25 min.

Transfer the interphase layer containing the infected schizonts and leukocytes quantitatively into one or more new Falcon tubes (the interphase layer should be removed until no more brown layer is left; this will correspond to about 10 ml/tube). Top up the tubes with medium to balance them (use residual media from the last centrifugation). Centrifuge at 2,000 rpm for 10 min to pellet parasites.

*(Rats to be recipients of electroporation product should now be placed at 37$^0$C).*

Remove supernatant with a pump leaving behind a little bit of medium. Mix very gently and make the volume to 100 μl with PBS, if needed. These parasites are ready to be electroporated. Do not put them on ice, continue right away with electroporation.

**Electroporation**

Add the 100 μl parasites to 300 μl of plasmid preparation which should be a maximum of 50 μl TE made to volume with cytomix. The DNA might be on ice or not. Mix well with P1000, transfer to a 0.4 cm electroporation cuvette (room temperature, not on ice) and electroporate using a single pulse (1.1 kV, 25 μF; time constant should be around 0.8 ms) in a Bio-Rad electroporator (no capacity extender, no pulse controller). Transfer cuvette to ice.

**Injection of rats and pyrimethamine injections**

Anaesthetise with ether (or equivalent e.g. halothane) one rat (previously left at 37$^0$C for 20 min) and introduce its tail in hot water for half a minute. Then inject the 0.4 ml product of each electroporation into the tail vein of the animal. After about 6 hr, prepare a slide from blood taken from the tail of the animal to make sure it is infected. Leave the animals for 24 hr before applying the selection pressure. The following day, preferably in the afternoon, a blood smear can be prepared to test for the developing parasitaemia. 1 ml of the pyrimethamine stock is injected i.p. in each rat, to make a 10 mg/kg final concentration. Second, third and fourth injections are applied the following days (the fourth injection can alternatively be done on day 5). If multiple rats are being analysed having received different constructs then different needles and syringes must be used to prevent contamination. It is possible to inject rats i.p. without resorting to anaesthesia, however, this is not recommended, unless you are well versed in handling large rodents.

**Monitoring the infection**

Slides from tail blood from rats should be prepared starting from 7 to 8 days after electroporation. As soon as animals present parasitaemias of 1 parasite/field (usually 10 days after electroporation) the blood must be transferred into mice. 4 droplets of tail blood taken with a 4 ml PBS/heparin filled syringe are obtained, well mixed and each 0.2 ml of this injected i.p./mouse (2 drops rat blood/mouse).

**Glycerol stocks for liquid $N_2$**

When the rat has a high parasitaemia, glycerol stocks of the parasites are prepared, from either heart blood or from tail blood: 1 ml of heart blood (taken in a heparinised syringe) is mixed with 1 ml of sterile 30% glycerol and aliquoted into two 0.5 ml cryotubes. After incubating several min at $4^0$C, the labelled cryotubes are introduced into liquid $N_2$. The blood of 2 heparin capillaries from the tail are mixed with 1 ml 30% glycerol and placed into two cryotubes (only necessary if you are afraid that the rat might die at an early stage).

**Infection in mice**

Pyrimethamine selection of infected mice is performed as for rats. Glycerol stocks of resistant parasites should also be obtained. About 1 ml of blood can be obtained from the heart of mice in 1 ml syringes filled with 0.1 ml heparin. This blood is then mixed with 30% glycerol (1:1) and frozen in $N_2$. Parasites obtained from resistant mice will have to be cloned for further analysis. This is performed by limiting dilution (inject 20 to 30 mice/cloning).

**DNA-RNA isolation**

The rat or mouse should have a parasitaemia of around 5 to 10% to give a worthwhile yield (about 100 μg DNA, 200 to 300 μg RNA). In a 50 ml tube mix 5 ml PBS with 0.2 ml of EDTA stock (5 mM final). Total blood (5 ml) is collected by heart puncture (without heparin) and immediately mixed 1:1 with 5 ml PBS-EDTA. Pre-wash a Plasmodipur leucocyte filter with about 10 ml of PBS. Pass blood over Plasmodipur filter to remove leukocytes. Wash filter with 40 ml PBS. Pellet RBC for 8 min at 1,500 rpm. Take off supernatant with pump. Lyse the RBC by resuspending the pellet in 50 ml of 1x RBC lysis buffer, shake strongly. Incubate on ice for 15 min. Spin for 8 min at 2,000 rpm to pellet the parasites. Take off supernatant with pump and use parasite pellet for chromosome gel (CHEF or FIGE) and DNA and/or RNA isolation.

## ESTABLISHMENT OF VERACITY OF DRUG RESISTANT PARASITES

### Transfer to mouse and drug challenge

Mice are infected by i.p. inoculation of dilute infected rat blood (additional tests are performed on the rat material - see below) and will develop a patent parasitaemia 4 to 5 days later. At this point the mice are subsequently challenged with pyrimethamine to ensure that the parasites that reappeared in the rat infection are truly drug resistant. This also expands the population of parasites for testing and later handling (see below). All of the tests outlined below can be carried out on the rat parental population and the subsequent populations raised in the mouse.

### Chromosome mapping of the integration event

If the integration event has occurred as expected in a site-specific manner via a mechanism that involves homologous recombination, then it ought to be possible to locate that event to a specific chromosome (or multiple chromosomes in the case of repeated disperse loci). To date random integration of foreign DNA into the malaria parasite genome has not been well documented, the vast majority of integration events are site specific (contrast *T. gondii*) and are directed by as little as 500 bp on either side of the selectable marker. Thus, experiments are only designed to investigate well-characterised loci whose chromosome location is generally known. Clamped homogenous electrophoretic field (CHEF) gel electrophoresisis a commonly used technique to resolve the relatively simple chromosomes of lower eukaryotes and works well for most malaria parasite genomes. Thus, integration into a specific parasite chromosome can be monitored by performing a Southern analysis of separated chromosomes of transfected parasites and hybridising the blot with a probe specific for the integration event. In *P. berghei* most of our constructs use the *T. gondii* DHFR-TS selectable marker linked to the 3' UTR region of the homologous gene from *P. berghei*. Therefore, we hybridise such blots with the 3' UTR region as this illuminates both the integration event but also provides an internal control through hybridisation to the homologous DHFR/TS gene that resides on chromosome 7 (of 14). Integration and episomal events are easily discriminated.

### PCR testing of parasites

Rat-derived parasite material (and the mouse later) can be tested for presence of the vector (in episomal transformations) and directly for integration in situations where there is enough ancillary information about the genetic locus under consideration, such that PCR primers that lie just outside the region of integration. This combination can be used to amplify the hybrid region created by the integration and confirm the presence and accuracy of the integration event. Design and location of the primer (within and relative to the construct) are dependent upon the form of integration being considered (single or double crossover). Generally we use a locus specific primer and either a pUC based primer (single crossover) or a selectable marker specific primer (double crossover). Marker specific primers can confirm the presence of episomes and integrants. For detection of the former this must be combined with plasmid rescue of the episomal DNA by transformation of *E. coli* and restriction analysis. It is essential to demonstrate that the episome is not rearranged. Controls include amplification from non-transfected wild-type parasite DNA. This excludes the possibility that the drug resistant parasites are naturally resistant (presumably with mutations in the wild-type DHFR-TS gene on chromosome 7) and existed within the parasite population prior to transfection. PCR analysis provides further proof that the experimental transfection has proceeded according to design and given integration at the desired chromosomal locus. This analysis and CHEF generally give enough confidence to proceed with parasite cloning whilst at the same time performing Southern analysis of restricted parasite DNA to prove fidelity of integration.

**Southern analysis of genomic DNA**

Integration of foreign DNA into the *Plasmodium* genome invariably occurs in a site-specific manner. Random integration is not frequently observed and when seen constitutes a minor experimental background noise. The site-specific nature of integration can be combined with the knowledge of the restriction maps of both the wild-type locus and the integration vector and used to create a prediction of the organisation of the rearranged genomic locus at the site of integration. Wild-type parasite and recombinant parasite population DNA can then be restricted with specific enzymes that will reveal fragments that are both diagnostic of the integration but also indicate the fidelity of the event.

**Phenotypic analysis of the transformed parasites**

This is usually only performed upon cloned parasites that have passed all the evaluation procedures detailed above and no conclusions should be drawn from phenotype evaluations performed on parasites which have not undergone this verification process. Clearly the phenotypic analysis is entirely related to the gene that one is seeking to introduce, modify or disrupt. As examples, we will describe experiments to make parasites that express green fluorescent protein, express the thymidine kinase gene of herpes simplex virus and knock out ribosomal RNA units.

Procedures:

**CHEF gel analysis of parasite chromosomes**

*The resolution of each pulsed field electrophoretic system varies, but most rodent malaria parasites chromosomes can be resolved on all systems.*

Prepare 100 ml of 1% agarose (low EEO) in 0.5xTBE (adjust volume back to original. Beware of evaporation). Place in 50°C water bath.

Tape up the glass plate that will support the gel. Place on levelled surface and set comb above the surface of the plate with a 2 mm gap between the two. Pour the molten gel smoothly onto plate and reserve a small amount that is placed back in the water bath. Allow to set. Once set place 30 ml 0.5xTBE on the surface of the gel and slowly and carefully remove the comb.

Blocks are pre-prepared (see Appendix 3), cut a 1mm slice of each block and place carefully in a well. Seal each well over with the remaining molten agarose.

Run parameters for 3 phase run:

Phases 1&3 Switch time: 120 sec; Duration: 24 hr; volts/cm: 4.2; Angle: 120°

Phase 2 Switch time: 130 sec; Duration: 24 hr; volts/cm: 3.9; Angle: 120°

*If the Bio-Rad DR3 machine is available the automatic setting selected by the microprocessor will adequately resolve all the chromosomes in a single 24 hr run.*

At the end of run, remove and stain in dilute EtBr solution for 20 min. Visualise on UV box and photograph. Treat with 0.1 M HCl for 10 min. Transfer to denaturing buffer for 30 min, neutralising buffer for 30 min. Set up to blot. Transfer over night.

**PCR analysis of chromosomal DNA to confirm transfection (and integration)**

This is performed on the infected rat blood as the parasitaemia becomes patent (about 7 days after transfection) to rapidly gain an impression of the success of the protocol. Parasite DNA is isolated from the blood as soon as the parasitaemia of the rat has reached 1%.

**Rapid DNA isolation from rat tail blood (Gentra kit)**

To 3 to 5 drops (=1 vol) of tail blood, add 10 μl 10 mM EDTA and 3 volumes of RBC lysis solution. Invert to mix and stand at room temperature for 10 min.

Centrifuge 20 sec in Eppendorf centrifuge, top speed and CAREFULLY remove supernatant with pipette. Vortex pellet free and add 1 vol Cell Lysis Solution. Pipette to mix. (Optional $37^0$C incubation, if cell lysis not homogenous and cool sample to room temperature.) Add 1/3 volume of protein precipitation solution.

Vortex vigorously for 20 sec and centrifuge in Eppendorf centrifuge (top speed) for 3 min. Remove supernatant by pipette and transfer to new tube. Add 0.8 volume isopropanol. Invert rapidly and repeatedly. Thin strands of DNA become visible and will eventually clump.

Centrifuge in Eppendorf for 1 min, discard supernatant and dry edges of tube on tissue. Wash with 70% EtOH. Spin and repeat wash. Air dry (15 min or in SpeedVac, if available). Add 100 μl of DNA hydration solution and incubate at $65^0$C for 1 hr with periodic mixing. Store at $4^0$C. Ready for PCR analysis. Use 5 to 10 μl/reaction.

**Southern analysis of recombinant parasite DNA**

Standard southern analysis of the drug resistant parasite DNA is performed based upon an expectation that accurate integration of the DNA construct has occurred. If so then a completely predictable pattern of DNA rearrangement can be visualised by appropriate restriction and hybridisation with specific DNA probes. Generally, we use both the selectable marker (it is not present within the wild-type parasite genome) and a probe representing the target locus.

## GAMETOCYTE ACTIVATION AND OOKINETE DEVELOPMENT IN VITRO

These techniques mimic the early stages of development of the transmitted malaria parasite (the gametocyte forms) in the midgut of the mosquito vector. They can be performed with purified gametocytes or they can be more easily performed with tail blood from a mouse that is highly infected. One can make use a transformed parasites that expresses green fluorescent protein and visualise these processes using fluorescence microscopy.

The crucial triggers of gametogenesis are pH (8.0 in the mosquito midgut), temperature (20°C) and midgut-contained activation factors (Billker et al, 1998). The sequence of events upon transmission of infected mouse blood is as follows:

1. Gametocytes are activated and emerge from the erythrocyte.
2. The female does little more but round up forming the female gamete.
3. The male undergoes three rounds of DNA replication yielding 8 gametes which are thin highly motile flagellar like forms.
4. Fertilisation occurs forming the zygote.
5. Meiosis and mitotic reduction take place.
6. The zygote gradually develops through characteristic morphs to form the motile banana-shaped ookinete. This remains infectious to mosquitoes.

Why are we interested in these forms?

1. Transmission blocking vaccines/drugs etc.
2. Gene knock out (ko) technology. Haploidy means we cannot ko essential bloodstage genes. However we can at this point in the life cycle, since this is downstream of the selection event. All events during the parasites' passage through the mosquito can be studied with genetically manipulated parasites. For example several sporozoite genes have been knocked out and their roles in parasite development and motility demonstrated (Ménard et al., 1997; Sultan et al., 1997).
3. Investigation of mosquito immune response. Infected blood meals only trigger an immune response if there are viable gametocytes present that fertilise and develop into ookinetes. These immune responses originate from a few specialised organs. An ability to look at the molecule(s) that might activate these responses will be a central technology to unravelling the processes of insect immunity.

Procedure:

Day 1. Infect mice with 0.2 ml of infected blood diluted in PBS (2 drops of 3 to 10% parasitaemic blood in 10 ml PBS).

Daily. Beginning on day 3, treat mice with pyrimethamine in drinking water (7 mg/ml in 0.5% lactic acid. 1:100 in drinking water or by injection as before), if dealing with transformed, drug resistant parasites.

Day 5. Begin daily monitoring of parasitaemia with Giemsa stains plus microscope observations.

Day 5 to 9. Thaw medium and FCS and place at 20$^{O}$C. Bleed out animal when parasitaemia/gametocytaemia is still on rise i.e. 1 to 3% (0.8 to 1.2 ml). Put in 10 ml culture medium with heparin (see drug test methods). Wash and pellet. During spin make 10 ml medium to 10% FCS, aliquot 1 ml portions into eppendorf tubes and replace at 20$^{O}$C. Take parasites gently resuspend by pipetting and make 1 ml aliquots to 1% cell suspension. Place for 10 min at 20$^{O}$C. Remove 20 μl aliquot (replace remainder at 20$^{O}$C) and examine in haemocytometer under light microscope (condenser in lowered position, 40x objective). Exflagellating male gametes/gametocytes should be visible. The remaining culture is incubated for 18 to 24 hr, examine with the haemocytometer once more and slides prepared for Giemsa staining. Furthermore, the culture can be examined by fluorescence microscopy throughout the period of development using the settings for fluorescein isothiocyanate (FITC) visualisation exciting at 488 nm.

## DRUG TESTING IN *P. BERGHEI*

We test numerous compounds for their ability to inhibit parasite growth in vitro and these compounds can serve as the basis for the development of additional selectable markers, if a suitable gene encoding an inactivator (or transporter) or resistant target protein is available. We test additional compounds to demonstrate the utility of the current selectable markers as well as demonstrate the applicability of other compounds to the development of alternative selection strategies.

### Drugs to be tested

**Aciclovir**: A structural homologue of ganciclovir establishes the natural resistance of the parasite to these suicide thymidine analogues and sets the stage for the subsequent determination of sensitivity of the recombinant parasites that hopefully are generated during the module. These are antiviral compounds that are specifically utilised by the thymidine kinase (tk) activity encoded by the tk gene of Herpes Simplex Virus (HSV). The drug is metabolised into a suicide substrate of DNA metabolism that is lethally incorporated into the elongating DNA strand preventing further nucleotide addition. Cells are normally resistant to this drug and become sensitive upon transformation with constructs expressing the HSV tk gene. The tests carried out here will establish the resistance of the untransformed parasite to aciclovir and hopefully its acquired sensitivity to the drug upon transformation of P. *berghei* with the appropriate DNA construct.

**WR99210** (Fidock et al., 1997) is a drug that can be used in the development of a second positive selectable marker. It acts upon dihydrofolate reductase (as does pyrimethamine) but in a manner independent of pyrimethamine. There is no reported case of resistance to WR99210 and the human DHFR enzyme is completely insensitive to the drug in the millimolar concentration range. The cDNA encoding the human gene has been successfully used to select transformed *P. falciparum*. This raises the possibility that the human enzyme could be used as a selectable marker in P. *berghei* transformation.

**Pyrimethamine**, the original selection drug shows that the parasites and systems are behaving normally.

Procedure:
Asynchronous parasites do not develop to schizonts if inhibited. Schizonts of *P. berghei* do not rupture in culture. Therefore, incubate in presence of drug and assess maturity compared to no drug controls (see Appendix 5 for a guide to parasite appearance). The parasites transformed with the HSV tk expression construct will be tested against three drugs and compared to wild-type untransformed parasites and parasites carrying the DHFR-TS selectable maker.

**Experimental requirements**
Infected mouse blood, parasitaemia 1 to 3%. RPMI 1640 medium with FCS (20%). Drug of choice. Gas: 10% $CO_2$; 5% $O_2$; 85% $N_2$, 24-well microtitre plates.

| Drug | Concentration range | Stock |
|---|---|---|
| Pyrimethamine | 100 µg/ml to 0.1 ng/ml | 5 mg/ml suspension in PBS/Tween |
| WR99210 | 300 nM to 3 pM | 10 mg/ml in DMSO |
| Aciclovir | 1 mM to 10 nM | 6 mM stock in medium |

**Experimental format**

Final volume is 1.5 ml per assay point (all in duplicate). Cell concentration at 1%.

Cells prepared in RPMI medium 750 μl per assay point. Drug prepared in same volume (at 2x conc). Mix in 24 well flat bottom microtitre plate. Incubate on with shaking.

**Preparation of cells**

Warm medium to 37$^{O}$C. Anaesthetise infected mouse (1 to 3% parasitaemia, infected 4 to 5 days earlier). Take blood by heart puncture and dispense into 10 ml medium containing 20 U heparin. Spin 1,500 rpm 7 min and aspirate, resuspend in appropriate volume of medium.

Prepare drugs and salt (dissolved or suspended) stocks in medium. Make 5 ml volumes at 2x concentration and also use these to prepare the dilutions. Rapidly mix (pH considerations) aliquots of cells and drug. Move to gassed incubation chamber at 37$^{O}$C. Leave static for 5 min. Switch on shaker. Leave overnight.

**Harvest of culture**

Remove plates from incubator. Cells will have settled. With 1 ml Pipettman, resuspend culture by gentle pipetting and remove 1 ml to a marked eppendorf tube. Spin 5 sec in eppendorf centrifuge at top speed. Rapidly, but carefully aspirate culture supernatant away to 2 to 3 mm above cell pellet. Carefully remove rest of supernatant with 20 μl Pipettman. Resuspend cells by pipetting. Place 1 to 2 μl on labelled glass slide and make smear. Stain and visually inspect light microscope (oil immersion 100x magnification). Compare development with that of no drug controls.

## PRACTICAL ANALYSIS OF GENE KNOCK OUTS IN *P. BERGHEI*

**Ribosomal RNA unit knock out**

This example investigates whether the stage specific rRNA genes can be disrupted using the recombination mechanisms described in the introduction. There are 4 gene units (A-D), each located on a separate chromosome. These units are formally divided into two groups (A, Asexual bloodstage and S, Sporozoite) based on sequence similarity and pattern of expression (Waters et al., 1997, Thompson et al, 1999). Vegetative growth in the bloodstages is the point at which selection is imposed for successful transformation (the S type units are virtually silent at this period). However, all four units are potential targets. Ultimately, we hope to learn the nature of the control of gene expression the parasite brings to bear on these genes and the biological reason underlying their maintenance. Furthermore, we would like to know the nature of the requirement of the parasite for these different genes at the structure-function level. A first step is to demonstrate that the parasite has no requirement for the S-type genes during asexual bloodstage growth and that they are essential for propagation in the vector. Therefore we have devised two experiments that target different combinations of the two s-type rRNA gene units.

**S-unit gene disruption**

The four rRNA gene units of *P. berghei* are distributed on different chromosomes. A construct pMD207 has been assembled that contains an incomplete copy of the C unit SSU gene flanked by Eco RI sites. Linearisation at a unique Spe I site creates a plasmid that should integrate by a single crossover mechanism. Single crossover integration differs from that of the double crossover mechanism in that it duplicates the target region contained on the integration vector. This mechanism can be exploited to disrupt or modify genes and it is in theory reversible.

Procedure:

Vector preparation is through Spe I digestion of pMD207 (Fig. 3). The mode of integration is single crossover that generates a 1 kb and 0.8 kb targeting fragments. The target region is the C unit SSU rRNA gene and integration effectively separates most of the structural region of the unit from the promoter. Recombinant parasites will be genotypically analysed by PCR and southern analysis of restricted genomic DNA and fractionated chromosomes of the recombinant parasites.

**PCR diagnosis**

| | |
|---|---|
| D wild type | L260R/L271R |
| C wild type | L270R/L271R |
| C Integrant | L270R/328A |
| D Integrant | L260R/328A |

Conditions: 1 cycle at $94^0$C, 2 min;
40 cycles at $94^0$C, 30 sec; $55^0$C, 30 sec; $72^0$C, for 4 min.

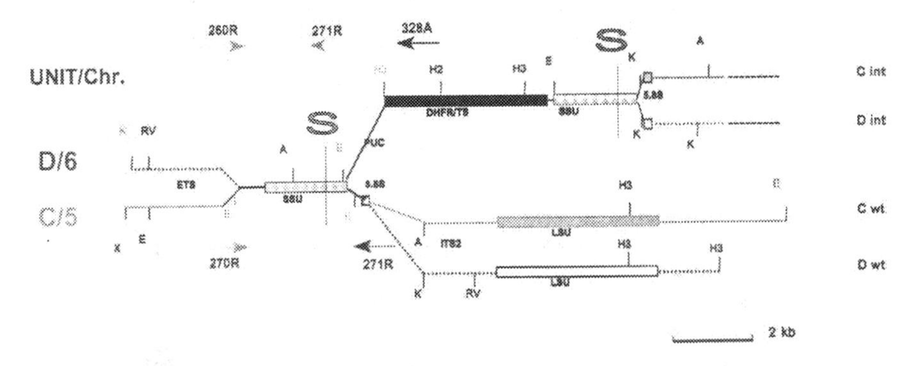


**Figure 3.** Integration of pMD207 into the S-type rRNA loci of *P. berghei.* The possible integration events (int. top) are illustrated alongside the wild-type loci (wt bottom). The point of linearisation of the vector at the Spe I site is indicated as S. Oligos for the specific amplification of the possible loci are illustrated in grey for the C locus and grey shade for the D locus. Black indicates common oligos. Restriction sites used in the diagnostic analysis of the integration are also indicated. The most informative is a Kpn I digest.

**Diagnostic Southern of recombinant parasites containing an inserted pMD207**

A Hind III/Kpn I digest of 3 μg parasite DNA probed with the S-type ETS probe will visualise a 7.1 kb fragment in mutant parasites where integration has occurred in the D unit (wt fragment about 20 kb). The Cint cut the same way gives 21 kb band (wt 18 kb) and cannot be easily discriminated. Therefore, an Eco RI digest probed in the same way will visualise a 3.0 kb mutant fragment corresponding to integration into the C unit (wt fragment about 14 kb). The Dint is obvious, but less informative (20 kb with a 44 kb wt band).

**D-unit rRNA gene ko**

It is also possible to target the individual rRNA units specifically using unit-specific sequences as targets. Note that only one of the target regions need be unique. It is desirable to carry out both the S-type universal knock out and that of the specific individual units as a confirmation of observed phenotype. This is especially important since to date it has not been possible to complement and restore a genetic mutation in *Plasmodium* due to a lack of independent selectable markers. The D rRNA gene unit is one member of the

S units and, therefore, is also not expressed during bloodstage development of parasites. It can, therefore, safely be targeted for disruption using unit specific regions that will prevent the possible spurious integration into the remaining three units (Fig. 4 ).

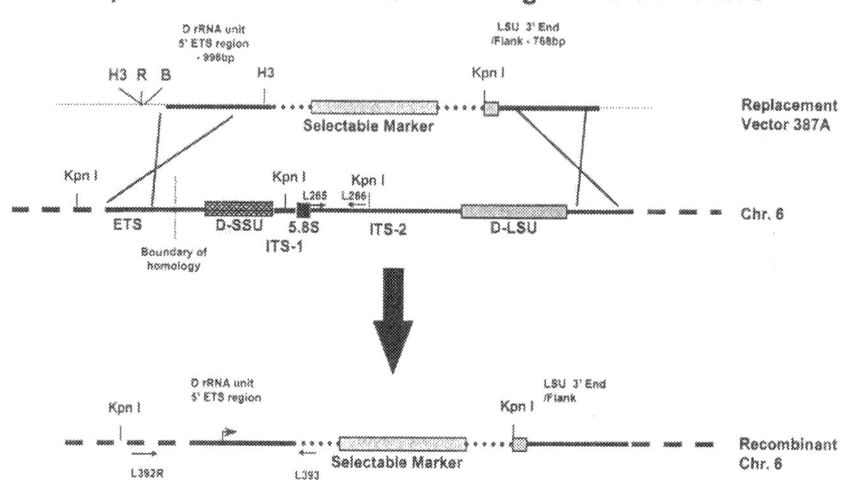


**Figure 4.** Integration into the D rRNA gene unit on chromosome 6 of *P. berghei.* Vector 387A (Top) integrates into parasite chromosome 6 by a double crossover replacement mechanism disrupting the unit which is expressed during parasite development in the mosquito vector.

Procedure:

Vector Preparation: Take 50 μg of vector 387A and perform sequential, separate restriction digests with: Bam HI and Eco RI. The mode of integration of this vector is double crossover into chromosome 6 of *P. berghei.* The vector 387A consists of the following targeting sequence elements that flank the selectable marker expression cassette. 5' target region: D unit ETS 1.0 kb that lies upstream of elements that are shared between the C and D units. 3' target region: 0.8 kb D LSU rRNA gene that contains specific and common elements. Successful integration into the parasite genome will delete most of the structural region of the unit (Fig. 4).

**Diagnostic PCR**

Integrant: Primers L392R/L393D, $94^{0}$C, 2 min [ $94^{0}$C, 30 sec; $55^{0}$C, 30 sec; $72^{0}$C, 1 min] 30 cycles. Product = 1 kb.

Wild-type: Primers L265R/L266R, $94^{0}$C, 2 min [$94^{0}$C, 30 sec; $60^{0}$C, 30 sec; $72^{0}$C, 1 min] 30 cycles. Product = 1.2 kb.

Diagnostic Southern analysis of recombinant parasite DNA is obtained by a Kpn I digest and gives 5.2 kb (wt) OR 7.5 kb (Dko) OR 9.7 kb (vector) bands. Digest 3 µg of parasite DNA with 30 units Kpn I overnight using standard conditions. Fractionate on 0.7% agarose gel, photograph and blot following standard protocol. Probe the blot with 3' UTR of DHFR/TS.

**Transgene expression in *P. berghei***

Using the expression vectors and precisely the same methodologies described above. We can create transgenic parasites that express either the green fluorescent protein (GFP) or the Herpes Simplex Virus thymidine kinase, both under control of the DHFR/TS promoter in vector pEXPRESS1 ($pD_BD_{Tm}D_{B:}D_{B.}$-.$D_B$) creating plasmids $pD_BD_{Tm}D_{B:}D_{B.}G.D_B$ (GFP) and $pD_BD_{Tm}D_{B:}D_B.T_k.D_B$ (hsvtk), respectively. These will give immediately observable phenotypes. GFP expressing parasites will glow green (de Koning-Ward et al., 1998; Sultan et al., 1999) under the fluorescence microscope and can be used to observe sexual development and fertilisation. Thymidine kinase should confer sensitivity to *P. berghei* to aciclovir, a suicide analogue of thymidine (Black et al., 1996).

## DISCUSSION AND CONCLUSIONS

The ability to genetically transform and manipulate malaria parasites is a relatively new technology that still requires much development. If one were to compare the technologies that can be applied to the trypanosomatids some of which are described in the accompanying chapter then clearly there is much that still can (and should) be developed. Some of the work described in this chapter attempts to address these needs, for example the use of thymidine kinase as a negative selectable marker. We require: more selectable markers (positive and negative); random insertion methods, cosmids for complementation; improved culture systems and model systems, inducible expression systems; and basic technologies like improvements to the efficiency of transformation. Many groups around the world are addressing these issues and there is no doubt that the ability to genetically manipulate malaria parasites will continue to improve over the next decade. Given that we will have access to at least one complete genome sequence for *Plasmodium* very shortly, the prospects for improved understanding of the biology of malaria are enormous. It is the challenge to the community to convert this knowledge into effective intervention strategies that can be employed in malarious areas of the world.

## ACKNOWLEDGEMENTS

This is a rapidly developing field and I am indebted to numerous colleagues who have provided information and developed techniques that have been incorporated into this module. In particular, I would like to thank Chris Janse who almost single-handedly developed the model and many of the concepts that enable transfection to be developed in *P. berghei.* I would like to also thank the members of the malaria group at Leiden, past and present, Robert Ménard, David Fidock and especially Kai Wengelnik and Annemarie van der Wel.

## REFERENCES

Billker, O., Lindo, V., Panico, M., Etienne, A.E., Paxton T, Dell, A., Rogers, M., Sinden, R.E., Morris, H.R. (1998) Identification of xanthurenic acid as the putative inducer of malaria development in the mosquito. Nature *392*, 289-292.

Black, M.E., Newcomb, T.G., Wilson, H.M., Loeb, L.A. (1996) Creation of drug-specific herpes simplex virus type 1 thymidine kinase mutants for gene therapy. Proc Natl Acad Sci *93,* 3525-3529.

Crabb, B.S.,, Cowman, A.F. (1996) Characterization of promoters and stable transfection by homologous and nonhomologous recombination in *Plasmodium falciparum.* Proc Natl Acad Sci (USA) *93*, 7289-7294.

Crabb, B.S., Cooke, B.M., Reeder, J.C., Waller, R.F., Caruana, S.R., Davern, K.M., Wickham, M.E., Brown, G.V., Coppel, R.L., Cowman, A.F. (1997) Targeted gene disruption shows that knobs enable malaria-infected red cells to cytoadhere under physiological shear stress. Cell *89*, 287-296.

Crabb, B.S., Triglia, T., Waterkeyn, J.G., Cowman, A.F. (1997) Stable transgene expression in *Plasmodium falciparum.* Mol Biochem Parasitol *90*, 131-144.

de Koning-Ward, T.F., Thomas, A.W., Waters, A.P. & Janse, C.J. (1998) Stable, drug-selected expression of Green Fluorescent Protein in *Plasmodium. berghei.* Mol. Biochem. Parasitol. *97*, 247-252.

Goonewardene, R., Daily, J., Kaslow, D., Sullivan, T.J., Duffy, P., Carter, R., Mendis, K., Wirth, D. (1993) Transfection of the malaria parasite and expression of firefly luciferase. Proc Natl Acad Sci (USA) *90*, 5234-5236.

Kocken, C.H.M.., van der Wel, A.M., Dubbeld, M.A., Narum, D.L., van de Rijke, F.M., van Gemert, G.-J., van der Linde, X., Bannister, L.H., Janse, C.J., Waters, A.P. & Thomas, A.W. (1998) Precise timing of expression of a *Plasmodium falciparum* derived transgene in *P. berghei* is a critical determinant of subsequent subcellular location. J Biol Chem *273*, 15119-15124.

Mènard, R., Sultan, A.A., Cortes, C., Altszuler, R., van Dijk, M.R., Janse, C.J., Waters, A.P., Nussenzweig, R.S., and Nussenzweig, V. (1997) Circumsporozoite protein is required for development of malaria sporozoites in mosquitoes. Nature, *385*, 336-340.

Mons B. (1986) Intra erythrocytic differentiation of Plasmodium berghei, an approach to develop a standard model system for gametocytogenesis in vivo and in vitro. *Acta Leidensia, 54*, 1-124.

Sultan, A.A., Thathy, V., Frevert, U., Robson, K,.J.H., Crisanti, A., Nussenzweig, V., Nussenzweig, R.S., and MÈnard, R. (1997) TRAP is necessary for gliding motility and infectivity of plasmodium sporozoites. Cell *90*, 511-522.

Sultan A.A., Thathy V., Nussenzweig V., Menard R. (1999) Green fluorescent protein as a marker in *Plasmodium berghei* transformation. Infect Immun *67*, 602-2606.

Thompson, J., van Spaendonk, R.M.L., Choudhuri, R., Sinden, R.E., Janse, C.J. & Waters, A.P. (1999) Heterogeneous ribosome populations are present in *Plasmodium berghei* during development in its vector. Mol. Micro. *31*, 253-260.

Tomas, A.M., van der Wel, A.M., Thomas, A.W., Janse, C.J. & Waters, A.P. (1998) "Transfection systems for animal models of malaria." Parasitol.Today *14*, 245-249.

Triglia T., Wang P., Sims P.F., Hyde J.E., Cowman A.F. (1998) Allelic exchange at the endogenous genomic locus in Plasmodium falciparum proves the role of dihydropteroate synthase in sulfadoxine-resistant malaria. EMBO J *17*, 3807-3815.

van der Wel, AM, Tomas, AM, Cocken, CHM, Malhotra, P, Janse, C.J., Waters, A.P. & Thomas, A.W. (1997) Transfection and *in vivo* selection of the primate malaria parasite, *Plasmodium knowlesi*. J. Exp. Med. *185*, 1499-1503.

van Dijk, M.R., Waters, A.P. and Janse, C.J. (1995). Stable transfection of malaria parasite blood stages. Science. *268*, 1358-1362.

van Dijk, M.R., Janse, C.J. and Waters, A.P. (1996). Expression of a *Plasmodium* gene introduced into subtelomeric regions of *P. berghei* chromosomes. Science. *271*, 662-665.

Waters, A.P., van Spaendonk, R.M.L., Thompson, J., Vervenne, R.A.W., Dirks, R.W. & Janse C.J. (1997) Species specific regulation and switching of transcription between stage specific ribosomal RNA genes in *Plasmodium berghei*. J. Biol. Chem. *272*, 3583-89.

Wu, Y., Sifri, C.D., Lei H.H., Su, X.Z., Wellems, T.E. (1995) Transfection of *Plasmodium falciparum* within human red blood cells. Proc Natl Acad Sci (USA) *92*, 973-977.

Wu Y., Kirkman L.A., Wellems, T.E. (1996) Transformation of *Plasmodium falciparum* malaria parasites by homologous integration of plasmids that confer resistance to pyrimethamine. Proc Natl Acad Sci (USA) *93*, 1130-1134.

## Appendix 1. Materials and solutions

Infection of mice: **Directly from stabilates**

Take 1 cryotube from $N_2$, thaw it between hands and inject, i.p., 4 mice, each with 0.125 ml. 5 days later, make a slide from the tail blood. When parasitaemias reach 1% dissolve 2 drops of tail blood into 10 ml PBS and inject 0.125 ml/mouse into new mice. These can be used to start the transfection protocol.

**From infected animal**

Rat. Cardiac puncture animal and dilute 10 drops per 1ml with PBS. Inject 1ml diluted blood i.p. infection is normally 3 to 10% 4 days later. Animal should be anaesthetised.

Mouse. 2 drops in 10ml PBS. Inject 0.2 ml infection patent 6 to 7 days later.

Cryo-preservation of parasites

Bleed out infected animal (heparin in syringe). Add equal volume 30% glycerol in PBS, mix. Aliquot into 0.5ml batches in Nunc cryotubes. $4^{O}C$ for 10 min. Place in liquid nitrogen. Thaw at RT and inject ip into mice/rats - whole aliquot for 2 rats.

Phenylhydrazine-HCl: stock solution

250 mg in 10 ml 0.9% NaCl, aliquot in 1.5 ml tubes, store at –20°C. Inject in mice 0.1 to 0.15 ml. Inject in rats 0.3 to 0.5 ml.

RPMI: RPMI 1640 medium (pH 7.2) with neomycin 50000 ug/l (=5 ml/l). Dissolve 10.41 g RPMI 1640 in 1 liter $H_2O$, add 1.75 g $NaHCO_3$ sterile filtrate, aliquot 150 ml and store at -20 C just before use add 20% FCS (heat inactivated)(20 ml FCS to 100 ml RPMI).

Neomycin stock: 10,000 ug/ml (Gibco)

Heparin stock:

1 ampoule of 0.2 ml = 5000 IU in 25 ml RPMI without FCS = 200 IU/ml (Liquemin, Roche)

Nycodenz: NycoPrep 1.150, Nycomed PharmaAS, Oslo density: 1.150g/ml, osmolarity: 290 Osm

Giemsa staining:
Fix 1 sec in methanol, dry under fan, 2 drops Giemsa (Merck) for 1 ml $H_2O$, 4ml per slide, stain for 25 min, fast staining: 4 drops Giemsa/ml, stain for 12 min, rinse with tap water, dry under fan.

Cytomix: 120 mM KCl, 0.15 mM $CaCl_2$, 10 mM $K_2HPO_4/KH_2PO_4$ pH7.6, 25 mM HEPES pH 7.6, 2 mM EGTA pH 7.6, 5 mM $MgCl_2$, pH adjusted with KOH.

Cuvette: BioRad gene Pulser Cuvette #165-2088,0.4 cm electrode gap, 50 use new sterile cuvette for each electroporation.

Pyrimethamine:
Sigma P-7771for rats: 20 mg in 10 ml $H_2O$, pyrimethamine is not really soluble in water, add 1 drop of Tween20 and sonicate in bath for 10 min inject 1 ml in 200 g rat = 10 mg/kg, mix suspension well before injection for mice: 25 mg in 10 ml DMSO, inject 0.1 ml in 25 g mice = 10 mg/kg.

Plasmodipur filter:
Euro-Diagnostica, Arnhem, NL. Fax: +31-26-3645111 lot 9507, 25 filters

10x RBC lysis buffer:
1.5 M $NH_4Cl$, 0.1 M $KHCO_3$, 0.01 M EDTA, pH 7.4, store at room temperature.

TNE buffer:
10 mM Tris HCl pH 7.5, 100 mM NaCl, 5 mM EDTA

EDTA: Stock solution of 250 mM in $H_2O$ (pH 8.0)

## Appendix 2. Oligonucleotides used in study

| | | |
|---|---|---|
| **328A** | TTTTCCCAGTCACGACGT | pUC forward primer |
| **L190A** | CGGGATCCATGCATAAACCGGTGTGTC | *T. gondii* dhfr forward primer |
| **L191A** | AGGG ATCC ATT GCTTCTGTATTTCCGC | *T. gondii* dhfr reverse primer |
| **L260R** | ATACTGTATAACAGGTAAGCTGTTATTGTG | DrRNA unit 5' ETS oligo forward |
| **L265R** | CATTAAACATATATGTTGTTCTCTC | D rRNA unit ITS2 forward |
| **L266R** | CCCAGGTTCCAGTCGCAATAG | D rRNA unit ITS2 reverse |
| **L270R** | GTGTAGTAACATCAGTTATTGTGTG | C rRNA unit 5' ETS oligo forward |
| **L271R** | CTTAGTGTTTTGTATTAATGACGATTTG | S-type ITS1 common oligo reverse |
| **L313D** | ACTCATTATATGACTTCATTTTAC | *P. berghei* dhfr 5' UTR reverse |
| **L392R** | AAATAGTCAATTAAAATCCTATGG | D rRNA unit ETS forward |
| **L393D** | ATAATTATATGTTATTTTATTTCCAC | *P. berghei* dhfr 5' UTR reverse |

## Appendix 3. Preparation of parasite blocks for pulsed-field electrophoresis

Mice are pre-treated with phenyl hydrazine to induce reticulocytosis and encourage a subsequent infection with high parasitaemia. Mice must be infected with the clone or parental population of interest by i.p. inoculation of either frozen stabilate or direct passage of blood from an infected animal. At a parasitaemia of 5 to 15%, blood is collected from 2 to 3 mice by heart-puncture under anaesthesia (0.8 to 1.2 ml blood per mouse). The blood is collected in 5 ml PBS containing 0.2 ml $Na_2$-EDTA stock-solution (250 mM $Na_2$.EDTA in water, pH 8.0. Leucocytes are removed from this suspension by passing it through a Plasmodipur leucocyte filter.

**Removal of leukocytes.** 'Pre-wash' the small leukocyte-filter (Plasmodipur, Eurodiagnostica) with 10 to 15 ml of PBS using a 20 ml syringe that is placed on top of the filter. The infected blood suspension (5 ml) – minimal dilution 1:1 with PBS) is passed through the filter using a 20 ml syringe. Elute with 15 ml PBS. Erythrocytes are pelleted by centrifugation (8 min; 1,800*g*). Supernatant is removed and the erythrocytes are lysed in erythrocyte lysis buffer.

**Lysis of infected erythrocytes.** Add 30 to 50 ml of cold erythrocyte-lysis buffer to the pelleted erythrocytes from 2 to 8 ml blood and mix thoroughly. Leave this suspension on ice for 5 to 10 min (the clarity of the suspension is an indicator of the state of the lysis procedure i.e. when fully clear lysis is complete). Pellet the parasites by centrifugation for 8 to 10 min at 450 *g*. Part of the parasite pellet (usually 60%) is stored in an eppendorf tube at –20°C for extraction of DNA (see below: DNA purification) for PCR, plasmid rescue and Southern analysis of restriction fragments. The remaining part of the parasite pellet is used for preparing agarose blocks containing chromosomes for separation of chromosomes using Pulsed Field Gel Electrophoresis.

### Preparing agarose blocks for separation of chromosomes by Pulsed Field Gel Electrophoresis

Mix part of the parasite pellet 1 to 1(1.5) with 1.5 to 2% low melting agarose (Sigma) at 37°C (water bath). Prepare small blocks using a mould or by carefully pouring the agarose suspension on a microscope slide in the area which is marked with a waterproof marker-pen. Let the agarose set at room temperature for a few minutes. Cut the blocks (5mm/5mm/2mm; length/width/height). Place the blocks in 5 to 10 ml SE buffer (0.5 M EDTA in 1% sarcosyl lauroyl sulphate solution; pH 8). Add 100 µl of 20 mg/ml stock of proteinase K to 10 ml SE buffer and keep the blocks overnight at 37°C (tube on a shaker). Store the blocks in the SE buffer at 4°C (can be stored for several years without loss of quality).

**Appendix 4.** Techniques are not described, where they are routine in any laboratory that uses even a minimum of molecular biology. Those who wish such detail are referred to Current Protocols in Molecular Biology (Eds. Ausubel, F.M. et al [1999] John Wiley & Sons, Inc) which gives comprehensive detail of all procedures.

**Appendix 5 (following page).** Diagram of the bloodstage development of *Plasmodium berghei.* (Reproduced with permission from the University of Leiden).

PLASMODIUM BERGHEI

| hpi | characteristics | LM morphology |
|---|---|---|
| 0-4 | invasion (reticulocytes) |  |
| 4-8 | intra cellular growth |  |
| 8-12 | susceptible to sexual commitment |  |
| 12-18 | sexuality determined |  |
| 18-20 | schizogony, sexuality manifest |  |
| 20-22 | sexual dimorphism not visible | 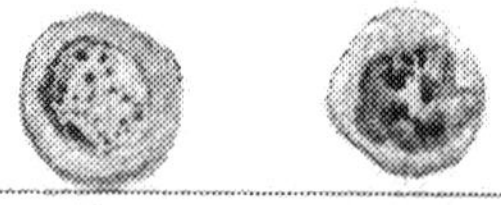 |
| 22-23 | schizogony completed | 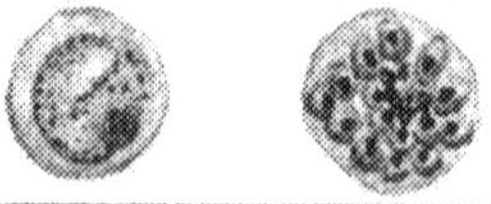 |
| 23-26 | sexual dimorphism visible | 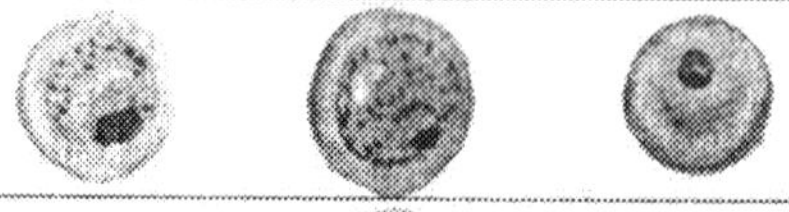 |
| 26-27 | gametocytogenesis completed | 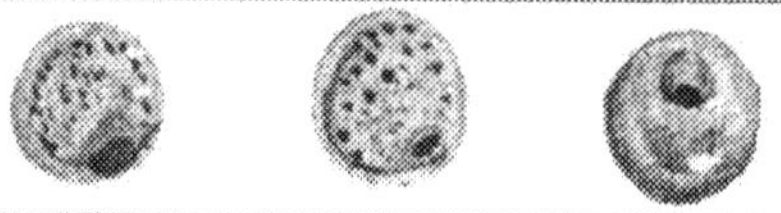 |
| 30-48 | stable gametocytaemia |  |
| 48-57 | degeneration of gametocytes |  |

From: B. Mons (1986), Acta Leidensia 54.

# 4

# MALARIA CHEMOTHERAPY: PARADIGMS FROM PYRIMIDINE METABOLISM

Pradipsinh K. Rathod
*Department of Biology, The Catholic University of America, 620 Michigan Av, NE, Washington, DC 20064*

## INTRODUCTION

Malaria remains one of the most important infectious diseases of the world. It afflicts over three hundred million people and kills about 2 million young children in Africa every year (Trigg and Kondrachine, 1998). Our ability to control the disease in tropical countries has been unimpressive. Public sanitation measures combined with antimalarial drugs have offered the only arsenal against this devastating disease. The problem has become even more acute with the widespread emergence of malarial parasites resistant to traditional drugs. Identification of efficacious new antimalarials continues to be relevant. The challenge is to do so economically.

If we are to identify new, high-quality targets for drug development, particularly from emerging data from the malaria genome sequencing efforts, it is important to know what it takes to develop potent, selective antimalarial agents. Our models for successful drug development need to be current, not oversimplified, and testable.

In this chapter, using de novo pyrimidine biosynthesis as a model, aspects of malaria drug development that are on solid footing are considered along with those that require significant illumination.

## STATUS OF MALARIA CHEMOTHERAPY

### Currently used drugs

Chloroquine appeared as an antimalarial agent in the 1940s and established itself as a first-line drug in many countries due to its potency against all forms of human malaria, its low cost, and its authorized use in children as well as pregnant women (Krogstad and De, 1998). In many countries, Fansidar (a combination of pyrimethamine and sulfadoxine) and the old standby quinine have served as second line drugs. Some other important agents include artemesinin, atovaquone, doxycycline, halofantrine, mefloquine, primaquine, proguanil, and quinidine, some of which offer significant risk from toxicity and others are only useful as a part of combination chemotherapy (Milhous and Kyle, 1998).

### Malaria drug resistance

Resistance to traditional established drugs continues to make malaria a serious global threat (Trigg and Kondrachine, 1998). Resistance against chloroquine first appeared in Southeast Asia and in South America in the late 1960s. Today, chloroquine resistance is prevalent in all tropical and many subtropical countries. In recent years, second line drugs, such as Fansidar have also failed in many parts of the world at a rapid rate. New drugs such as Mefloquine and Halofantrine have had a very short useful life, especially in Southeast Asia. There is growing evidence that parasites in some regions of the world, such as Southeast Asia, are developing resistance at a very high rate (White, 1992; Rathod et al., 1997). With increased travel between countries, and mass migrations across borders, drug-resistance in malarial parasites is no longer a localized threat.

### Nature of drug resistance

Resistance to antimetabolites, that have specific protein drug targets appears primarily through point mutations that alter the binding of the drug to the target enzyme. Such changes have been demonstrated for pyrimethamine, sulfadoxine (Cowman, 1998), and atovaquone (Vaidya, 1998). Additional mechanisms may include alterations in folate utilization (Milhous et al., 1985; Wang et al., 1999). Chloroquine, mefloquine, and halofantrine have complex mechanisms of action (Sullivan et al., 1998) and resistance is usually associated with enhanced efflux or partitioning of these drugs (Krogstad and De, 1998).

## ANTIMETABOLITES FOR CHEMOTHERAPY

### The need for new antimalarials

In response to malaria drug resistance, it is necessary to develop new effective antimalarial agents at a faster pace. Since we now know a lot about metabolic pathways and enzymes in the parasite and the host (http://sites.huji.ac.il/malaria/), antimetabolites can be source of drugs.

### Recognisable paradigms for drug development

Identification of new antimalarials amongst antimetabolites should be facilitated by adhering to some traditional criteria for selective chemotherapy.

*Essential enzymes*. Above all, it is considered necessary to inhibit parasite enzymes that are essential to the survival of the organism. The ability to integrate plasmids at specific loci in malaria now makes it possible to directly test whether a candidate gene is non-essential and if a protein is a target of an established drug (Van Dijk et al., 1995; Wu et al., 1996; Cowman, 1998).

*Different enzymes*. When the host has an analogous essential enzyme, one searches for parasite enzymes that are different from the host enzymes. This time-old model for selective chemotherapy is based on some classic antimalarials such as pyrimethamine that are useful at least in part, because they bind the parasite enzyme 1,000 times more tightly than the host enzyme. It is important to emphasise that while host-parasite differences in target enzymes can be important in selective chemotherapy, it is not the only way to get efficacy.

*Different pathways and processes*. In addition to studying individual target enzymes in detail, it is equally important to identify target-related biochemical steps, pathways, or processes that are important for the parasite but not the host. Parasites frequently lose whole metabolic pathways and thereby become completely dependent on an alternate route or a transport step that the host is not relying on for survival. Pyrimidine metabolism in malaria offers such an example (see differential rescue below).

*Potent new inhibitors*. Once an enzyme target has been identified, knowledge about the structure of the target, about the chemical mechanism of the enzyme, combined with some combinatorial chemistry and high-throughput screens ought to reveal tight-binding inhibitors that kill parasites.

*Potent old inhibitors.* New inhibitors are not always necessary. As one understands more about the details of a physiological process, existing antiproliferative agents, that previously had not been tested as antimalarials, begin to look appealing (see thymidylate synthase inhibitors below). Turning to existing compounds has the advantage of "piggy-backing" on prior pre-clinical and clinical experiences in other systems and can be expected to reduce the costs of developing new antimalarials.

**A need for new components in paradigms for drug development**

Our inability to find good selective antimalarials on a regular basis, plus the total lack of cellular toxicity of some very potent enzyme inhibitors, indicates that there is probably much more to successful chemotherapy than simply identifying important target enzymes and tight-binding inhibitors.

**Components of newer paradigms for drug development**

*Transport and delivery.* The inhibitor in question has to find its way to the target. If the transport routes for a drug or a prodrug are unique to the parasite and are missing in the host, this can facilitate selectivity. In addition to simply binding a drug less tightly, a host cell can avoid toxicity, if the inhibitor fails to enter the host cell, if it is readily pumped out of the host, or if the host cell (but not the parasite) can render a drug non-toxic by selective utilisation of nutrients (see pyrimidine salvage below).

*Metabolic responses.* Even if the parasite enzyme target has been inhibited successfully, the cell does not have to die. Substrate concentrations can build up until the inhibitor is displaced. In vitro, inhibitors with even nanomolar $K_i$ values can be utterly poor inhibitors, when substrate concentrations are in the saturating range. On the other hand, even if the enzyme is not completely inhibited, the accumulation of some substrates and depletion of product, in itself, can result in toxicity to the parasite. Nucleotide imbalances, whether caused by chemotherapeutic agents or through inborn errors of metabolism, are notorious for the cellular pathology that they cause (Kunz et al., 1994). For these reasons, some steps in some metabolic pathways can be much better suited as drug targets than other steps (see dUMP build up below).

*Cellular responses.* When treated with an inhibitor, a cell can avoid death by simply synthesising more of the target protein. Most cells have elaborate regulatory mechanisms that are exquisitely sensitive to metabolic needs. Cells can respond to depletion of specific reaction substrates or products by altering transcription and/or translation of the inhibited enzyme (Chu et al., 1991). If there is a host-parasite difference in cellular response to a drug, this can end up playing a role in selective chemotherapy. Similarly, a target protein that is undergoing slow turnover in the parasite, but not in the host, may be selectively vulnerable to inactivators (Phillips et al., 1987).

*Genetic responses*. Successful killing of a few parasite cells in a 96-well plate is not the same as eliminating large pools of parasites in a village or in the whole world. It is large genetic variations in large parasite populations that fuel emergence of drug resistance (Gassis and Rathod, 1996). If a drug can be compromised by the action of one or two point mutations, it has no chance of curing global parasite populations (see section on initiation of drug resistance below). Drugs such as chloroquine that have had high utility in the world must have presented parasite populations with large barriers to drug resistance.

*Population differences*. Finally, there are recent indications that parasites in some parts of the world may be able to generate resistance at a higher frequency than traditional parasites (Rathod et al., 1997). It is conceivable that repeated exposure of malarial parasite populations to new and different drugs over many years not only selected for individual drug-resistance traits but selected for the ability to initiate drug resistance through higher rates of mutations. In such a scenario, antimetabolites that appear successful in one part of the world may not fare so well in a different part of the world.

## TARGETING PYRIMIDINE METABOLISM

Study of pyrimidine biosynthesis remains important in developing realistic paradigms for malaria chemotherapy and for testing the utility of these paradigms. Firstly, previously successful antimalarial agents, such as pyrimethamine, cycloguanil, and sulfadoxine, seem to cause toxicity by indirectly inhibiting de novo pyrimidine metabolism The inhibitors probably cause parasite cell death through depletion of methylenetetrahydrofolate pools, which are necessary for thymidylate synthase (TS) reaction. Loss of TS function leads to nucleotide imbalances, DNA strand fragmentation, and cell death (see details below). Secondly, since malarial parasites lack enzymes for the salvage of pyrimidines, the whole de novo pyrimidine biosynthesis pathway in the blood stage form of the malarial parasites is a potential target for selective chemotherapy.

Atovaquone, a naphthoquinone derivative has often been implicated in de novo pyrimidine biosynthesis as an inhibitor of dihydroorotate dehydrogenase. At non-physiological concentrations, it can deplete pyrimidine nucleoside triphosphate pools (Seymour et al., 1994). However, at physiologically relevant concentrations its primary action does not appear to involve depletion of orotate pools (Gassis and Rathod, 1996) but to collapse mitochondrial membrane potential, which can trigger cell death (Vaidya, 1998).

**Pyrimidine biosynthesis in malaria infected erythrocytes is odd**

Most of what we know about the metabolic pathways of malaria has been learned from studying the blood-stage form of the parasite. In recent years, this is also the form of the parasite that is most accessible, because it can be manipulated in vitro.

Mature erythrocytes are devoid of nucleic acids and of free pyrimidine nucleotides. However, human red blood cells are rich in purines such as ATP. The metabolic machinery of malarial parasites reflects this environment. Malarial parasites efficiently salvage purines and are devoid of enzymes for de novo purine biosynthesis (Reyes et al., 1982; Sherman, 1998). On the other hand, they possess a robust de novo pyrimidine biosynthesis apparatus and lack enzymes for the salvage of pyrimidines (Reyes et al., 1982; Sherman, 1998).

**An opportunity and a challenge**

A combination of a potent inhibitor of de novo pyrimidine biosynthesis inhibitor combined with nucleosides ought to be efficacious, because proliferating human cells utilize preformed pyrimidines efficiently and malaria parasites do not. Failure to identify antimalarials directed at pyrimidine pathway should force us to conclude that our current paradigms for drug development are grossly inadequate.

**Which de novo pyrimidine enzymes to target?**

Of all the enzymes related to the de novo pyrimidine biosynthesis pathway, the two that draw special attention are dihydrofolate reductase (DHFR) and thymidylate synthase (TS). These enzymes historically have proven to be good targets for chemotherapy. DHFR is the target of anticancer drugs such as methotrexate, antibacterial agents such as trimethoprime, and antiprotozoan agents such as pyrimethamine and cycloguanil. TS is the target of anticancer agents such as 5-fluorouracil and ZD1694.

In malarial parasites (Bzik, et al., 1987), as in all other protozoan parasites (Coderre et al., 1983), DHFR and TS are part of a single polypeptide that is synthesized from a single RNA molecule. The amino end of each protein folds into a DHFR domain and the carboxyl end of two polypeptides assemble to form a TS homodimer. Clearly, the bifunctional status of malarial DHFR-TS represents a very significant difference in host-parasite biochemistry but, until recently, it has not been clear how one would exploit this difference for selective chemotherapy.

**Why are DHFR and TS inhibitors such potent cell killers?**
Normally, DNA contains deoxycytidine (dC) and thymidine (dT) as the pyrimidine nucleosides. However, at very low frequency, deoxyuridine (dU) can form in DNA molecules as a result of spontaneous hydrolysis of dC residues (Friedberg et al., 1995). The resulting dU in DNA has to be promptly excised and the DNA has to be repaired to avoid mutagenesis during replication. In addition, dU can also appear in DNA molecules as a result of the DNA polymerase accidentally using dUTP as a substrate instead of dTTP. To avoid this problem, normally, cells go to extraordinary efforts to prevent accumulation of dUTP. Cells contain a dUTPase to quickly degrade any dUTP that is formed (Friedberg, et al., 1995).

In mammalian cells, inhibition of TS activity immediately leads to decreases in TMP and TTP levels and increases in dUMP and dUTP levels. This increase in dUTP overwhelms the standard proofreading and repair mechanisms of the cell and causes excessive incorporation of dU residues into DNA. This, in turn, triggers massive DNA strand fragmentation and cell death (Yoshioka et al., 1987). Inhibition of DHFR in mammalian cells, indirectly, leads to the same type of nucleotide imbalances, strand fragmentations, and cell death (Ingraham et al., 1986). Therefore, part of the reason that DHFR and TS are such good targets for chemotherapy is that accumulation of the substrate (dUMP) and depletion of product (dTMP), in itself, is sufficient to trigger cell death. It is not even necessary to inhibit TS to 100% in order to kill a cell (Houghton et al., 1989).

## INHIBITING MALARIA DHFR WITH SELECTIVITY

A comparison of amino acid sequences between malarial DHFR domain and human DHFR reveals a mere 27% identity (Bzik et al., 1987). This sequence diversity effects the active site. It is known that pyrimethamine binds malarial DHFR as much as 1,000 times more tightly than to human DHFR (Ferone, 1970). The importance of this differential host-parasite binding is underscored by sequencing and kinetic characterization of mutant DHFR isolated from malarial strains that are pyrimethamine resistant and by transfection studies where integration of mutant DHFR-TS sequences into pyrimethamine-sensitive malaria results in resistance to pyrimethamine (Sirawaraporn et al., 1997; Wu et al., 1996).

More recent data on translation control of DHFR-TS suggest that additional factors related to the species-specific cellular responses to DHFR and TS inhibitors may play an important role in the selective antimalarial activity of agents directed at this target (Zhang and Rathod, unpublished data; see discussion below).

## INHIBITING MALARIAL TS WITH SELECTIVITY

Unlike DHFR, TS in living organisms is a very conserved enzyme. A comparison of amino acid coding sequence reveals about 56% identity between human TS and malarial TS domain. Furthermore, unlike the family of DHFR enzymes, which show very different sensitivities to DHFR inhibitors, all TS enzymes have essentially identical kinetic properties and bind TS inhibitors with equal avidity (Hekmat-Nejad and Rathod, 1996; Jiang et al., 2000). So how can one inhibit malarial TS with selectivity? TS catalyses a reaction involving two substrates, dUMP and methylene-tetrahydrofolate (MTHF). There exist substrate analogs of both dUMP and MTHF which inhibit TS enzymes at nanomolar concentrations.

**Selectivity through differential uptake and metabolism of prodrugs**
5-Fluoro-2'-deoxyuridylate (5-FdUMP) binds malarial as well as human TS with a $K_i$ of about 1 nM (Hekmat-Nejad and Rathod, 1996). 5-FdUMP cannot be used as a drug itself because it does not permeate cells. Mammalian cells are extremely vulnerable to 5-fluorouracil and to 5-fluoro-2'-deoxyuridine, because these preformed pyrimidines can be converted to the toxic 5-FdUMP by salvage pathways. Malarial parasites are inherently resistant to these

compounds because they lack the enzymes to activate these molecules to nucleotides (Reyes et al., 1982; Rathod et al., 1989).

However, 5-fluoroorotate is a potent and selective inhibitor of malaria proliferation. It shows an $IC_{50}$ of about 5 nM against all malarial strains tested (Rathod et al., 1989). It doesn't show toxicity to mammalian cells until one uses about 1,000 times higher concentrations. 5-fluoroorotate can cure malaria in mice, whether delivered intraperitoneally or orally (Gomez and Rathod, 1990). 5-Fluoroorotate, after entering malarial parasites, gets metabolised to 5-FdUMP and inactivates TS (Rathod et al., 1992).
The potent antimalarial activity of 5-fluoroorotate is likely due to the selective and efficient transport of this molecule through the tubovessicular membrane (TVM) network of infected erythrocytes (Laur et al., 1997) and the easy activation of this molecule to nucleotides (Rathod and Reyes, 1983). Mutants resistant to 5-fluoroorotate have decreased ability to take up exogenous 5-fluoroorotate and orotate (Rathod et al., 1994).

**Selectivity through differential rescue of mammalian cells**
Folate-based TS inhibitors like D1694 and 1843U89 arrest mammalian cells with an $IC_{50}$ of about 1 to 10 nM (Jackman et al., 1991; Duch et al., 1993). This toxicity can be completely reversed with 10 μM thymidine. The complete lack of toxicity of these folate-based inhibitors in the presence of thymidine underscores the specific and selective activity of the compounds against TS. The compounds do not interfere with any other important function in the cell. Since malarial parasites cannot utilise exogenous thymidine, a combination of a folate-based TS inhibitor and thymidine was expected to inhibit malarial parasites with absolutely no toxicity to host cells (Rathod and Reshmi, 1994).

Recently, 1843U89, even without polyglutamylation, was shown to be a potent folate-based inhibitor of purified malarial TS (Jiang et al., 2000). The binding was noncompetitive with respect to methylenetetrahydrofolate and had a $K_i$ of 1 nM. The compound also had potent antimalarial activity in vitro. *Plasmodium falciparum* cells in culture were inhibited by 1843U89 with an $IC_{50}$ of about 70 nM. The compound was effective against drug-sensitive as well as drug-resistant clones of *P. falciparum*. As predicted by the biochemistry of the parasite, the potent inhibition of parasite proliferation by 1843U89 could not be reversed with 10 μM thymidine (Jiang et al., 2000). In contrast, in the presence of 10 μM thymidine, mammalian cells were unaffected by 1843U89 even at concentrations as high as 0.1 mM. This greater than 10,000 fold in vitro therapeutic window between malarial and mammalian cells is as good as it gets for antimetabolites. On this basis, folate-based TS inhibitors may offer a powerful additional tool to combat drug-resistant malaria.

**An opportunity to interfere with protein-protein interactions**

In *P. falciparum*, the DHFR and TS activities that are conferred by a single 70 kDa bifunctional polypeptide (DHFR-TS) assemble into a functional 140 kDa homodimer. In mammals, the two enzymes are smaller distinct molecules encoded on different genes. A 27 kDa amino acid domain of malarial DHFR-TS is sufficient to provide DHFR activity but, until recently, the structural requirements for TS function had not been established.

Although the 3'-end of DHFR-TS had high homology to TS sequences from other species, expression of this protein fragment failed to yield active TS enzyme and it failed to complement TS$^-$ *E. coli.* (Shallom et al., 1999). Unexpectedly, even partial 5'-deletion of full-length DHFR-TS gene abolished TS function on the 3'-end. Thus, it was hypothesized that the amino end of the bifunctional parasite protein played an important role in TS function. When the 27 kDa amino domain (DHFR) was provided in trans, a previously inactive 40 kDa carboxyl-domain from malarial DHFR-TS regained its TS function. Physical characterization of the "split enzymes" revealed that the 27 kDa and the 40 kDa fragments of DHFR-TS had reassembled into a 140 kDa hybrid complex (Shallom et al., 1999). Therefore, in malarial DHFR-TS, there are physical interactions between the DHFR domain and the TS domain and these interactions are necessary to obtain a catalytically active TS. Interference with these essential protein-protein interactions could lead to new selective strategies to treat malaria resistant to traditional DHFR-TS inhibitors.

It has been argued that electrostatic channelling plays an important role in the kinetics of parasite DHFR-TS: In *Leishmania* and *Toxoplasma*, as much as 80% of DHF may move directly from TS to DHFR without equilibrating with bulk solvent (Meek et al., 1985; Trujillo et al., 1996). However, the importance of this channelling to malaria pharmacology is probably negligible. The active sites of purified DHFR-TS can accept substrates from bulk solvent, they are fully active when the other site is inhibited, and, in intact parasites, human DHFR can replace the functions of an inhibited malaria DHFR even when the external DHFR is not setup for channelling (Fiddock and Wellems, 1997).

**Malarial serine hydroxymethyltransferase**

Three enzymes are involved in methylenetetrahydrofolate recycling in malaria: TS, DHFR, and serine hydroxymethyltransferase (SHMT). Unlike DHFR-TS, properties of malarial SHMT have remained a mystery for a long time. Recently, the gene for this enzyme was cloned, sequenced, and the SHMT protein from *P. falciparum* was expressed in functional form (Alfadhli and Rathod, 2000). The genomic sequence had 1,485 bp including a 159 bp intron near the 5'-end of the gene. The open reading frames coded for a 442 amino acid protein with 38 to 47% identity to SHMT sequences from other species. The function of this sequence was established through transformation of malarial SHMT coding sequence (minus the intron) in *glyA* mutants of *E. coli*. Expression of malarial SHMT relieved glycine auxotrophy in these mutants and permitted assay of SHMT catalytic activity in bacterial cell lysates.

Thus, it may be possible to attack malarial TS indirectly not just by inhibiting malarial DHFR and attacking de novo folate biosynthesis, but also by attacking SHMT. The malarial enzyme and the heterologous expression system described above will be useful for screening of SHMT inhibitors and for fine tuning the additional broad strategies for selectively inhibiting malarial TS indirectly.

## CELLULAR RESPONSE TO DHFR-TS INHIBITORS

**How do mammalian cells respond to inhibitors of DHFR and TS?**

During cancer chemotherapy, treatment of mammalian cells with inhibitors of DHFR or TS is associated with large accumulation of dead target protein (Chu et al., 1991). Careful analysis of this phenomena has revealed that inhibition of TS or DHFR triggers production of more TS or DHFR protein, respectively. This induction of protein synthesis appears to be due to de-repression of translation. Mammalian DHFR binds its own RNA and mammalian TS binds its own RNA (Chu et al., 1993). This controls the amount of DHFR or TS that is made at a given time. The pharmacological relevance comes from the fact that when the enzyme is inhibited by a drug, the RNA is no longer associated with the protein and is free to synthesise more of the target protein. As a result of such a mechanism operating in mammalian cells, larger quantities of DHFR and TS inhibitors are needed to kill a mammalian cell, because as protein binds incoming drug, the cell responds by making more target protein. Recently, it has been argued that the post-transcriptional increase in TS levels in response to TS inhibitors is primarily due to stabilisation of the target protein and not an increase in protein synthesis (Kitchens et al., 1999). However, this study has its own ambiguities. It may be relying on assumptions that may not be valid (e.g. the rate-limiting step in TS synthesis was expected to be the loading of the ribosomes, it is assumed that a single binding site on the TS RNA is solely responsible for all the translational control of TS).

**Is there translational regulation of DHFR-TS in malarial parasites?**

Since post-transcriptional regulation of TS is seen in *E. coli*, mice, and humans, it is likely that malarial parasites may also use such a mechanism. Whether malarial parasites use translational control mechanisms to regulate DHFR-TS or not has profound implications on what type of drugs are likely to be effective antimalarials. Given the bifunctional nature of malarial DHFR-TS and not the host, there are opportunities for differential host-parasite responses to antimetabolites. If malarial bifunctional DHFR-TS involves protein-RNA contact, which inhibitors of DHFR or TS would successfully release RNA from protein and, thereby, promote translation of more DHFR-TS protein? This question is important because compounds which can inhibit DHFR action or TS action without inducing synthesis of more target protein are likely to be effective at very low concentrations. This type of analysis also has bearing on whether drug combinations will be synergistic or antagonistic. For instance, if both the DHFR and TS active sites are essential for binding RNA tightly, inhibitor of one or the other active site will work as antimalarials at low concentrations because the other active site will continue to repress translation. However, in this example, a combination of DHFR and TS inhibitors would promote protein synthesis and be antagonistic. Preliminary data shows that malarial DHFR-TS protein binds its own RNA tightly and selectively (Zhang and Rathod, unpublished data).

**Translation control of DHFR-TS may explain some mysteries in malaria**

Malarial parasites have as few as five hundred molecules of DHFR-TS per cell which is about 1,000 times lower than what is seen in mammalian cells (Zhang and Rathod, unpublished data). This low-level expression of malarial DHFR-TS in the parasite may be related to the difficulties in overexpressing malarial DHFR-TS in heterologous systems such as *E. coli*. It has generally been assumed that the later probably has to do with the odd codon usage of malarial genes (Sirawaraporn et al., 1997). While this may be partially true, tight binding of malarial DHFR-TS to RNA would also limit its expression in malaria as well as in heterologous expression systems.

It is not fully understood why compounds such as WR99210 inhibit malarial, but not host cell proliferation. Fidock and Wellems (1997) have clearly demonstrated that transfection of human DHFR into the malarial parasite confers 4,000-fold resistance to WR99210, leaving little doubt that DHFR is the target of WR99210. However, kinetic studies show that WR99210 is a potent inhibitor of both malarial and host DHFR (Zhang and Rathod, unpublished results). The selective activity of WR99210 may have a lot to do with the differences in cellular response of mammalian versus malarial parasites to the antifolate. Additionally, the potent and selective activity of pyrimethamine may have to do with more than just tighter binding of this compound to malarial DHFR.

While it is clear that some mutations in the DHFR-TS region confer higher $K_i$ values for DHFR inhibitors, as one analyses parasites with many mutations in the DHFR region, there isn't a perfect one-to-one correlation between $K_i$ values and the degree of resistance conferred in the parasite (Sirwarapom et al., 1997). This is true even after one removes the "folate effect" by measuring inhibition of parasite proliferation in folate-free media. It is possible that some point mutations in RNA and in protein result in weaken translational inhibition and increased target production, which can contribute to resistance.

## INITIATION OF DRUG RESISTANCE IN MALARIA

### Population sizes of parasites and emergence of resistance

Treating population-based infectious diseases like malaria with chemotherapeutic agents is fundamentally different than treating patient-centered diseases, such as cancer or heart ailment. When a drug fails in cancer treatment, the resistant cells do not pass on to a new patient. In cancer chemotherapy, drug resistance in each patient has to initiate as a de novo event. In sharp contrast, when a malaria patient acquires resistance to a drug, that drug resistance trait can be propagated through a vector to the parasite populations at large and the drug is lost indefinitely.

> *For these reasons, every drug developed for malaria chemotherapy has to be effective not just against one individual or a village but against*

*parasites in all the people in the world for a sustained amount of time.*

**Sources of genetic variation**

The ability of DNA polymerases to discriminate between nucleoside triphosphate substrates, their ability to correct mistakes through proofreading mechanism, and the ability of DNA repair enzymes to correct errors in newly synthesised strands, together, determine the fidelity of DNA replication (Friedberg et al., 1995). For most normal cells the error rate is usually about $10^{-10}$ per base pair per replication (Table 1). It is such error rates that contribute to genetic diversity and initiation of drug resistance. If a drug can be compromised by a specific type of point mutation, to find one cell that has the appropriate nucleotide substitution at a given base in the parasite genome, one would have to start with a population of at least $10^{10}$ cells.

**Table 1**
**Comparing standard frequencies of mutations to parasite population sizes**

| RATES PER GENERATION | | PARASITE POPULATION SIZES | |
|---|---|---|---|
| | | Mosquito Bite | $10^{2}$ parasites |
| Loss of function | $10^{-7}$ | | |
| 1 nucleotide change | $10^{-10}$ | One patient | $10^{11}$ parasites |
| | | A million patients | $10^{17}$ parasites |
| 2 simultaneous changes | $10^{-20}$ | The world | $10^{19}$ parasites |
| | | 10 year use of a drug | $10^{22}$ parasites |

**Parasite population sizes**

A typical patient with serious malaria can have $10^{11}$ parasites in the body. If the mutation rate in malaria is similar to other "standard" organisms, no matter how selective and how safe a drug appears in preclinical trials and if the effectiveness of an antimalarial agent can be compromised by one point mutation, the drug will never be able to cure a single patient, leave alone a global population of parasites.

If a drug can be compromised by a loss of function mutation (whose frequencies can be as high as $10^{-6}$ and $10^{-7}$), the drug may not even be able to eliminate all parasites in a simple, small-animal model (Gomez and Rathod, 1990; Vaidya, 1998).

Even for drugs or some drug combinations that require three to four point mutations to be relegated as useless (Cowman, 1998), if there are incremental benefits to the parasite from single point mutations (Sirawaraporn, 1997; Reynolds and Roos, 1998; Cowman, 1998), it is likely that the drug would be compromised in a relatively short time.

> *New antimalarial agents that have staying power will be those that require many point mutations simultaneously before they are compromised.*

**Drug combinations**

In many discussions on drug combinations, attention is focused on the importance of synergistic combinations (Canfied et al., 1995). In an individual patient, synergy can be very useful, because it decreases potential toxic effects. However, on a population scale, the true powers of drug combinations come from the fact that the frequency of resistance to two compounds that work through independent mechanisms is a multiple of the frequencies of resistance to each compound.

In an experimental system, two non-synergistic compounds, 5-fluoroorotate and atovaquone, could eliminate exponentially larger populations of parasites compared to each compound alone (Gassis and Rathod, 1996). Drug combinations should not be dismissed or discounted, because they lack synergistic effects against small parasite populations.

**Variations in frequency of drug resistance in malaria**

Recent history teaches us that drug-resistance to previously successful antimalarial agents arose from specific parts of the world (Trigg and Kondrachine, 1998). Traditional explanation for such observations could be that parasites in places like Southeast Asia simply have had greater opportunities for developing resistance because parasites in that part of the world have been exposed to more new drugs than anywhere else. One could also argue that physicians or patients in that part of the world have not used drugs as recommended. However, another plausible possibility is that, parasites in these regions have developed generic mechanisms for initiating drug resistance at a more rapid rate than parasites in other parts of the world.

To test this hypothesis, different *P. falciparum* clones were treated with two new antimalarial agents: 5-fluoroorotate and atovaquone (Rathod et al., 1997). All parasite populations were equally susceptible in small numbers.

However, when large populations of these clones were challenged with either of the two compounds, significant variations in frequencies of resistance became apparent. On one extreme, clone D6 from West Africa, which was sensitive to all traditional antimalarial agents, failed to develop resistance under simple non-mutagenic conditions in vitro. In sharp contrast, the Indochina clone W2, which was known to be resistant to all traditional antimalarial drugs, independently acquired resistance to both new compounds as much as a 1,000 times more frequently than D6. Additional clones which were resistant to some (but not all) traditional antimalarial agents acquired resistance to atovaquone at high frequency, but not to 5-fluoroorotate. These findings were unexpected and surprising based on current views of the evolution of drug resistance in *P. falciparum* populations. Such new phenotypes, named Accelerated Resistance to Multiple Drugs (ARMD), raise important questions about the genetic and biochemical mechanisms related to the initiation of drug resistance in malarial parasites. Some potential mechanisms underlying ARMD phenotypes have public health implications that are ominous.

## CONCLUSIONS

Experience with inhibitors of de novo pyrimidine biosynthesis is teaching us that:

(i) it is possible to understand the molecular basis of successful chemotherapy against malaria,
(ii) it is possible to make successful predictions on what types of compounds would be efficacious against malaria infections, and
(iii) it is possible to understand fundamental new rules that govern initiation of drug resistance in malaria.

At the same time, it is also becoming clear that our old, simple generic models and paradigms for drug development may be inadequate for the efficient development of a new generation of antimalarials. As our understanding of cell function has matured in the last fifty years, our models for successful chemotherapy need to keep pace. The new emerging tools for malaria functional genomics (Hayward et al., 2000; Wellems et al., 1999) will undoubtedly offer unexpected views of cellular processes that underlie selective chemotherapy and our models and paradigms will continue to improve. Studies on inhibitors of pyrimidine metabolism in malaria, hopefully, will continue to help in this process.

## ACKNOWLEDGEMENT

This review has focused primarily on research from the author's laboratory. Due to the breadth of the topics covered and to space limitations, all relevant manuscripts could not be cited. PKR has been supported by grants from the National Institute of Allergy and Infectious Diseases and by a New Initiatives in Malaria Research Award from the Burroughs Wellcome Fund.

## REFERENCES

Alfadhli, S., and Rathod, P. K. (2000). *Plasmodium falciparum*: Cloning sequencing and functional expression of serine hydroxymethyltransferase. Exp. Parasitol., submitted.

Bzik, D. J., Li, W. B., Horii, T., and Inselburg, J. (1987). Molecular cloning and sequence analysis of the *Plasmodium falciparum* dihydrofolate reductase-thymidylate synthase gene. Proc. Natl. Acad. Sci. (USA) *84*, 8360-8364.

Canfield, C. J., Pudney, M., and Gutteridge, W. E. (1995). Interactions of atovaquone with other antimalarial drugs against Plasmodium falciparum in vitro. Exp Parasitol. *80*, 373-381.

Chu, E., Koeller, D.M., Casey, J.L., Drake, J.C., Chabner, B.A., Elwood, P.C., Zinn, S., and Allegra, C.J. (1995). Autoregulation of human thymidylate synthase messenger RNA translation by thymidylate synthase. Proc Natl Acad Sci (U S A). *88*, 8977-8981.

Chu, E., Takimoto, C.H., Voeller, D., Grem, J.L., and Allegra, C.J. (1993). Specific binding of human dihydrofolate reductase protein to dihydrofolate reductase messenger RNA in vitro. Biochemistry *32*, 4756-4760.

Coderre, J. A., Beverley, S. M., Schimke, R. T., and Santi, D. V. (1983). Overproduction of a bifunctional thymidylate synthetase-dihydrofolate reductase and DNA amplification in methotrexate-resistant *Leishmania tropica*. Proc. Natl. Acad. Sci. (USA) *80*, 2132-2136.

Cowman, A. F. (1998). The molecular basis of resistance to the sulfones, sulfonamides, and dihydrofolate reductase inhibitors, p. 317-330 *In* I. W. Sherman (ed.), Malaria: Parasite Biology, Pathogenesis, and Protection, ASM Press, Washington, D.C.

Diggens, S. M., Gutteridge, W. E., and Trigg, P. I.. (1970). Altered dihydrofolate reductase associated with a pyrimethamine-resistant *Plasmodium berghei berghei* produced in a single step. Nature *228*, 579-580.

Duch, D.S., Banks, S., Dev, I. K., Dickerson, S. H., Ferone, R., Heath, L. S., Humphreys, J., Knick, V., Pendergast, W., Singer, S., et al. (1993). Biochemical and cellular pharmacology of 1843U89, a novel benzoquinazoline inhibitor of thymidylate synthase. Cancer Res. 53: 810-818.

Ferone, R. (1970). Dihydrofolate reductase from pyrimethamine-resistant *Plasmodium berghei*. J Biol Chem *245*, 850-854.

Fidock, D. A., and Wellems, T. E. (1997) Transformation with human dihydrofolate reductase renders malaria parasites insensitive to WR99210 but does not affect the intrinsic activity of proguanil. Proc Natl Acad Sci (U S A) *94*, 10931-10936.

Friedberg, E.C., Walker, G. C., and Siede, W. 1995. "DNA repair and mutagenesis", 2nd Ed., ASM press, Washington DC.

Gassis, S and Rathod, P. K. (1996). Frequency of drug resistance in *Plasmodium falciparum*: a nonsynergistic combination of 5-fluoroorotate and atovaquone suppresses in vitro resistance. Antimicrob. Agents Chemother. *40*, 914-919.

Gomez, Z. M., and Rathod, P. K. (1990). Antimalarial activity of a combination of 5-fluoroorotate and uridine in mice. Antimicrob. Agents Chemother. *34*, 1371-1375.

Hayword, R., DeRisi, J., Alfadhli, S., Kaslow, D., Brown, P., and Rathod, P.K. (2000). "Shotgun DNA microarrays and stage-specific gene expression in *Plasmodium falciparum* malaria", Mol. Microbiol., in press.

Hekmat-Nejad, M., Lee, P-C.and Rathod, P. K. (1997a). *Plasmodium falciparum*: Direct cloning and expression of pyrimethamine sensitive and pyrimethamine resistant dihydrofolate reductase domains. Exp. Parasitol. *85*, 303-305.

Hekmat-Nejad, M., Lee, P-C.and Rathod, P. K. (1997b) *Plasmodium falciparum*: Kinetics of interactions of WR99210 with pyrimethamine-sensitive and pyrimethamine-resistant dihydrofolate reductase domains. Exp. Parasitol., *87*, 222-228.

Hekmat-Nejad, M., and Rathod, P. K. (1996). Kinetics of *Plasmodium falciparum* thymidylate synthase: interactions with high-affinity metabolites of 5-fluoroorotate and D1694. Antimicrob. Agents Chemother. *40*, 1628-1632.

Houghton, P. J., Germain, G. S., Hazelton, B. J., Pennington, J. W., and Houghton, J. A. (1989). Mutants of human colon adenocarcinoma, selected for thymidylate synthase deficiency. Proc. Natl. Acad. Sci. (USA). *86*, 1377-1381.

Ingraham, H. A., Dickey, L., and Goulian., M. (1986). DNA fragmentation and cytotoxicity from increased cellular deoxyuridylate. Biochemistry *25*, 3225-3230.

Jackman, A.L., Taylor, G. A., Gibson, W., Kimbell, R., Brown, M., Calvert, A. H., Judson, I. R., and Hughes, L. R. (1991). ICI D1694, a quinazoline antifolate thymidylate synthase inhibitor that is a potent inhibitor of L1210 tumor cell growth in vitro and in vivo: a new agent for clinical study. Cancer Res. *51*, 5579-5586.

Jiang, L., Lee, P.-C., White, J., and Rathod, P. K. (2000) "Potent and selective activity of a combination of thymidine and 1843U89, a folate-based thymidylate synthase inhibitor, against *Plasmodium falciparum*" , Antimicrob. Agents Chemother., in press.

Kitchens, M. E., Forsthoefel, A. M., Rafique, Z., Spencer, H. T., Berger, F. G. (1999) Ligand-mediated induction of thymidylate synthase occurs by enzyme stabilization. Implications for autoregulation of translation. J Biol Chem *274*, 12544-12547.

Krogstad, D. J., and De, D. (1998). Chloroquine: Modes of action and resistance and the activity of chloroquine analogs. p. 331-339. *In* I. W. Sherman (ed.), Malaria: Parasite Biology, Pathogenesis, and Protection, ASM Press, Washington, D.C.

Kunz, B. A., Kohalmi, S. E., Kunkel, T. A., Mathews, C. K., McIntosh, E. M., and Reidy, J. A.. (1994). International Commission for Protection Against Environmental Mutagens and Carcinogens. Deoxyribonucleoside triphosphate levels: a critical factor in the maintenance of genetic stability. Mutat. Res. *318*, 1-64.

Laur, S. A. , Rathod, P. K., Ghori, N. and Haldar, K. (1997). "A membrane network for nutrient transport in red cells infected with the malaria parasite", Science *276*, 1122-1125.

Meek, T. D., Garvey, E. P., and Santi, D. V. (1985) Purification and characterization of the bifunctional thymidylate synthetase-dihydrofolate reductase from methotrexate-resistant Leishmania tropica. Biochemistry *24*, 678-686.

Milhous, W. K. and Kyle, D. E. (1998). In troduction to the modes of action of and mechanisms of resistance to antimalarials, p. 303-316. *In* I. W. Sherman (ed.), Malaria: Parasite Biology, Pathogenesis, and Protection, ASM Press, Washington, D.C.

Milhous, W.K., Weatherly, N.F., Bowdre, J.H., Desjardins, R.E.. (1985). In vitro activities of and mechanisms of resistance to antifol antimalarial drugs. Antimicrob Agents Chemother. *27*, 525-30.

Phillips, M.A., Coffino, P., Wang, C.C. (1987). Cloning and sequencing of the ornithine decarboxylase gene from *Trypanosoma brucei*. Implications for enzyme turnover and selective difluoromethylornithine inhibition. J Biol Chem *262*, 8721-8727.

Rathod, P. K., and Gomez, Z. M. (1991). *Plasmodium yoelii*: oral delivery of 5-fluoroorotate to treat malaria in mice. Exp. Parasitol. *73*, 512-514.

Rathod, P. K., Khatri, A., Hubbert, T., Milhous, W. K.. (1989). Selective activity of 5-fluoroorotic acid against Plasmodium falciparum in vitro. Antimicrob. Agents Chemother. *33*, 1090-1094.

Rathod, P. K., Khosla, M., Gassis, S., Young, R. D., and Lutz, C. (1994). Selection and characterization of 5-fluoroorotate-resistant *Plasmodium falciparum*. Antimicrob. Agents Chemother. *38*, 2871-2876.

Rathod, P. K. Leffers, N. P., and Young, R. D. (1992). Molecular targets of 5-fluoroorotate in the human malaria parasite, *Plasmodium falciparum*. Antimicrob. Agents Chemother. *36*, 704-711.

Rathod, P. K., McErlean, T., and Lee, P. C.. (1997). Variations in frequencies of drug resistance in *Plasmodium falciparum*. Proc. Natl. Acad. Sci. (USA) *94*, 9389-9393.

Rathod, P. K., and Reshmi, S. (1994). Susceptibility of *Plasmodium falciparum* to a combination of thymidine and ICI D1694, a quinazoline antifolate directed at thymidylate synthase. Antimicrob. Agents Chemother. *38*, 476-480.

Rathod, P. K. and Reyes, P. 1983. "Orotidylate metabolizing enzymes of the human malarial parasite, *Plasmodium falciparum*, differ from host cell enzymes". J. Biol. Chem. *258*, 2852-2855.

Reyes, P., Rathod, P. K., Sanchez, D. J., Mrema, J. E., Rieckmann, K. H., and Heidrich, H. G. (1982). Enzymes of purine and pyrimidine metabolism from the human malaria parasite, *Plasmodium falciparum*. Mol. Biochem. Parasitol. *5*, 275-290.

Reynolds, M. G. and Roos, D. S. (1998). A biochemical and genetic model for parasite resistance to antifolates. *Toxoplasma gondii* provides insights into pyrimethamine and cycloguanil resistance in *Plasmodium falciparum*. J. Biol. Chem. *273*, 3461-3469.

Seymour, K. K., Lyons, S. D., Phillips, L., Rieckmann, K. H., and R. I. Christopherson, R. I. (1994). Cytotoxic effects of inhibitors of *de novo* pyrimidine biosynthesis upon *Plasmodium falciparum*. Biochemistry *33*, 5268-5274.

Shallom, S., Zhang, K., Jiang, L., and Rathod, P. K. (1999). Essential protein-protein interactions between *Plasmodium falciparum* thymidylate synthase and dihydrofolate reductase domains. J. Biol. Chem., in press.

Sherman, I. W. 1998. Purine and pyrimidine metabolism of asexual stages, p.177-184 *In* I. W. Sherman (ed.), Malaria: Parasite Biology, Pathogenesis, and Protection, ASM Press, Washington, D.C.

Sirawaraporn, W., Sathitkul, T., Sirawaraporn, R., Yuthavong, Y., Santi, D.V. (1997) Antifolate-resistant mutants of *Plasmodium falciparum* dihydrofolate reductase. Proc Natl Acad Sci (U S A) *94*, 1124-1129.

Sullivan, D.J. Jr, Matile, H., Ridley, R.G., Goldberg, D.E. (1998). A common mechanism for blockade of heme polymerization by antimalarial quinolines. J. Biol. Chem. *273*, 31103-31107.

Srivastava, I. K., and Vaidya, A. B. (1999). A mechanism for the synergistic antimalarial action of atovaquone and proguanil. Antimicrob. Agents Chemother. *43*, 1334-1339.

Trigg, P. I., and A. V. Kondrachine. (1998). The current global malaria situation, p.11-22 *In* I. W. Sherman (ed.), Malaria: Parasite Biology, Pathogenesis, and Protection, ASM Press, Washington, D.C.

Trujillo, M., Donald, R. G., Roos, D. S., Greene, P. J., and Santi, D. V. (1996) Heterologous expression and characterization of the bifunctional dihydrofolate reductase-thymidylate synthase enzyme of Toxoplasma gondii. Biochemistry *35*, 6366-6374.

Vaidya, A. B. (1998) Mitochondrial physiology as a target for atovaquone and other antimalarials. p. 355-368. *In* I. W. Sherman (ed.), Malaria: Parasite Biology, Pathogenesis, and Protection, ASM Press, Washington, D.C.

van Dijk, M.R., Waters, A.P., Janse, C.J. (1995). Stable transfection of malaria parasite blood stages. Science *268*, 1358-1362.

Wang, P., Brobey, R.K., Horii, T., Sims, P.F., Hyde, J.E. (1999) Utilization of exogenous folate in the human malaria parasite *Plasmodium falciparum* and its critical role in antifolate drug synergy. Mol Microbiol. *32*, 1254-1262.

Wellems, T. E., Su, X-Z., Ferdig, M., and Fidock, D. A. (1999) Genome projects, genetic analysis and the changing landscape of malaria research. *Curr Opin Microbiol* **2**: 415-419

White, N. J. (1992). Antimalarial drug resistance: the pace quickens. J. Antimicrob. Chemother. *30*, 571-585.

Wu, Y, Kirkman, L. A., and Wellems, T. E. (1996). Transformation of *Plasmodium falciparum* malaria parasites by homologous integration of plasmids that confer resistance to pyrimethamine. Proc. Natl. Acad. Sci. (USA). *93*, 1130-1134.

Yoshioka, A., Tanaka, S., Hiraoka, O., Koyama, Y., Hirota, Y., Ayusawa, D., Seno, T., Garrett, C., and Wataya, Y. (1987). Deoxyribonucleoside triphosphate imbalance. 5-Fluorodeoxyuridine-induced DNA double strand breaks in mouse FM3A cells and the mechanism of cell death. J. Biol. Chem. *262,* 8235-8241.

# 5

# POLYAMINE AND GLUTATHIONE BIOSYNTHETIC ENZYMES FROM *TRYPANOSOMA BRUCEI* AND *TRYPANOSOMA CRUZI*

Lisa N. Kinch, Deirdre L. Brekken, Margaret A. Phillips
*University of Texas Southwestern Medical School, Dallas, TX 75235-9041*

## OVERVIEW AND BACKGROUND INFORMATION

### The polyamine and glutathione biosynthetic pathways

The polyamines putrescine and spermidine are ubiquitous cell growth factors that are synthesized from ornithine and S-adenosylmethionine (Fig. 1). Inhibition of the polyamine biosynthetic enzymes, or the knockout of genes encoding these enzymes, causes cell growth arrest in both prokaryotic and eukaryotic cells, e.g. *E. coli*; (Tabor and Tabor, 1984), mammalian cells (Svensson and Persson, 1996), yeast (Cohn et al., 1980), and *Trypanosoma brucei* (Li et al., 1996). A number of inhibitors of polyamine biosynthesis have been demonstrated to be effective anti-trypanosomal agents (Wang, 1995), identifying these enzymes as drug targets for the treatment of trypanosomatid infections. In addition to the common pathway found in almost all cell types, protozoa from the family Trypanosomatidae synthesize a unique cofactor that is a conjugate of spermidine and glutathione (Fig. 1). This cofactor, trypanothione, is required to maintain redox balance in the cell (Fairlamb and Le Quesne, 1997).

### Ornithine decarboxylase (ODC) as a drug target

Ornithine decarboxylase catalyses the first committed step in the biosynthesis of polyamines (Fig. 1). A number of mechanism-based inhibitors of ODC have been designed and tested as anti-trypanosomal agents (Bitonti et al.,

1985). The best characterized of these inhibitors, α-difluoromethylornithine (DFMO; eflornithine), was demonstrated to cure mice infected with the African trypanosome, *Trypanosoma brucei* and was subsequently found to cure both early and late stage Gambian trypanosomiasis in man (Wang, 1995). DFMO was approved for the treatment of this disease in 1990. DFMO depletes the parasites of ODC activity and of putrescine, spermidine and trypanothione. Cell growth arrest is the final consequence. This growth arrest is fully reversible by putrescine supplementation (Bacchi and Yarlett, 1993). Further, the ODC knockout mutant of *T. brucei* is dependent on putrescine for growth (Li et al., 1996). These studies demonstrate that ODC is the sole target of DFMO action.

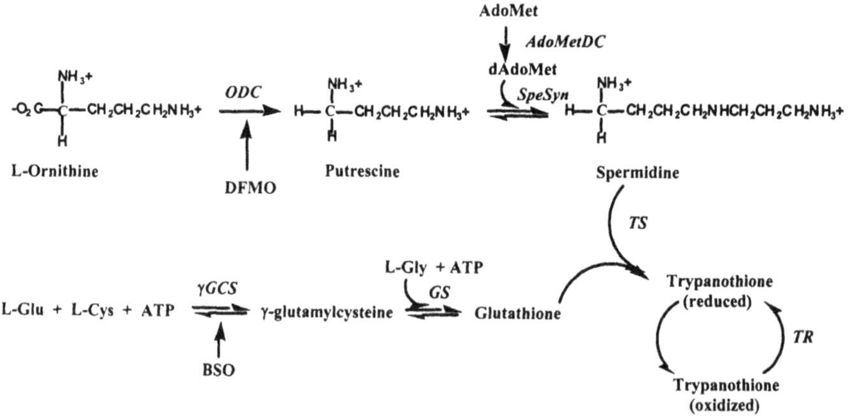

**Figure 1.** Polyamine and glutathione metabolism in trypanosomes. ODC, ornithine decarboxylase; AdoMetDC, S-adenosylmethionine decarboxylase; γGCS, γ-Glutamylcysteine synthetase; GS, glutathione synthetase; TS, glutathionyl-spermidine and trypanothione synthetases; TR, trypanothione reductase; SpeSyn, spermidine synthetase; DFMO, difluoromethylornithine; BSO, buthionine sulfoximine.

The basis for the selective toxicity of DFMO on the parasite is not well established (Wang, 1995). Selective inhibitor binding to ODC is not a factor, however, multiple metabolic differences have been identified. First, mammalian ODC is short lived with an intracellular half-life of 15 to 60 min, while *T. brucei* ODC is stable (Wang, 1995). These results suggest that while both enzymes are inhibited, the mammalian enzyme can be rapidly resynthesized, which may allow the cell to overcome the DFMO inhibition. Secondly, DFMO depletes the *T. brucei* cells of trypanothione hindering their ability to maintain cellular redox balance (Fairlamb and Le Quesne, 1997). This novel role for spermidine may make the cells more sensitive to polyamine depletion than mammalian cells. Finally, in *T. brucei* cells, DFMO treatment elevates the levels of AdoMet. Increased AdoMet levels cause hypermethylation of DNA, an outcome that potentially may impact cell growth rates (Bacchi et al., 1993, Goldberg et al., 1999). In contrast, mammalian cells tightly regulate AdoMet synthetase and DFMO treatment does not increase AdoMet levels.

Despite the success of DFMO against *T. b. gambiense* and its low toxicity in humans, the pharmacology of DFMO is relatively poor; large doses of drug must be administered by IV for an extended course (150 mg/kg, dosed every 6 hr for 14 days; Kuzoe, 1993). Additionally, clinical isolates of *T. b. rhodesiense* are naturally resistant to DFMO, although the mechanism of resistance is unclear. Differences between *T. b. gambiense* and *T. b. rhodesiense* have been found in ODC turnover (Iten et al., 1997), DFMO uptake (Bacchi et al., 1993) and in AdoMet levels (Bacchi et al., 1993). The problems of drug resistance and poor pharmacology demonstrate that there is a need for novel ODC inhibitors that can overcome these shortcomings.

Finally, DFMO is ineffective against the intracellular parasite *T. cruzi*. A series of recent experiments demonstrate that *T. cruzi* cells lack ODC as well as arginine and lysine decarboxylase activities (Fairlamb and Le Quesne, 1997). In contrast to *T. brucei*, *T. cruzi* cells have fast, high affinity transporters for putrescine and cadaverine and they have access to substantial intracellular pools of these amines, which are not found in serum. Thus, *T. cruzi* relies on scavenged putrescine and cadaverine to maintain its polyamine pools (Fairlamb and Le Quesne, 1997).

**AdoMetDC as a drug target**

AdoMetDC catalyses the decarboxylation of AdoMet to produce the substrate that serves as the aminopropyl donor for the formation of spermidine from putrescine (Fig. 1). Several very promising in vivo trials have been undertaken demonstrating that AdoMetDC inhibitors can cure *T. brucei* infections in mice, including a synthetic adenosine analogue that is a mechanism based inhibitor of AdoMetDC (MDL73811 (5'-[Z-4-amino-2butenyl)methylamino]-5'-dexyadenosine; Bacchi et al., 1992). A novel AdoMet transporter has been identified in *T. brucei*, but not in mammalian cells, suggesting that uptake of AdoMet analogs into the parasite but not the host could form the basis for selective toxicity (Goldberg et al., 1999). Unlike ODC, AdoMetDC has been identified in all trypanosomatids studied to date. Several studies have been reported that demonstrate AdoMetDC inhibitors have efficacy against intracellular parasites. For example, MDL73811 decreases the ability of *T. cruzi* to infect and to multiply in rat heart myoblasts (Yakubu et al., 1993). Inhibitors of AdoMetDC were also shown to arrest *Plasmodium falciparum* growth at the trophozoite stage of the erythrocytic cycle (Das et al., 1997).

**γGCS as a drug target**

Major differences in the utilization of spermidine and glutathione have been found between trypanosomatids and mammalian cells. Mammalian cells rely on the antioxidant glutathione to protect against oxidative injury by peroxides or free radicals and to detoxify xenobiotics (Griffith and Mulcahy, 1999). In place of glutathione, trypanosomes utilize trypanothione (Fairlamb and Le Quesne, 1997). Trypanothione is synthesized from glutathione and spermidine (Fig. 1). The reduced form of trypanothione is maintained by trypanothione reductase (TR), a homologue of mammalian glutathione reductase (Fairlamb and Le Quesne, 1997). The first step in the biosynthesis of glutathione is catalysed by γ-glutamylcysteine synthetase (γGCS; Fig. 1), which has been demonstrated to be the rate-limiting enzyme in the biosynthesis of glutathione in mammalian cells (Griffith and Mulcahy 1999) and of trypanothione in *L. tarentolae* (Grondin et al., 1997).

An enzyme-activated inhibitor of γGCS, buthionine sulfoximine (BSO), was shown to cure or prolong survival of mice infected with *T. brucei* (Arrick et al., 1981). This study implicated γGCS as a potential drug target. The effectiveness and selectivity of BSO against *T. brucei* infection suggests that trypanosomes are more sensitive to glutathione depletion than mammalian cells. This conclusion is supported by the findings that knockout of TR is lethal in *T. brucei* (Clayton, personal communication) and that deletion of a single allele of TR in *L. donovani* decreased their ability to survive in macrophages (Dumas et al., 1997). Trypanosomes have been reported to lack catalase (Mehlotra, 1996), which in mammals breaks down hydrogen peroxide in the peroxisome. This finding may account for the detrimental effect of glutathione depletion (Meshnick et al., 1977). *T. cruzi*, *L donovani* and *P. falciparum* also have limited ability to detoxify hydrogen peroxide (Mehlotra, 1996). Recent studies suggest that *P. falciparum* is unable to scavenge glutathione from red blood cells and must synthesize glutathione de novo (Atamna and Ginsburg, 1997). In vitro cultures of *P. falciparum* are killed by BSO (Geary et al., 1985) and by inhibitors of glutathione reductase (Schirmer et al., 1995). Thus, inhibitors of glutathione biosynthesis may also have efficacy against intracellular parasites.

Inhibitors of γGCS may also have a role in the treatment of drug resistant parasites or in enhancing the potency of existing chemotherapy. Up-regulation of glutathione is often associated with resistance of cancer cells to many chemotherapeutic agents (Schroder et al., 1996), and in clinical trials co-treatment with BSO restores sensitivity to patients with drug resistant cancer (Schroder et al., 1996). γGCS is amplified in *L. tarentolae* cells resistant to antimonials and arsenicals, resulting in increased pools of trypanothione and glutathione (Grondin et al., 1997). Similarly, chloroquine resistance in *P. berghei* correlates with increased levels of both glutathione and ABC transporters (Dubois et al., 1995). It is not known if γGCS activity is up-regulated in these resistant cells, however BSO was able to partially restore drug sensitivity. Finally, BSO potentiates the toxic effects of nifurtimox and benznidazole on *T. cruzi* epimastigotes in culture (Moncada et al., 1989).

## EXPERIMENTAL APPROACHES

### Structural, functional and mechanistic analysis of *T. brucei* ODC

Eukaryotic ODC is a pyridoxal 5'-phosphate (PLP) dependent enzyme that is active only as a dimer (Phillips, 1999). The extensive family of eukaryotic enzymes (Prosite # PS00878) are related to bacterial arginine decarboxylase, but not to bacterial ornithine decarboxylase (Grishin et al., 1995). Each monomer is composed of two domains and the N-terminal domain folds into a β/α barrel. The X-ray structures of mouse (Kern et al., 1999), human (Almrud et al., 1999) and *T. brucei* (Grishin et al., 1999) ODC have all recently become available. The *T. brucei* ODC structure solved in our lab is bound to the irreversible inhibitor, DFMO and provides insight into the nature of the substrate binding site and the catalytic function of residues found in the active site. The two identical active sites are composed of residues from the N-terminal domain of one monomer and the C-terminal domain (labelled B) of the other (Fig. 2). PLP is bound primarily by the N-terminal β/α barrel domain, while DFMO is bound in the subunit interface.

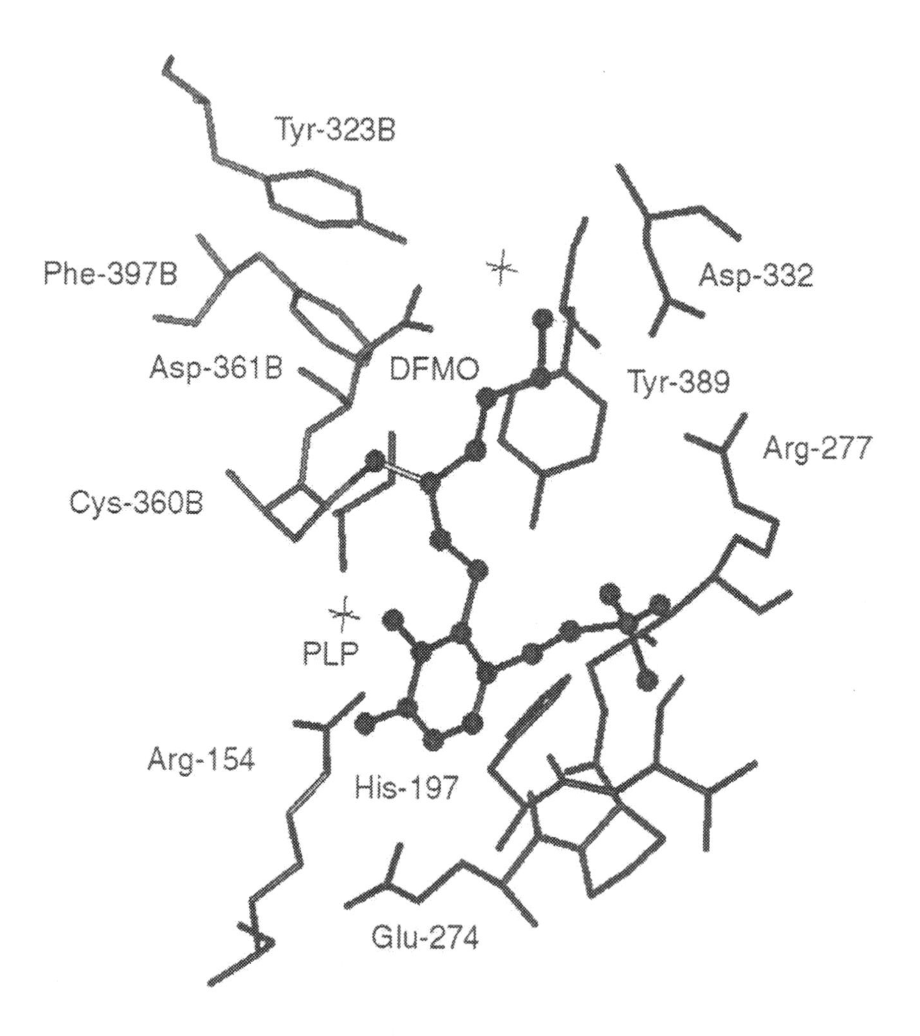


**Figure 2.** The X-ray structure of *T. brucei* ODC determined at 2.0 Å resolution. The active site is displayed. PLP and DFMO are displayed in ball and stick and select water molecules are displayed as stars. The active site is formed at the subunit interface and residues contributed to the active site from the N-terminal domain of one subunit are labelled by residue and number only, those from the C-terminal domain of the second subunit are additionally labelled B.

We have delineated the roles of a number of the active site residues in catalysis, substrate binding and cofactor binding through the use of site-directed mutagenesis, steady-state and presteady-state kinetic and spectral analysis and through the study of cofactor analogs. The reaction mechanism of wild-type *T. brucei* ODC was analysed by multi-wavelength UV/vis stopped-flow spectroscopy (Brooks and Phillips, 1997). The absorption spectrum of the PLP cofactor bound to the active site site is a sensitive indicator of the electronic and tautomeric state of the cofactor (Fig. 3). Analysis of the presteady-state phases of the reaction by stopped-flow spectroscopy allowed determination of the rates of formation and break down of key intermediates. The minimal model for the reaction with L-Orn is rapid formation of the Schiff base species, followed by two first order steps, decarboxylation and product release ($k_{decarb} = 20\ s^{-1}$ and $k_{off.Put} = 1\ s^{-1}$ at 4°C), where the latter step is rate determining (Fig. 4A; Brooks and Phillips, 1997). A fast step, which may be protonation of $C_{\alpha}$ is likely to separate these steps. Formation and hydrolysis of the Schiff base species is catalysed by Lys-69 (Osterman et al., 1999), the residue that forms a Schiff base to PLP in the absence of substrate (Poulin et al., 1992). Mutation of Lys-69 to Ala or Arg reduces the $k_{cat}$ by $10^4$-fold. The absorption spectrum of K69R ODC is strikingly different from the wild-type spectrum and lacks the 420 nm band. Addition of putrescine to the enzyme restores the 420 nm band, consistent with the reaction of an amine with the free aldehyde of PLP. Time dependent spectral analysis of this change was used to monitor the rates of Schiff base formation and hydrolysis for the mutant enzyme. The mutation drastically slows the rate of these steps by $10^3$-$10^4$-fold compared to wild-type ODC (Osterman et al., 1999). In addition, the rate of the decarboxylation step, measured by single turnover analysis, is decreased by $10^4$-fold, suggesting that Lys-69 also plays a role in the proper positioning of the $CO_2$ group for efficient bond hydrolysis (Osterman et al., 1999).

The efficiency of decarboxylation is also enhanced by the presence of Glu-274 in the active site (Osterman et al. 1995). Glu-274 forms a hydrogen bond to the pyridine nitrogen of PLP (Fig. 2). Mutation of this residue to Ala decreases the steady-state $k_{cat}$ by 50-fold. However, activity can be restored to near wild-type levels by replacing the PLP cofactor with N-methylPLP, a cofactor analogue that fixes the positive charge on the pyridine nitrogen. These studies demonstrate that the role of Glu-274 is to maintain the ring in the protonated state, thereby enhancing the electron withdrawing capacity of the cofactor. Generation of the carbanion intermediate upon decarboxylation is thus facilitated.

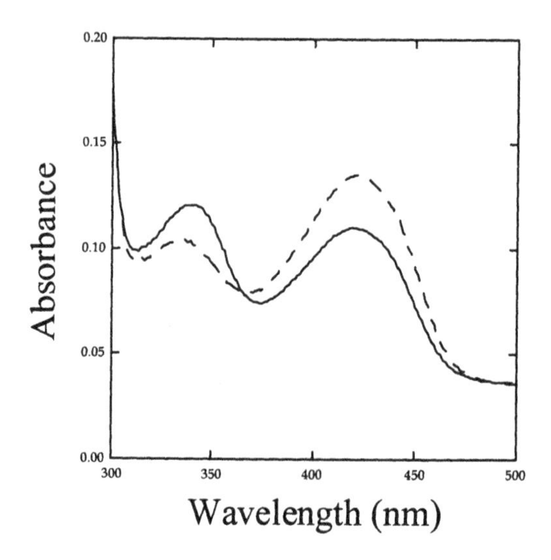


**Figure 3.** Absorbance spectra of *T. brucei* ODC. Absorbance spectra of ODC (22 μM) before (solid line) and after mixing with 10 mM Orn (dashed line) using a 1 cm cell, pH 7.3.

High affinity PLP binding to the active site is dependent on Arg-277. This residue forms a salt bridge with the 5'-phosphate of PLP (Fig. 2). Mutation of Arg-277 to Ala, increases the $K_m$ for PLP by 270-fold (Osterman et al., 1997), however the $k_{cat}$ at neutral pH is unaffected. Similar effects on PLP binding affinity were also observed, confirming that the primary role of this residue is to promote high affinity PLP binding, while it does not have a direct role in the reaction chemistry.

The substrate binding site is formed at the subunit interface (Fig. 2). DFMO is bound in the subunit interface forming a Schiff base with PLP and a covalent bond to Cys-360. Studies on the chemistry of the inactivation suggested that inactivation would occur after decarboxylation via nucleophilic attack by Cys-360 (Fig. 4B). The structural analysis confirms this result (Grishin et al., 1999). The δ-amino group is positioned between two aspartic acid residues, each from an opposite subunit (Asp-361 and Asp-332). Mutation of Asp-361 to Ala increases the $K_m$ for L-Orn by 2000-fold, while having little effect on $k_{cat}$ (Osterman et al., 1995). These studies confirm the importance of Asp-361 for the energetics of substrate binding, similar studies of Asp-332 have not been done. The requirement that residues from both subunits participate in the active site has been biochemically demonstrated by the finding that activity can be restored to the inactive mutant enzymes, K69A and C360A ODC, by formation of mixed mutant heterodimers (Tobias and Kahana, 1993; Coleman et al., 1994; Osterman et al., 1994).

**Fig. 4A. Mechanism of decarboxylation by Ornithine decarboxylase**

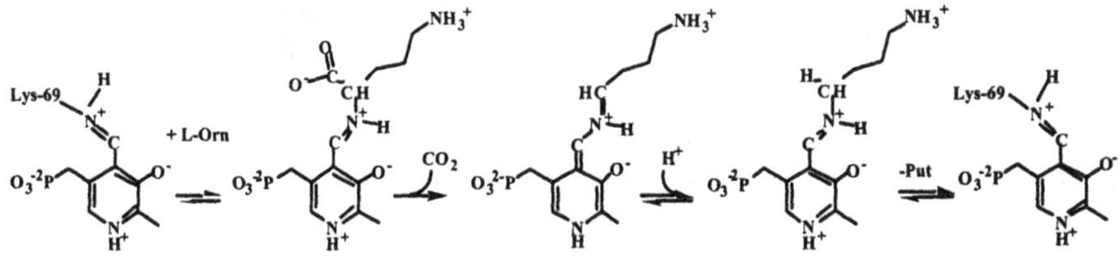

**Fig. 4B. Mechanism of inactivation by α-difluoromethylornithine**

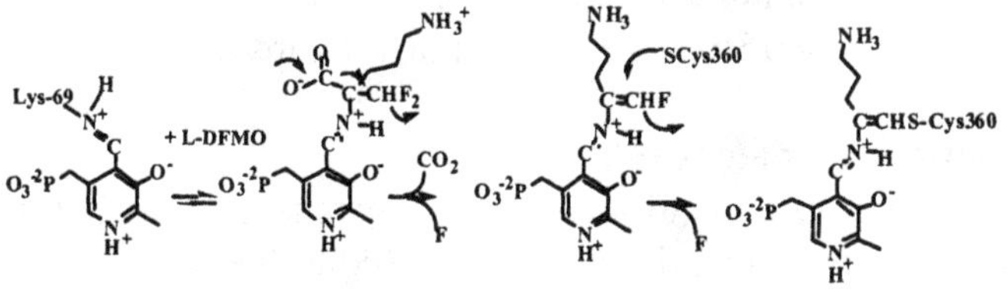

## ODC protein expression and purification

*T. brucei* ODC is expressed as a fusion protein with an N-terminal extension, which includes a $His_6$-tag and the TEV protease site (MHHHHHHAENLYFQGA) from the $T_7$ promoter in *E. coli* BL21/DE3 cells (Osterman et al., 1994). The enzyme is expressed to levels of 25 to 50 mg/L of *E. coli* culture if grown in shaker flasks. As much as 2 g of enzyme can be obtained from a 12 L fermentor preparation (Brooks and Phillips, 1997). ODC is purified by $Ni^{+2}$-agarose column chromatography, followed by gel filtration on a Pharmacia Superdex 200 (16 x 60) column. ODC for crystallization studies is further purified by the removal of the $His_6$-tag by TEV protease immobilized on Glutathione-agarose in $A_h$ buffer (Grishin et al., 1996). ODC recovered in the flow through (tag-minus) is further purified by gel filtration.

Procedures:

**Cell growth and harvest**

1. Transform competent BL21/DE3 cells with expression vector.
2. Inoculate LB broth + Ampicillin (0.1 mg/ml) with 1/40 dilution of an overnight culture.
3. Grow cells to 0.8 to 1 $OD_{600}$ at 37°C.
4. Cool cells to ≤ 30°C and grow 4 to 6 hr.
5. Spin down cells at 4,000 x G for 20 min.
6. Resuspend cells in 1/30th the original volume of $A_t$ buffer plus protease inhibitors (1x PicI, 1x Pic II and 1 mM phenylmethylsulfonylfloride (PMSF))
7. Add 1/50 vol. of lysozyme (50 mg/ml stock). Incubate on ice for 30 min.
8. Freeze in liquid Nitrogen (At this step, the frozen protein can be left in the -80°C freezer overnight or can immediately be thawed for the next step).
9. Sonicate cells until the suspension is no longer viscous.
10. Centrifuge the suspension in a Ti45 rotor for 1 hr at 8°C and 35,000 rpm, or for small preps in a microfuge for 10 min. Discard the pellet.

**$Ni^{+2}$-agarose column chromatography**

1. Apply the supernatant to an equilibrated $Ni^{+2}$-agarose column (Quiagen). The column volume should be about 1/400th the starting culture volume (15 ml for 6 L prep; 0.25 to 0.5 ml for 50 ml) and is equilibrated in $A_t$ buffer. The flow rate for a 15 ml column should be approximately 3 ml/min.

2. Wash the column with 20 volumes of $A_t$ Buffer, 10 volumes of $A_t$ plus 1 M NaCl, and 10 volumes of $A_h$ buffer.

3. Starting with 100% $A_h$ buffer elute the protein with the following gradient: 0 to 20% $B_h$ in 10 min, 20% $B_h$ for 15 min; 20 to 100% $B_h$ in 60 min. *T. brucei* ODC elutes between 50 to 70% $B_h$.

$A_t$ Buffer: 50 mM Tris pH 8.0, 100 mM NaCl, 5 mM ß-mercaptoethanol, 20 μM PLP, 0.15% Brij.

$A_h$ Buffer: 20 mM Hepes pH 7.0, 100 mM NaCl, 5 mM ß-mercaptoethanol, 20 μM PLP, 0.15% Brij.

$B_h$ Buffer: $A_h$ buffer, 200 mM Imidazole pH 7.0.

Protease inhibitors: Pic I (1000 x): 1mg/ml Leupeptin, 2 mg/ml antipain, 10 mg/ml Benzamide in water. PicII (1000 x): 1 mg/ml chymostatin, 1 mg/ml pepstatin in DMSO.

*Inhibitors and buffer components are purchased from Sigma.*

**Gel filtration column chromatography**

Protein is concentrated to 20 to 40 mg/ml in a centriprep 10 concentrator (Amicon) and is applied (volume 1 to 2 ml) to a Pharmacia Superdex 200 (16 x 60) column equilibrated in $A_h$ buffer in which 1 mM dithiothreitol (DTT) is substituted for β-mercaptoethanol. The peak fraction is collected and the concentration of ODC is determined by measuring the OD at 280 nm ($\varepsilon$ = 0.85 $OD(mg/ml)^{-1}cm^{-1}$). Protein purity is verified by SDS-PAGE analysis using 10 to 12 % polyacrylamide gel.

**ODC activity assays and the effect of inhibitors**

There are two assays for ODC activity that we routinely use in the lab: a NADH spectral assay where the production of $CO_2$ is coupled to the oxidation of NADH and the $^{14}CO_2$-trapping assay, where released labelled $CO_2$ is trapped on filter paper soaked in base and quantitated by scintillation counting (Phillips, 1999). The $^{14}CO_2$-trapping assay has been used traditionally to monitor ODC activity and it has the strength of being a sensitive assay that can be used for both steady-state and presteady-state analysis. However, we find that, unless we are analysing a very inactive mutant or doing single turnover experiments, the spectral assay is greatly preferred because the reaction can be followed continuously over time, making determination of the steady-state phase of the reaction easy. Additionally, this assay allows data to be collected rapidly and does not require radioactive material. Both assays can be adapted to microtiter plates for high throughput analysis.

It is essential to confirm for any of the assays that a linear dependence of the initial rate (v) on ODC concentration is observed. Substrate concentration should be used at saturation for specific activity measurements. We use 10 mM L-Orn for the spectral assay and 1 mM for the $^{14}CO_2$-trapping assay (for wild-type *T. brucei* ODC, $K_m$ = 0.2 mM, $k_{cat}$ = 15 $s^{-1}$). Because the $^{14}C$-labeled L-Orn will have low specific activity at high L-Orn concentrations, this assay does not allow for as broad a concentration range of substrate to be used in the spectral assay. For mutant enzymes with elevated $K_m$'s the spectral assay is therefore much more suitable. For the determination of $K_m$

and $k_{cat}$, L-Orn is varied over a 10-fold range with equal data points acquired above and below $K_m$. The initial rates are fitted to the Michaelis-Menten equation to determine the steady-state kinetic parameters (Segel, 1975).

**Steady-state spectral assay for ODC activity**

ODC activity can be measured by coupling the production of 1 mol of $CO_2$ to the oxidation of 1 mol of NADH as follows:

$$\text{L-Ornithine} \xrightarrow{\text{ODC}} \text{putrescine} + CO_2\,(HCO_3^-)$$

$$HCO_3^- + \text{PEP} \xrightarrow{\text{PEPC}} \text{OAA} + H_2PO_4$$

$$\text{OAA} + \text{NADH} \longrightarrow \text{Malate} + NAD^+ + H^+$$

PEPC, phosphoenol pyruvate carboxylase; MDH, malate dehydrogenase; OAA, oxaloacetate.

Reagents:

$CO_2$ detection kit (Sigma 132-B), L-Orn, DTT, PLP, disposable cuvettes (Fisher). All reagents are purchased from Sigma, unless specified.

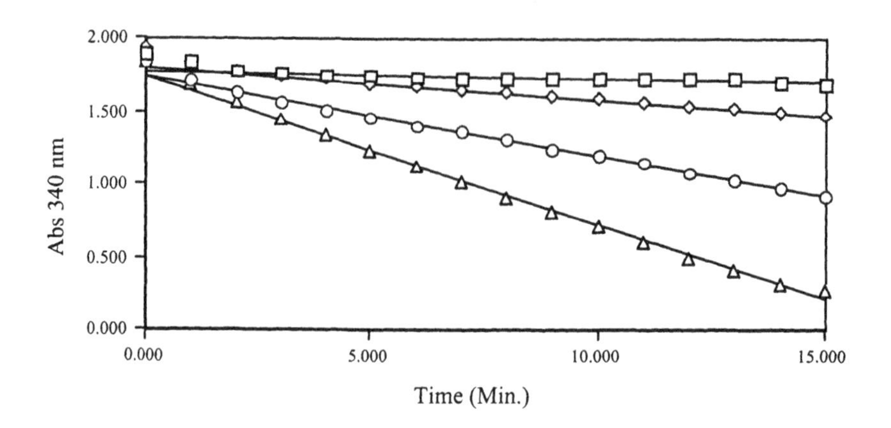


**Figure 5**. Spectral assay of ODC activity. Oxidation of NADH is followed at 340 nm. Squares, no ODC; diamonds, 0.25 μg/ml (0.022 OD/min) ODC; circles, 0.5

μg/ml (0.054 OD/min) ODC; triangles, 1.0 μg/ml (0.10 OD/min) ODC. L-Orn = 1 mM.

Procedure:

1. Dissolve Reagent A (Sigma) in the recommended volume of water that has been freshly boiled and cooled to < 50°C. Should be made fresh daily. Alternatively reagent A can be prepared as follows: 2.2 mM phosphoenol pyruvate, 0.32 mM yeast NADH, 10 mM $MgCl_2$, 50 mM Tris-HCl pH 8.0, 0.1 mg/ml BSA.
2. Dissolve Reagent B in the recommended volume of $CO_2$ diluent. Reagent B can also be prepared as follows: 5 U plant phosphoenol pyruvate carboxylase (Sigma P2023), 30 U porcine heart malate dehydrogenase (Sigma M2634), 10 mM $MgCl_2$, 2% glycerol, 20 mM MOPS pH 7.0, volume 2 ml.
3. Prepare stocks of: 1 M L-Orn pH 7.5; 1 M DTT pH 7.5; 20 mM PLP pH 7.5.
4. Make reagent C: 100 μl 1/100 x L-Orn stock, 25 μl 1 M DTT, 25 μl 20 mM PLP, 825 μl water. Make fresh daily.
5. Prepare the final assay mix as follows: Place disposable 1 ml cuvettes in the sample holder preheated to 37°C. Add 50 μl reagent B, 400 μl reagent A (pre-warmed to 37°C), 50 μl reagent C, 5 to 20 μl ODC stock to a final concentration of 0.25 to 3 μg/ml. Total volume 0.5 ml.
6. Mix by stirring or gently shaking. Do not introduce air.
7. Collect data at 340 nm. for 10 to 20 min at 37°C. The absorbance will decrease from 2 to 2.5 OD towards zero (Fig. 5). Determine the initial rate (v) of the reaction after steady-state is reached. The first 1 to 3 min of assay often must be discarded because they are influenced by pre-dissolved $CO_2$ in the buffer. Absorbance (A) is related to the concentration (c) of a solute by Beer's law, $A = \varepsilon lc$. The extinction coefficient ($\varepsilon$) for NADH is 6.22 OD (μmoles/ml)$^{-1}$ cm$^{-1}$. The path length (l) of the standard cuvette is 1 cm. The monomer molecular weight of *T. brucei* ODC is 49,000 gram/mole.

**$^{14}CO_2$ steady- state assay for ODC activity**

ODC activity can be measured by following the release of $^{14}CO_2$ trapped on filter paper soaked in base (Osterman et al., 1999).

Reagents:
1-$^{14}CO_2$-L ornithine (Amersham CFA491); Barium Hydroxide (Sigma); Trichloroacidic acid (TCA; Sigma); L-Orn (Sigma); 16 x 125 mm glass test tubes (Fisher 14-961-30); Rubber sleeve stoppers (Fisher 14-126BB); 18 gauge needles with 3 to 5 ml plastic syringes; whatman filter paper. DTT, PLP, Hepes and BSA are all purchased from Sigma.

Procedure:
Reaction mix final concentrations: 20 mM Hepes pH 7.5, 1 mg/ml BSA, 2 mM DTT, 0.05 mM PLP, 0.13 to 0.25 μCi 1-$^{14}CO_2$-ornithine (typically 60 mCi/mmol, 50 μCi/ml), 1 mM L-Orn, 0.04 to 0.2 μg ODC. Final volume 0.25 ml.

Add reaction components to a test tube sitting on ice. Cut strips of whatman filter paper (1 cm x 4 cm cut so that there is an extra hook that can be placed outside the test tube) and treat the paper with 0.075 ml saturated Barium Hydroxide. Add enzyme to the reaction mixture and hang a treated filter paper in the test tube such that the hook is left outside and the main part of the strip is hanging inside the test tube, cap with a sleeve stopper and incubate at the desired temperature for 5 to 60 min. The assay is typically run at 37°C, but this assay can be employed at any temperature. Stop the reaction by injecting 1 ml 40% TCA through the cap. Incubate 30 min. Remove filter paper and place it in a scintillation vial, add scintillant and count. The efficiency of counting should be tested by running a reaction to completion. Typical efficiencies are 50 to 70%. Cpm (counts per minute) are related to moles of released $CO_2$ based on the specific activity of the label, which must be calculated for each substrate concentration and for each reagent, and based on the relationship that $2.22 \times 10^6$ dpm/μCi.

**Experimental set-up for inhibitor analysis: DFMO and putrescine**
**Analysis of reversible competitive inhibitors of ODC**
Putrescine is a competitive inhibitor of ODC ($K_I = 0.6$ mM). Initial rates (v) are measured for a matrix of substrate and inhibitor concentrations. Typically we collect 6 substrate concentrations per each of 4 inhibitor concentrations in addition to the no inhibitor data set. The steady-state kinetic parameters ($k_{cat}$, $K_M$ and $K_I$) are than determined by fitting the data to the steady-state rate equation describing a reversible competitive inhibitor (Copeland, 1996).

**Analysis of the kinetics of DFMO inactivation.**
DFMO is a suicide inhibitor of ODC. The mechanism of DFMO inactivation has been studied biochemically (Poulin et al., 1992). DFMO forms a Schiff base with PLP and is decarboxylated eliminating the first floride molecule to form an activated species. This species is attacked by the nucleophile Cys-360 in the active site, eliminating the second floride to form a covalent bond between inhibitor and enzyme (Fig. 4B). The X-ray structure of DFMO in complex with *T. brucei* ODC confirms this mechanism of inactivation (Fig. 2). Kinetic analysis of a time-dependent inhibitor, such as a suicide inhibitor, is performed as described (Copeland, 1996). ODC (0.25 to 2 μM) is incubated with different concentrations of inhibitor (for DFMO the range should be 0.025 to 0.25 mM). Aliquots are removed after varying incubation times (t) and diluted (100-fold) into assay mixture to determine the amount of enzyme activity remaining at time (t). Enzyme activity decays via a single exponential process and the observed rate constant for this decay ($k_{obs}$) is determined for each inhibitor concentration. This rate constant ($k_{obs}$) is related to the apparent $K_I$ and to the rate constant for inactivation as described (Copeland, 1996).

**ODC crystallization**
Crystallization conditions were identified in Hampton crystallization screens (Riverside, CA) as described (Grishin et al., 1996, Grishin et al., 1999). The crystals were obtained at 16°C by vapour diffusion, mixing equal volumes of ODC (20mg/ml ODC without the $His_6$-tag in 10 mM HEPES-NaOH pH 7.2, 50 mM NaCl, 0.5 mM EDTA, 0.015% Brij35, 10mM DTT) and well solution (16% PEG 3350, 0.2 M Na ($CH_3COO$), 100 mM HEPES, pH 7.5). This procedure results in mostly twinned and stacked crystals, which are used for micro-seeding into the solution of the above content. Plate shape single crystals grew close to the surface of the drop within a few days. The crystals belong to space group $P2_1$ (*a*=66.8Å, *b*=154.5Å, *c*=77.1Å, β=90.58°) with 4 molecules per asymmetric unit and approximately 32% solvent.

**Site-directed mutagenesis**
Mutant enzymes were created by the method of Kunkel (Kunkel, 1985) or using the Quick Change kit (Stratagene #200518).

**Mechanistic analysis of *T. cruzi* AdoMetDC**
AdoMetDC catalyses the pyruvoyl-dependent decarboxylation of S-adenosylmethionine (AdoMet) (Stanley and Shantz, 1994). Pyruvoyl-dependent enzymes generate their cofactor through an auto-proteolytic

cleavage into α and β subunits, leaving the pyruvate covalently attached to the N-terminus of the α-subunit. The mechanism of decarboxylation by pyruvate dependent enzymes begins with formation of a Schiff base between substrate and the pyruvate cofactor. The resulting imine promotes decarboxylation by serving as an electron sink for the carbanion produced upon removal of the carboxyl group. Following decarboxylation, the α-carbon is protonated, and the product is released by subsequent hydrolysis of the Schiff base (Fig. 6).

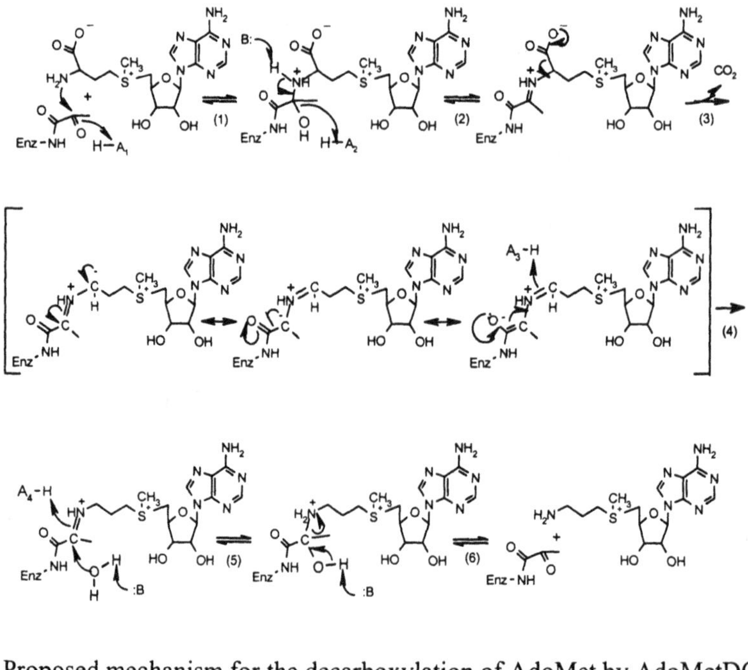

**Figure 6.** Proposed mechanism for the decarboxylation of AdoMet by AdoMetDC.

The X-ray crystal structure of the human AdoMetDC enzyme has recently been solved (Ekstrom et al., 1999). The enzyme is a dimer of α–β subunits, with each (αβ) monomer comprising two anti-parallel β sheets flanked by α helixes. Although the structure does not contain any bound substrate or inhibitors, the active site pocket is indicated by the presence of the pyruvate cofactor. Previous biochemical analysis has suggested the involvement of various active site residues in the catalytic mechanism of the enzyme. Mutation of several conserved residues including Glu-8, Glu-11, Cys-82, His-127, Glu-133, Ser-229 and His-243 significantly reduces AdoMetDC activity (Pegg et al., 1998; Xiong et al., 1999), with all but Glu-133 and His-127 found in proximity to the active site. The residue Cys-82 has been implicated in protonation of the α-carbon prior to Schiff base hydrolysis (Xiong et al., 1999). However, the specific roles of each of the other residues in the reaction mechanism remain to be determined.

Although AdometDCs from different sources perform this same pyruvoyl-dependent reaction, their requirements for accomplishing the reaction differ considerably. The *E. coli* enzyme requires a divalent metal ion for activity (Markham et al., 1983), while putrescine activates the yeast and mammalian enzymes (Pegg et al., 1998). Other SamDCs (*Tetrahymena*, some plants) do not exhibit activation with either divalent metal ions or putrescine (Poso et al., 1976). Additionally, the kinetics of activation appears to differ in these various systems. Addition of increasing amounts of a divalent cation to the *E. coli* enzyme increases $V_{max}$ while it decreases $K_m$, while putrescine exerts its effects on the yeast enzyme by lowering $K_m$ for substrate.

Our lab has characterized the *T. cruzi* AdoMetDC kinetics of activation by putrescine and has screened a large array of amines for activation (Kinch et al., 1999). The steady state kinetics fit to a model of a non-essential activator (Segel, 1975) with a $k_{cat}$ of 0.007 $s^{-1}$ and $K_d$ of 0.05 mM. The presence of putrescine increases the $k_{cat}$ by 9 fold to 0.06 $s^{-1}$, and increases the $K_d$ by 2 fold to 0.01 mM, with a model derived $K_d$ for putrescine of 3 mM. The parasite enzyme is 10 to 50-fold less active than the human enzyme, while the apparent $K_d$ for the putrescine activator is 50-fold higher than for the mammalian enzyme. Finally, the diamine cadaverine activates the *T. cruzi* enzyme to the same extent as putrescine, while this is not the case for the human enzyme. To address the low specific activity of the *T. cruzi* enzyme, the active site was titrated with a suicide inhibitor, MDL73811. This titration showed that the amount of active enzyme correlated with the amount of enzyme estimated to be present in the reaction mixtures.

**AdoMetDC expression and purification**

*T. cruzi* AdoMetDC is expressed with a $His_6$-tag and purified over a $Ni^{+2}$-agarose column, followed by resource Q column chromatography. The procedure for the $Ni^{+2}$-agaorosoe column is identical to that described for ODC accept for the $A_h$ and $B_h$ buffer composition.

$A_h$ Buffer: 50 mM HEPES, pH 8.0, 50 mM NaCl, 10 mM Putrescine, 2.5 mM β-mercaptoethanol.
$A_Q$ Buffer, 50 mM HEPES, pH 8.0, 50 mM NaCl, 5 mM DTT.
$B_h$ Buffer: $A_h$ Buffer plus 200 mM Imidazole pH 7.0.
$B_Q$ Buffer: 50 mM HEPES, pH 8.0, 500 mM NaCl, 5mM DTT.

Procedure for Mono Q column chromatography:

1. Equilibrate Resource Q column (10 ml) at a flow rate of 4 ml/min with 5 column volumes $A_Q$ Buffer, followed by 10 column volumes $B_Q$ Buffer, and then another 5 column volumes $A_Q$ Buffer.
2. Apply pooled $Ni^{+2}$-agarose column chromatography fractions to the Resource Q column using a 50 ml superloop (2 ml/min).
3. Elute at a flow rate of 2 ml/min using a gradient of low ($A_Q$ Buffer) to high ($B_Q$ Buffer) salt: 0 to 40% $B_Q$ Buffer in 15 min (3 column volumes), then 40% to 100% $B_Q$ Buffer in 40 min (8 column volumes). Collect 4 ml fractions.
4. Check purity by SDS-PAGE.
5. Concentrate the protein (about 50 mg/ml) in Centriprep 10 (Amicon) and determine concentration for storage (-80°C).

**AdoMetDC activity assays**

Currently, a radioactive assay is the only method used for the study of AdoMetDC activity. For the analysis of steady-state rates, the labelled $CO_2$ product is trapped on filters with the same procedure as the ODC assay. It is important to establish a linear dependence on both the enzyme concentration and the time course of the reaction. To study the apparent activation of AdoMetDC by putrescine, decarboxylation rates are measured at various concentrations of putrescine and saturating concentrations of AdoMet substrate (0.5 mM).

Reagents:

The same supplies as used for the steady-state $^{14}CO_2$ ODC assay except for the substrate: S-Adenosyl-L-[carboxyl-$^{14}$C] methionine (Amersham Pharmacia CFA477). S-adenosyl-L-methionine, HCl, Sodium Bicarbonate, and putrescine are all purchased from Sigma.

**Steady-state assay**

Reaction Mixture: 50 mM Hepes pH 8.0, 50 mM NaCl, 1 mg/ml BSA, 2 mM DTT, 0 to 25 mM Putrescine, 0.1 μCi $^{14}CO_2$-AdoMet (56 mCi/mmol, 25 μCi/ml), 10 to 500 μM AdoMet, 2 to 8 μg AdoMetDC, Final volume 0.2 ml.

The remaining procedure is identical to that described for ODC.

**Active site titration with MDL73811**

Procedure:

MDL73811 (0.9 to 4 μM) is incubated with AdoMetDC (2.5 μM) for 2 or 4 hr at 37°C in buffer (50 mM Hepes pH 8.0, 50 mM NaCl, 10 mM Putrescine, and 5 mM DTT). The fraction of steady-state activity remaining after inhibitor incubation is compared to activity of enzyme incubated without inhibitor to determine the amount of active AdoMetDC.

**Kinetic analysis of *T. brucei* γGCS**

The gene encoding the 77 kdal *T. brucei* γ-GCS was cloned utilizing degenerate oligonucleotides designed to conserved regions of the γ-GCS sequences from rat and yeast (Lueder and Phillips, 1996). Subsequently we have also cloned the gene from *T. cruzi* utilizing the same strategy (Genbank accession number AF095637). The deduced amino acid sequence encoded by the *T. cruzi* γGCS gene shares 68% sequence identity with the *T. brucei* enzyme. We characterized the full kinetic profile for the *T. brucei* catalytic subunit expressed and purified from a heterologous bacterial system (Brekken and Phillips, 1998). The data is best fit to the equation describing a random ter-reactant mechanism (Segel, 1975), where $K_{glu}$=2.6 mM, $K_{aba}$=5.4 mM and $K_{ATP}$=1.4 mM. The model also predicts that the binding affinities of the substrates are not always independent of each other. For example, the binding of ATP or Glu to the *T. brucei* enzyme increases the binding affinity for the other by 12-fold (compare $K_{ATP}$ to $\beta K_{ATP}$). Glu and aminobutyrate (Aba) have a negative interaction energy and ATP and Aba do not interact.

The mammalian enzymes are composed of two subunits: a regulatory and catalytic subunit (termed the holoenzyme). The holoenzyme of rat γGCS is reported to have a 12-fold lower $K_m^{app}$ for Glu (the $K_m$ measured at saturating concentrations of the other two substrates) than the catalytic subunit alone (Griffith and Mulcahy, 1999). In our preliminary kinetic studies we found that the *T. brucei* enzyme had a $K_m^{app}$ for Glu that was lower than for the mammalian holoenzyme, suggesting that it may not require a regulatory subunit for efficient catalysis (Lueder and Phillips, 1996). There is yet no evidence for a regulatory subunit outside of the mammalian kingdom. γGCS purified from the worm *A. suum* is a single 70 kdal subunit (Hussein and Walter 1995) and no evidence for a second subunit of the *T. brucei* enzyme could be found by immunoprecipitation of the catalytic subunit from procyclic *T. brucei* 427 cell extracts (unpublished observation).

**γGCS protein expression and purification**

*T. brucei* and *T. cruzi* γGCS are expressed as C-terminal $His_8$-tag fusion proteins from the T7 promoter (Brekken and Phillips, 1998) as described for ODC except that the buffers are modified by the removal of PLP and Brij and by the addition of 5 mM $MgCl_2$. Further DTT inhibits the enzyme activity and 1 mM β-mercaptoethanol is used during all steps of the purification instead.

**Steady-state spectral assay for γGCS activity**

γ-GCS activity is followed at $37^0$C using a spectrophotometric assay which couples ADP production to NADH oxidation as described (Lueder and Phillips, 1996). Buffer (100 mM Tris-HCl, pH 8.0, 150 mM KCl, 20 mM $MgCl_2$, 2 mM phosphenolpyruvate, 0.27 mM NADH) is mixed with type III rabbit muscle pyruvate kinase (5 units of 350 to 600 units/mg redissolved lyophilized powder (Sigma P9136)), type II rabbit muscle lactic acid dehydrogenase (10 units of 800 to 1200 units/mg ammonium sulfate suspension (Sigma L2500)) and γ-GCS substrates to a final reaction volume of 0.5 ml. The assay is initiated by the addition of γ-GCS. L-α-aminobutyric acid (L-Aba) is used in place of L-Cys unless specified. For specific activity measurements the concentrations of ATP, L-Aba and L-Glu used are 5 mM, 100 mM and 10 mM, respectively. γ-GCS concentration can range from 0.1 to 0.5 μM as determined by measuring the $OD_{280}$. The extinction coefficient for *T. brucei* γ-GCS was determined to be 1.35 (mg/ml)/OD. Data to assess the kinetic mechanism is collected for a complete matrix of rates as a function of substrate concentration (L-Glu, 0.1 to 8 mM; ATP, 0.04 to 2 mM and; L-Aba, 2.5 to 100 mM) such that for any given concentration of any one substrate the rates were measured over the entire range of the other two substrates.

## DISCUSSION AND CONCLUSION

The design of new chemotherapeutic agents to combat parasitic diseases is essential to the continued management of morbidity and mortality in disease endemic regions. Drug resistance and toxicity make much of the existing chemotherapy undesirable. The first step in this process is to identify potential protein targets for new chemotherapy. Once an enzyme has been identified as a valid chemotherapeutic target, detailed biochemical and structural analysis of the enzyme provides information about the enzyme active site that will aid in inhibitor design. In addition to the structures of ODC and AdoMetDC, the X-ray structures of a number of parasite enzymes that are drug targets have now been solved (reviewed in Hunter, 1997), opening up the possibility to utilize enzyme structure to aid in the process of inhibitor design. Many computational methods are available to utilize structural information towards this end, particularly when coupled with a detailed understanding of what residues in the active site are important for the energetics of ligand binding and for enzymatic catalysis (Blundel, 1997). The combination of site-directed mutagenesis with kinetic analysis of the reaction can be used to address these latter questions, while a detailed understanding of the enzyme mechanism provides the basis for the design of mechanism-based inhibitors. Finally, the advent of combinatorial chemistry techniques, in combination with high-throughput methods for enzyme assay, provides another powerful avenue for the generation of new chemical structures that can serve as lead compounds for the development of new anti-parasitic agents.

## REFERENCES

Almrud, J.J., Oliveira, M.A., Grishin, N.V., Phillips, M.A. and Hackert, M.L. (1999). Crystal structure of human ornithine decarboxylase at 2.1 Å resolution: structural perspectives of antizyme binding. submitted.

Arrick, B.A., Griffith, O.W. and Cerami, A. (1981). Inhibition of glutathione synthesis as a chemotherapeutic strategy for trypanosomiasis. J. Exp. Med. *153*, 720-725.

Atamna, H. and Ginsburg, H. (1997). The malaria parasite supplies glutathione to its host cell: Investigation of glutatione transport and metabolism in human erythrocytes infected with P. falciparum. Eur. J. Biochem. *250*, 670-679.

Bacchi, C.J., Garofalo, J. Ciminelli, M. Rattendi, D., Goldberg, B., McCann, P.P. and Yarlett, N. (1993). Resistance to DL-α-difluoromethylornithine by clinical isolates of

Trypanosoma brucei rhodesiense. Role of S-adenosylmethionine. Biochemical Pharmacol. *46*, 471-481.

Bacchi, CJ, Nathan, H.C., Yarlett, N., Goldberg, B., McCann, P.P., Bitonti, A.J. and Sjoerdsma, A. (1992). Cure of murine Trypanosoma brucei rhodesiense infections with an S-adenosylmethionine decarboxylase inhibitor. Antimicrobial agents and Chemotherapy. *36*, 2736-40.

Bacchi, C.J. and Yarlett, N. (1993). Effects of antagonists of polyamine metabolism of African trypanosomes. Acta Tropica. *54*, 225-236.

Blundell, TL (1996) Structure-based drug design. *Nature,* 384 supp., 23-26.

Bitonti, A.J., Bacchi, C.J., McCann, P.P. and Sjoerdsma, A. (1985). Catalytic irreversible inhibition of Trypanosoma brucei brucei ornithine decarboxylase by substrate and product analogs and their effects on murine trypanosomiasis. Biochemical Pharmacology. *34*, 1773-1777.

Brekken, D.L. and Phillips, M.A. (1998). Trypanosoma brucei γ-glutamylcysteine synthetase: characterization of the kinetic mechanism and the role of Cys-319 in cystamine inactivation. J. Biol. Chem. *273*, 26317-26322.

Brooks, H.B. and Phillips, M.A. (1997). Characterization of the reaction mechanism of Trypanosoma brucei ornithine decarboxylase by multiwavelength stopped-flow spectroscopy. Biochemistry. *36*, 15147-15155.

Cohn, M.S., Tabor, C.W. and Tabor, H. (1980). Regulatory mutations affecting ornithine decarboxylase activity in Saccharomyces cerevisiae. J. Bacterol. *142*, 791-799.

Coleman, C.S., Stanley, B.A., Viswanath, R. and Pegg, A.E. (1994). Rapid exchange of subunits of mammalian ornithine decarboxylase. J Biol Chem. *269*, 3155-3158.

Copeland, R.A. (1996). Enzymes: a practical introduction to structure, mechanism, and data analysis. New York, Wiley-VCH.

Das, B., Gupta, R. and Madhubala, R. (1997). Combined action of inhibitors of S-adenosylmethionine decarboxylase with an antimalarial drug, chloroquine, on Plasmodium falciparum. J. Euk. Microbiol. *44*, 12-17.

Dubois, V.L., Platel, D.F.N., Pauly, G. and Tribouley-Duret, J. (1995). Plasmodium berghei: Implication of intracellular glutathione and its related enzyme in chloroquine resistance in vivo. Exp. Parasitol. *81*, 117-124.

Dumas, C., et al. (1997). Disruption of the trypanothione reductase gene of Leishmania decreases its ability to survive oxidative stress in macrophages. EMBO J. *16*, 2590-2598.

Ekstrom, J.L., Mathews, I.I., Stanley, B.A., Pegg, A.E. and Ealick, S.E. (1999). The crystal structure of human S-adenosylmethionine decarboxylase at 2.25 Å resolution reveals a novel fold. Structure. *7*, 583-595.

Fairlamb, A.H. and Le Quesne, S.A. (1997). Polyamine metabolism in Trypanosomes. *Trypanosomiasis and Leishmaniasis*. G. Hide, J. Mottram, G. Coombs and P. Homlmes, CAB International: 149-161.

Geary, T.G., Divo, A.A., Bonanni, L.C. and Jensen, J.B. (1985). Nutritional requirements of Plasmodium falciparum in culture. III. Further observations on essential nutrients and antimetabolites. J. Protozool. *32*, 608-613.

Goldberg, B., Rattendi, D., Lloyd, D., Yarlett, N. and Bacchi, C.J. (1999). Kinetics of S-adenosylmethionine cellular transport and protein methylation in Trypanosma brucei brucei and Trypanosoma brucei rhodesiense. Archives of Biochemistry and Biophysics. *364*, 13-18.

Griffith, O.W. and Mulcahy, R.T. (1999). The enzymes of glutathione synthesis: g-glutamylcysteine synthetase. Advances in Enzymology and Related Areas of Molecular Biology. D. Purich. New York, John Wiley & Sons, Inc. **73:** 209-267.

Grishin, N.V., Osterman, A.L., Brooks, H.B., Phillips, M.A. and Goldsmith, E.J. (1999). The X-ray structure of ornithine decarboxylase from *Trypanosoma brucei*: the native structure and the structure in complex with α-difluoromethylornithine. Biochemistry, in press.

Grishin, NV, Osterman, A.L., Goldsmith, E.J. and Phillips, M.A. (1996). Crystallization and preliminary x-ray studies of ornithine decarboxylase from Trypanosoma brucei. Proteins. *24*, 272-273.

Grishin, NV, Phillips, M.A. and Goldsmith, E.J. (1995). Modeling of the spatial structure of eukaryotic ornithine decarboxylases. Protein Science. *4*, 1291-1304.

Grondin, K., Haimeur, A., Mukhopadhyay, R., Rosen, B.P. and Ouellette, M. (1997). Co-amplification of the γ-glutamylcysteine synthetase gene gsh1 and of the ABC transporter gene pgpA in arsenite-resistant Leishmania tarentolae. EMBO J. *16*, 3057-3065.

Hiromi, K. (1979). Kinetics of Fast Reactions. New York, Halsted Press.

Hunter, W.N. (1997). A structure-based approach to drug discovery; crystallography and implications for the development of antiparasite drugs. Parasitology. *114*, S17-S29.

Hussein, A.S. and Walter, R.D. (1995). Purification and charactization of g-glutamylcysteine synthetase from Ascaris suum. Mol. and Biochem. Parasitol. *72*, 57-64.

Iten, M., Mett, H., Evans, A., Enyaru, J.C.K., Brun, R. and Kaminsky, R. (1997). Alternations in ornithine decarboxylase characteristics account for tolerance of Trypanosoma brucei rhodeisiense to D,L-α-difluoromethylornithine. Antimicrobial agents and chemotherapy. *41*, 1922-1925.

Kern, A.D., Oliveira, M.A., Coffino, P. and Hackert, M. (1999). Structure of mammalian ornithine decarboxylase at 1.6 Å resolution: Stereochemical implications of PLP-dependent amino acid decarboxylase. Structure. *7*, 567-581.

Kinch, L.N., Scott, J., Ullman, B. and Phillips, M.A. (1999). S-adenosylmethionine from *Trypanosoma cruzi*: cloning, expression and kinetic characterization of the recombinant enzyme. Mol Biochem Parasitol. *101*, 1-11.

Kunkel, T.A. (1985). Rapid and efficient site-specific mutagenesis without phenotypic selection. Proc. Natl. Acad. Sci. U.S.A. *82*, 488.

Kuzoe, F.A.S. (1993). Current situation of African trypanosomiasis. Acta Trop. *54*, 153-162.

Li, F., Hua, S.B., Wang, C.C. and Gottesdiener, K.M. (1996). Procyclic Trypanosoma brucei cell lines deficient in ornithine decarboxylase activity. Mol. Biochem. Parasitol. *78*, 227-236.

Lueder, D.V. and Phillips, M.A. (1996). Characterization of Trypanosoma brucei γ-glutamylcysteine synthetase, an essential enzyme in the biosynthesis of trypanothione. J. Biol. Chem. *271*, 17485-17490.

Markham, G.D., Tabor, C.W. and Tabor, H. (1983). S-adenosylmethionine decarboxylase (Escherichia Coli). In: Tabor H, Tabor CW, eds. Methods in Enzymology. 94th ed. Academic Press,. 228-230.

Mehlotra, R.K. (1996). Antioxidant defense mechanisms in parasitic protozoa. Clinical Reviews in Microbiology. *22*, 295-314.

Meshnick, S.R., Chang, K.P. and Cerami, A. (1977). Biochem. Pharmacol. *26*, 1923-1928.

Moncada, C., Repetto, Y. and Aldunate, J., Letelier, M.E. and Morello, A. (1989). Role of glutathione in the suseptibility of Trypanosoma cruzi to drugs. Comparative Biochemistry and Physiology-C: Comparative Pharmacology and Toxicology. *94*, 87-91.

Osterman, A.L., Brooks, H.B., Jackson, L., Abbott, J.J. and Phillips, M.A. (1999). Lys-69 plays a key role in catalysis by *T. brucei* ornithine decarboxylase through acceleration of the substrate binding, decarboxylation and product release steps. Biochemistry. *in press*,

Osterman, A.L., Brooks, H.B., Riso, J.. and Phillips, M.A. (1997). The role of Arg-277 in the binding of pyridoxal 5'-phosphate to *Trypanosoma brucei* ornithine decarboxlyase. Biochemistry. *36*, 4558-4567.

Osterman, A.L., Grishin, N.V., Kinch, L.N. and Phillips, M.A. (1994). Formation of functional cross-species heterodimers of ornithine decarboxylase. Biochemistry. *33*, 13662-13667.

Osterman, A.L., Kinch, L.N., Grishin, N.V. and Phillips, M.A. (1995). Acidic residues important for substrate binding and cofactor reactivity in eukaryotic ornithine decarboxylase identified by alanine scanning mutagenesis. J. Biol. Chem. *270*, 11797-11802.

Osterman, A.L., Lueder, D.V., Quick, M., Myers, D., Canagarajah, B.J. and Phillips, M.A. (1995). Domain organization and a protease-sensitive loop in eukaryotic ornithine decarboxylase. Biochemistry. *34*, 13431-13436.

Pegg, A.E., Xiong, H., Feith, D.J. and Shantz, L.M. (1998). S-adenosylmethionine decarboxylase: structure, function and regulation by polyamines. Biochem Soc Trans. *26*, 580-586.

Phillips, M.A. (1999). Ornithine decaboxylase. in *The encyclopedia of molecular biology*. T. Creighton. New York, John Wiley & Sons.

Poso, H., Hannonen, P., Himberg, J. and Janne, J. (1976). Adenosylmethionine decarboxylase from various organisms: relation of the putrescine activation of the enzyme to the ability of the organism to synthesize spermine. Biochem Biophys Res Commun 68, 227-234.

Poulin, R., Lu, L., Ackermann, B., Bey, P. and Pegg, A.E. (1992). Mechanism of the irreversible inactivation of mouse ornithine decarboxylase by α-difluoromethylornithine: characterization of sequences at the inhibitor and coenzyme binding sites. J Biol Chem. *267*, 150-158.

Schirmer, R.H., Muller, J.G. and Krauth-Siegel, R.L. (1995). Disulfide-reductase inhibitors as chemotherapeutic agents: the design of drugs for trypanosomiasis and malaria. Angew. Chem. Int. Ed. Engl. *34*, 141.

Schroder, C.P., Godwin, A.K., O'Dwyer, P.J., Tew, K.D., Hamilton, T.C. and Ozols, R.F. (1996). Glutathione and drug resistance. Cancer Investigation. *14*, 158-168.

Segel, I.H. (1975). Enzyme kinetics, behavior and analysis of rapid equilibrium and steady-state enzyme systems. New York, John Wiley & Sons, Inc.

Stanley, B.A. and Shantz, L.M. (1994). S-adenosylmethionine decarboxylase structure-function relationships. Biochem Soc Trans. *22*, 863-869.

Svensson, F. and Persson, L. (1996). Regulation of ornithine decarboxlyase and S-adenosylmethionine decarboxlyase in a polyamine auxotrophic cell line. Mol. and Cell. Biochem. *162*, 113-119.

Tabor, C.W. and Tabor, H. (1984). Polyamines. Ann Rev Biochem. *53*, 749-790.

Tobias, K.E. and Kahana, C. (1993). Intersubunit location of the active site of mammalian ornithine decarboxylase as determined by hybridization of site-directed mutants. Biochemistry. *32*, 5842-5847.

Wang, C.C. (1995). Molecular mechanisms and therapeutic approaches to the treatment of African Trypanosomiasis. Annu. Rev. Pharmacol. Toxicol. *35*, 93-127.

Xiong, H., Stanley, B.A. and Pegg, A.E. (1999). Role of Cysteine-82 in the catalytic mechanism of human S-adenosylmethionine decarboxylase. Biochemistry. *38*, 2462-2470.

Yakubu, M.A., Majumder, S. and Kierszenbaum, F. (1993). Inhibition of S-adenosyl-L-methionine (adomet) decarboxylase by the decarboxylated adomet analog 5'{[(Z)-4-amino-2-butenyl]methylamino}-5'-deoxyadenosine (MDL73811) decreases the capacities of Trypanosoma cruzi to infect and multiply within a mammalian host cell. J. Parasitol. *79*, 525-532.

# 6

# ACETYLCHOLINESTERASES OF GASTROINTESTINAL NEMATODES

Murray E. Selkirk and Ayman S. Hussein
*Department of Biochemistry, Imperial College of Science, Technology and Medicine, London SW7 2AY, United Kingdom*

## OVERVIEW AND BACKGROUND INFORMATION

**Cholinesterases**

Vertebrate cholinesterases (ChEs) are broadly classified into two families based on their substrate specificity. Acetylcholinesterases (AChEs) terminate transmission of neuronal impulses by rapid hydrolysis of acetylcholine (ACh), and are therefore primarily associated with synaptic contacts in nerves and muscle (Fig. 1).

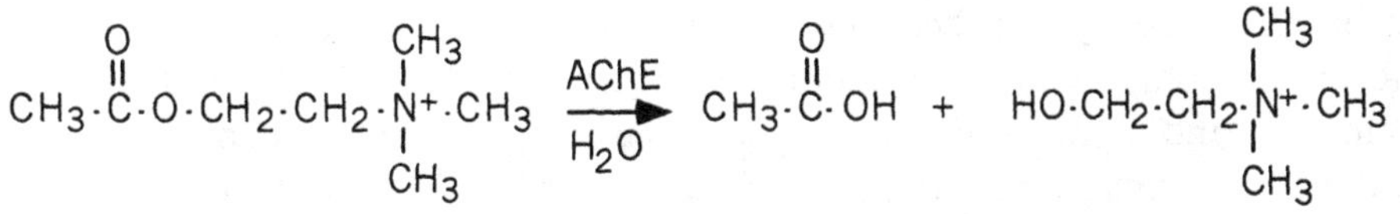

$$CH_3\text{-}\overset{\overset{O}{\|}}{C}\text{-}O\text{-}CH_2\text{-}CH_2\text{-}\underset{\underset{CH_3}{|}}{\overset{\overset{CH_3}{|}}{N^+}}\text{-}CH_3 \xrightarrow[H_2O]{AChE} CH_3\text{-}\overset{\overset{O}{\|}}{C}\text{-}OH + HO\text{-}CH_2\text{-}CH_2\text{-}\underset{\underset{CH_3}{|}}{\overset{\overset{CH_3}{|}}{N^+}}\text{-}CH_3$$

**Figure 1**. Hydrolysis of acetylcholine to acetate and choline.

Butyrylcholinesterases (BuChEs) hydrolyse other esters such as butyrylcholine (BuCh), although their function remains obscure (Taylor, 1991; Massoulié et al., 1993). In mammals, the liver represents a major site for synthesis of BuChE, which circulates at high concentration in the plasma, where it has been postulated to detoxify plant esters. This is not the whole story, however, as both enzymes are expressed at alternative sites, such as haematopoeitic cells and the developing nervous system, in which a variety of roles such as regulation of differentiation and morphogenesis have been proposed (Greenfield, 1991; Soreq et al., 1994). AChE hydrolyses ACh much more rapidly than BuCh, although BuChEs generally demonstrate appreciable activity against the smaller substrate. Both classes of enzymes show optimal activity against choline esters, and can be distinguished from other esterases by their sensitivity to the natural alkaloid physostigmine (eserine). AChEs are potently inhibited by the bis quaternary ligand 1,5-bis(4-allyldimethylammoniumphenyl) pentan-3-one dibromide (BW284C51), whereas BuChEs are highly sensitive to the organophosphate tetramonoisopropylpyrophosphortetramide (iso-OMPA) (Austin and Berry, 1953). In contrast to the latter enzymes, AChEs are further distinguished by marked inhibition at high concentrations of substrate, although the precise molecular basis for this is still unresolved (Radic et al., 1993).

**Structural diversity**

Cholinesterases exist in multiple molecular forms distinguished by their subunit interactions and hydrodynamic properties (Taylor, 1991; Massoulié et al., 1993). Vertebrates posess a single gene for AChE, and generate distinct catalytic subunits via alternative splicing. This process results in H and T subunits, although a third alternatively spliced transcript has been identified in several species (Giles, 1997). These subunits are assembled into asymmetric (A) or globular (G) forms, the latter consisting of monomers ($G_1$), dimers ($G_2$) and tetramers ($G_4$) of a catalytic subunit. Globular forms may be further subdivided into hydrophilic or amphiphilic species, and the latter associate with cell membranes either via a non-catalytic subunit bearing covalently attached fatty acids or glycolipid anchors. Assymmetric forms are composed of one to three tetramers (A4, A8 and A12) linked to Q subunits, and seem to be primarily restricted to vertebrates, associated with the basal lamina of neuromuscular junctions (Massoulié et al., 1993; Giles, 1997). Invertebrates appear to exclusively express globular forms of AChE, which also display considerable heterogeneity with respect to subunit interactions and membrane association (Toutant, 1989). Invertebrate AChEs invariably show considerable activity against BuCh, and it is generally considered that the evolution of an enzyme which is devoid of activity against this substrate coincided with the appearance of the vertebrates (Sanders et al., 1996). The secreted AChEs of parasitic nematodes provide an exception to this observation, but this will be explored later in the article.

**Enzyme structure and mechanism of action**

AChEs are serine esterases which react with their natural substrate at close to diffusion-controlled rates (Bazelyansky et al., 1986). Nucleophilic attack of the carbonyl carbon of ACh generates choline and an acylated form of the enzyme. This is followed by hydrolysis of the acyl-enzyme intermediate to liberate acetic acid (Wilson et al., 1950; Fig. 2).

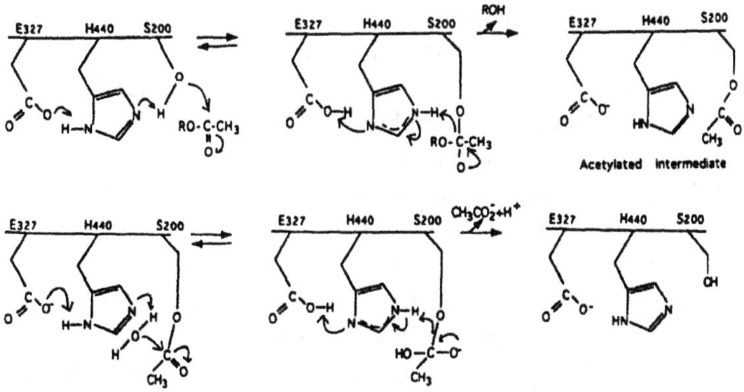

**Figure 2.** Mechanism of hydrolysis of acetylcholine by acetylcholinesterase

The three-dimensional structure of AChE isolated from the electric ray *Torpedo californica* was elucidated in 1991 (Sussman et al., 1991). It consists of a central 12 stranded mixed beta sheet surrounded by 14 alpha helices, and a similar structure (the alpha/beta hydrolase fold) is shared by a group of lipases, carboxypeptidases and adhesion molecules (Cygler et al., 1993). A remarkable feature of the enzyme is the position of the catalytic triad (Ser-200, His-440, Glu-327), which is located at the base of a narrow gorge extending approximately halfway (20Å) into the enzyme (Sussman et al., 1991). The enzyme has a strong electrostatic dipole aligned with the gorge leading to the active site, so that a positively charged substrate such as ACh can be attracted to the active site by the electrostatic field (Ripoll et al., 1993). The gorge is lined with 14 aromatic residues, thought to shield ACh from direct interaction with the negatively charged residues which contribute to the dipole, and possibly contribute to substrate guidance through the affinity of quaternary ammonium compounds for aromatic rings (Ripoll et al., 1993). Orientation of ACh in the active site is effected by interaction ot the quaternary nitrogen of choline with Trp-84 (Sussman et al., 1991). The structure of the *Torpedo* enzyme provides a template on which to compare other ChEs, and references to residue numbers are made as a rule with respect to *T. californica* AChE.

**Basis for inhibition**

An extensive literature has been compiled on inhibition of AChEs, due in part to a) the development of nerve gases targetting the mammalian enzyme, b) compounds utilised as insecticides or for control of ecto- and endo-parasites, and c) clinical applications of anti-cholinesterases, most recently their use to treat Alzheimer's disease (Giacobini, 1998). Compounds which mediate irreversible inhibition (eg organophosphates) form covalent adducts with the active site serine, and the acute toxicity of these compounds is due to the stability of the phosphoryl-enzyme complex. Reversible inhibitors may act in a competitive manner, effecting blockade of substrate at the active site (eg. edrophonium), or non-competitive, binding to a region of the enzyme that has been termed the 'peripheral' site, thought to reside near the lip of the active site gorge (eg. propidium). The bis quaternary ligands decamethonium and BW284C51 bind across both active and peripheral sites. A number of natural inhibitors of AChE have been isolated, notably from venomous snakes. Fasciculins, 61 amino acid peptides from mamba venoms, are the only known peptide inhibitors of AChE. They act in a highly selective manner on vertebrate and eel AChEs, but are weak inhibitors of avian and invertebrate AChEs or BuChEs. The crystal structure of the enzyme-inhibitor complex illustrated interaction of fasciculin with the peripheral site, sterically occluding access of substrate to the catalytic site, with potential additive allosteric effects (Bourne et al., 1995). The calabar bean (seeds of an African climbing plant) is a natural source of the alkaloid physostigmine, which carbamylates the active site serine residue of ChEs and has a long history of use in medicinal extracts (Taylor and Radic, 1994).

**Nematode cholinesterases**

In nematodes, as in many organisms, ACh is the major excitatory neurotransmitter which regulates motor functions (Rand, 1997). Where examined, distinct forms of AChE appear to be encoded by separate genes, in contrast to the situation in vertebrates and insects. *Caenorhabditis elegans* was long thought to possess three genes encoding kinetically distinct classes of AChE (Johnson et al., 1988), but recent data indicate the existence of four genes (Grauso et al., 1998). The two major classes, A and B, encoded by *ace-1* on chromosome X and *ace-2* on chromosome I are required for normal motility but appear to have overlapping functions. Homozygous mutants in *ace-1*, *ace-2* or *ace-3* have no visible phenotype, whereas *ace-1⁻/ace-2⁻* mutants are severely uncoordinated, and the triple mutation is lethal. Class C AChE, encoded by *ace-3* on chromosome II accounts for less than 5% of the total activity in the worm, and is characterised by a low $K_m$ for ACh and an unusual insensitivity to eserine (Kolson and Russell, 1985). The existence of a fourth type of enzyme, class D, representing less than 0.1% of the total AChE activity, had been suggested, but in the absence of any mutants was largely ignored until a recent study identified two *ace* sequences closely linked at the *ace-3* locus. These have been temporarily named *ace-x* and *ace-y*, as it is not yet clear which gene accounts for class C AChE (Grauso et al., 1998). Recent studies indicate that *ace-1* is expressed in the musculature of the body wall, anal sphincter, vulva and pharynx, in addition to cephalic sensory neurons (Culetto et al., 1999). In the closely related nematode *Steinernema carpocapsae*, *ace-1* encodes a hydrophilic catalytic subunit which is assembled into an amphiphilic tetramer via disulphide bonding to a hydrophobic (non-catalytic) subunit, and *ace-2* encodes an amphiphilic catalytic subunit which assembles into a glycosyl phosphatidylinositol-linked amphiphilic dimer (Arpagaus et al., 1992).

In addition to this well-defined role for AChE, many parasitic nematodes differ from free-living species in synthesizing AChEs in specialised secretory glands and expelling the enzymes to the external environment (Ogilvie et al., 1973). This unusual behaviour is exhibited primarily by parasites which inhabit the alimentary tract of their host. Much speculation on the physiological function of these enzymes has been made, and can be summarised as regulation of a) intestinal peristalsis or local spasm, b) intestinal secretory processes, and c) lymphoid/myeloid cell functions. No definitive conclusion has been reached however, and as discussion of this subject lies outside the boundaries of this article the reader is referred to a number of reviews (Rhoads, 1984; Lee, 1996; Selkirk et al., 1998).

**Nematode acetylcholinesterases as a drug target**

Many anthelmintics are targetted towards neuromuscular function in nematodes. The avermectins promote opening of glutamate-gated chloride channels, exerting a major effect via paralysis of pharyngeal pumping. Drugs such as levamisole and pyrantel are nicotinic acetylcholine receptor agonists, and act at neuromuscular junctions causing spastic paralysis, whereas piperazine is a GABA (gamma-amino-butyric acid) receptor agonist and causes flaccid paralysis. Inhibition of AChE leads to accumulation of ACh at neuromuscular junctions and also leads to paralysis via sustained contraction (Martin, 1997).

Numerous AChE inhibitors have been used to control animal-parasitic and plant-parasitic nematodes, although these have been restricted to organophosphates and carbamates which, as previously discussed, act by phosphorylating or carbamylating the active site serine residue of the enzyme. In general, these compounds are non-selective towards nematode AChEs and thus have a low therapeutic index. Variable levels of toxicosis may result from treatment of animals with organophosphates, and they also represent a significant health hazard to handlers of preparations for control of ectoparasites. Environmental contamination is a major concern, as in addition to control of plant-parasitic nematodes, both classes of compounds have also been widely used as insecticides. Thus, although these compounds are still in use, they are being progressively withdrawn from the market.

Nevertheless, as AChEs are validated targets for nematicidal drugs, it is reasonable to ask whether compounds with selective activity against parasite enzymes might be identified or designed. One might predict that a fruitful binding site for such compounds would lie at the entrance to the active site gorge, where they might block access of substrate or inhibit catalysis via allosteric effects. The conservation of aromatic residues lining the gorge in the nematode enzymes sequenced to date implies that this structure will be relatively invariant, although differences have been observed in residues implicated in inhibition by peripheral site ligands to suggest that this may be feasible (Arpagaus et al., 1994; Hussein et al., 1999a). This will become clearer following structural analysis of nematode AChEs, and this objective is facilitated by the high levels of expression obtained in *Pichia pastoris* (Hussein et al., 1999a). We are currently in the process of characterising the full complement of secreted and somatic AChEs expressed by *Nippostrongylus brasiliensis* as a model parasitic nematode of the gastrointestinal tract. Although a major objective in our laboratory lies in understanding the function of the secreted enzymes, development of inhibitors of both classes of enzymes is desirable, and could help to address the functional role of secreted AChEs in addition to providing a lead for new nematicides. The procedures described are thus aimed at defining the properties of these enzymes.

## EXPERIMENTAL APPROACHES

### Assays for cholinesterase activity

**Spectrophotometric assay**

The most commonly used assay for cholinesterase activity was devised by Ellman (Ellman et al., 1961). The assay is accurate yet simple, and typically uses acetyl- or butyrylthiocholine (ASCh or BuSCh) as substrate. The rate of hydrolysis of ASCh approximates to that of ACh (Ellman et al., 1961), although for comparative purposes, a radiometric assay employing [$^3$H] acetylcholine may be used (Johnson and Russell, 1975). In principle, the thiocholine released by hydrolysis reacts with 5:5-dithiobis-2-nitrobenzoate (DTNB) to give the yellow anion of 5-thio-2-nitrobenzoic acid. Each mole of anion produced represents hydrolysis of 1 mole of substrate, and the rate of production of the coloured ion can be recorded at 412 nm on a spectrophotometer. The assay is also easily adapted to a 96 well ELISA plate reader, allowing multiple measurements to be made at the same time. Control reactions should be carried out for non-enzymatic hydrolysis of substrate (by omitting the source of enzyme), and for the presence of thiols in the enzyme sample (by omitting ASCh). The former consideration becomes significant when assaying at high substrate concentrations. The second consideration is important when measuring AChE activity in tissue extracts, but in our experience is negligible when using parasite secreted products as a source of enzyme activity.

Reagents:

Substrates: Acetylthiocholine iodide, *S*-Butyrylthiocholine iodide, Propionylthiocholine iodide; inhibitors: Eserine, BW284C51, iso-OMPA; and DTNB are all purchased from Sigma.

Standardised procedure for plate reader:

1. Add enzyme to phosphate buffer and 1 mM DTNB in a final volume of 240 µl.
2. Start the reaction by addition of 10 µl substrate (e.g. ASCh) to a final concentration of 1 mM.
3. Monitor periodically or read after 5 min.

i.e. Each reaction should contain:

| | |
|---|---|
| Enzyme | 10 µl |
| 100 mM $NaPO_4$ pH 7.0 | 220 µl |

| | |
|---|---|
| 25 mM DTNB | 10 μl |
| 25 mM ASCh | 10 μl |

The reaction may be stopped with a suitable inhibitor (e.g. eserine at 10 μM), in which case the volume of phosphate buffer should be reduced accordingly to accommodate this, and the inhibitor should also be added to control reactions. However, it is generally advantageous to monitor the reaction continuously.

The change in absorbance per minute (ΔA) is measured. This can be converted to nmoles of substrate hydrolysed per min by taking into consideration the extinction coefficient (a measure of the amount of light absorbed by a substance in solution) of the coloured anion, which has been determined as $1.36 \times 10^4$ $M^{-1}$ $cm^{-1}$ (M = moles/litre). Thus, for a reaction volume of 1 ml in a spectrophotometer in which the cuvette has a standard light path of 1 cm, the rate of hydrolysis of substrate = $\Delta A \times 100/1.36$ nmoles $min^{-1}$.

Many plate readers will now automatically correct for the light path, but for a reaction volume of 250 μl with a depth of 0.9 cm, then the rate of hydrolysis of substrate = $\Delta A \times 22.5/1.36$ nmoles $min^{-1}$.

*One unit of AChE is generally defined as that which hydrolyses 1 micromole of substrate per minute at 20°C.*

**In situ assays for cholinesterase activity**

A number of cytochemical techniques have been developed to detect cholinesterase activity. These have been extensively applied in gel-based assays and localisation of enzyme activity in tissue sections, and can be used in combination with specific inhibitors to discriminate between AChEs and BuChEs (Silver, 1974). The method we prefer and routinely use is referred to as the 'direct colouring' method (Karnovsky and Roots, 1964). This is because the sites of enzyme activity are stained during the incubation itself rather than as a result of subsequent development, and therefore the reaction can be monitored and terminated as appropriate. The end-product is finer and more precisely localised than with other procedures. Thiocholine released by enzymatic activity reduces ferricyanide to ferrocyanide which reacts with copper to generate an insoluble precipitate of copper ferrocyanide. The gel-based assay can be used in order to discriminate between different isoforms of AChE based on their migration in a non-denaturing system. Thus, *N.*

*brasiliensis* secretes three forms of AChE designated A, B and C (Ogilvie et al., 1973) in addition to at least one form of the enzyme which is not secreted but is present in somatic extracts. All of these isoforms can be identified by their distinct electrophoretic mobilities (Hussein, 1999b).

**Procedure for gel-based assay**

1. Pour 7.5% polyacrylamide gels (without a stack) omitting SDS. Make up running buffer and loading buffer also without SDS. Load samples without boiling. Approximately 0.005 units of activity (ie. 5 nmoles of substrate hydrolysed per minute) is generally sufficient. Electrophorese until the dye is about 1 centimeter from the bottom of the gel. Take off and rinse in distilled water for 15 min.

2. Place the gel in 65 ml 0.1 M $NaPO_4$ pH 6.0. Add 50 mg ASCh and incubate at room temp (gently rocking) for 10 min. Add the following in order and mix before adding the subsequent reagent. It is important to follow this sequential addition of reagents as chelation of copper ions with citrate prevents the direct reaction of copper with ferricyanide:

| | |
|---|---|
| 5 ml | 0.1 M sodium citrate |
| 10 ml | 30 mM copper sulphate |
| 10 ml | distilled water |
| 10 ml | 5 mM potassium ferricyanide |

Incubate until a brown precipitate is clearly formed at the site of enzyme migration. Rinse in water and fix in 10% acetic acid. In order to discriminate between enzyme activities, pre-incubate the gel for 30 minutes in either $10^{-4}$ M BW284C51 (specific inhibitor of AChE) or $10^{-4}$ M iso-OMPA (specific inhibitor of BuChE) prior to addition of substrate at step 2.

**Procedure for localisation of ChEs in tissues**

The same procedure can be used to localise ChEs in tissues, and has been adapted to both optical and electron microscopy. It was this technique which first demonstrated a non-neuronal origin for nematode secreted AChEs (Lee, 1970; Ogilvie et al., 1973; McLaren et al., 1974). Cholinesterases are relatively resistant to fixation, and we generally fix sections in 4% paraformaldehyde at room temperature for 10 min (although this may be extended for a couple of hours). Wash them three times in phosphate-buffered saline and incubate in substrate buffer as outlined above for 1 to 2 hr at room temperature. After incubation, sections are washed several times in water, dehydrated and mounted. A detailed discussion of different histochemical techniques for detection of ChEs is provided by Silver (Silver, 1974).

**Determination of quaternary association: sucrose density centrifugation**
As outlined above, ChEs exist in multiple molecular forms distinguished by their subunit interactions and hydrodynamic properties. Although an old and relatively crude technique, sedimentation analysis has traditionally been the method used to define these forms, and newly discovered enzymes are routinely assigned sedimentation coefficients or S values. This is performed in sucrose gradients, and inclusion of non-denaturing detergents such as Triton X-100 or Brij 96 allows the identification of enzymes with hydrophobic domains, as in the absence of detergent such proteins tend to form aggregates. AChEs with hydrophobic domains therefore exhibit a characteristic shift in their S value in the presence of detergent, whereas hydrophilic forms are unaffected. This definition of hydrophobic versus hydrophilic forms is not strictly correlated with the requirements for solubilisation (a typical protocol for isolation of AChEs involves sequential solubilisation of tissue in buffers containing low salt, detergent and high salt, (Johnson and Russell, 1983), as aqueous extracts often contain proteins which remain in solution due to interaction with amphiphiles. Moreover, hydrophilic enzymes may only be maximally liberated by extraction with detergents due to their localisation in intracellular vesicles (the secretory AChEs of parasitic nematodes being a case in point). Interaction of hydrophobic AChEs with detergent may be abrogated by pre-treatment with phophatidylinositol-phospholipase C, or limited digestion with proteases, indicating that this interaction is mediated by a specific hydrophobic domain, often in the form of a glycolipid anchor (Toutant, 1989). These enzymes are thus more accurately defined as amphiphilic, and the combination of sucrose density gradient centrifugation with enzymatic digestion can yield valuable information on the properties of the AChE under investigation.

Reagents:
Detergents (Triton X-100, Brij 96, sodium deoxycholate, CTAB), marker enzymes (*E. coli* alkaline phosphatase, bovine liver catalase, *E. coli* β–galactosidase) and general reagents are all purchased from Sigma. Ultra-clear centrifuge tubes (14 x 89 mm) for the SW41 rotor are from Beckman.

Procedure:
A typical protocol for sedimentation analysis uses 2 to 20% sucrose gradients (e.g. 10 ml, with a 0.5 ml cushion of 50% sucrose) made in phosphate buffered saline (PBS) in the presence or absence of 1% Triton X-100 or 0.5% Brij 96 (10-oleylether).

1. Make 2% and 20% solutions of sucrose in PBS ($^{+}/_{-}$ 1% Triton X-100). Pour gradients with 5 ml of each solution over a 0.5 ml cushion of 50% sucrose. Weigh each tube carefully and balance with 2% sucrose.
2. Layer protein samples (up to 200 μl, with *E. coli* alkaline phosphatase, bovine liver catalase and *E. coli* β-galactosidase included as internal standards) onto the gradients, weigh again, balance with PBS and centrifuge for 16 hr at 36,000 rpm at 4°C in an SW41 rotor (170,000 x *g*). Carefully collect 0.25 ml fractions and assay for AChE, alkaline phosphatase (6.1S), catalase (11.3S), and β-galactosidase activities (16S). Use 5 to 10 U alkaline phosphatase, 1 to 2 mg catalase and 50 U β-galactosidase per gradient.
3. Alkaline Phosphatase assay: Prepare substrate buffer: 100 mM Tris-HCl, 100 mM NaCl, 50 mM $MgCl_2$, pH 9.5. To 20 ml of this add 165 μl BCIP (5-bromo-4-chloro-3-indolyl phosphate; 20 mg/ml in water) and 88 μl nitro blue tetrazolium (NBT, 75 mg $ml^{-1}$ in 70% *N,N*-dimethyl formamide). Add 10 μl of each fraction to 200 μl substrate buffer and measure absorbance at 405 nm.
4. Catalase assay: Add 200 μl of 30% hydrogen peroxide to 100 ml 50 mM $NaPO_4$ pH 7.0. Add 20 μl of each fraction to 1 ml hydrogen peroxide solution and assay the decrease in absorbance at 240 nm.
5. Beta-galactosidase assay: Prepare substrate buffer by adding 100 μl X-gal (5-bromo-4-chloro-3-indolyl-β-D-galactopyranoside, 20 mg $ml^{-1}$ in *N,N*- dimethylformamide) to 10 ml 50 mM phosphate buffer pH 7.0. Add 20 μl of each fraction to 200 μl substrate buffer, incubate for 1 hr at 37°C and measure the absorbance at 550 nm.
6. Plot the migration of each enzyme and calculate the sedimentation coefficient of AChE from the standards.

**Charge-shift electrophoresis**
An alternative method to screen for amphiphilic forms of AChE is provided by charge-shift electrophoresis (Toutant, 1986), based on the observation that positively- or negatively-charged detergents affect the migration of hydrophobic or amphiphilic proteins in non-denaturing systems (Helenius and Simons, 1977).

Procedure:

1. Cast 3 sets of 7.5% acrylamide non-denaturing gels as above, the first with the addition of 0.5% Triton X-100 (non-ionic detergent), the second with 0.5% Triton X-100 and 0.25% sodium deoxycholate (DOC; anionic detergent) and the third with 0.5% Triton X-100 and 0.1% cetyltrimethylammonium bromide (CTAB; cationic detergent).
2. Add 0.5% Triton X-100, 0.5% Triton X-100 + 0.25% DOC, and 0.5% Triton X-100 + 0.1% CTAB to 3 sets of running buffer, and electrophorese in parallel. Dilute the samples 1:1 in each running buffer plus glycerol, load and electrophorese until the dye migrates approximately halfway. In each gel include a sample of hydrophilic AChE (hAChE). Wash and stain for AChE activity as above.
3. Determine the migration of samples in each buffer system by a ratio R of the distance migrated in comparison to the hAChE. A shift in R indicates interaction with detergent.

**Purification of native enzymes**

Athough AChEs can be purified by conventional methodologies, several ligands have been synthesised by different laboratories or are commercially available which can be used for affinity chromatography, greatly simplifying the purification procedure. Synthesis of several of these ligands is described by Dudai and Silman (Dudai and Silman, 1974), and we have used one of these, 1-methyl-9-[*N*-(ε-aminocaproyl)-β-aminopropylamino]-acridinium bromide hydrobromide (MAC) to purify secreted AChEs of *N. brasiliensis* (Grigg et al., 1997). The utility of this ligand is based on the observation that *N*-methyl-acridinium is a potent inhibitor of AChE, and the MAC ligand was synthesised with an elongated spacer chain which could be covalently linked to Sepharose to yield an affinity matrix which binds AChE even at high ionic strength (Dudai and Silman, 1974). More conveniently, procainamide-Sepharose or edrophonium-Sepharose can be used, and the latter ligand has also been successfully used to purify nematode secreted enzymes in a one-step procedure (Pritchard et al., 1991; Griffiths and Pritchard, 1994).

Reagents:

Edrophonium chloride and procainamide are purchased from Sigma, and can be conjugated to Sepharose 4B via standard procedures (Pharmacia).

Procedure:

1. Dialyse parasite secreted products or extracts into low ionic strength buffer (e.g. 20 mM sodium phosphate pH 7.4).

2. Apply to the edrophonium-Sepharose 4B afffinity column at a flow rate of approximately 0.5 ml/min.
3. Wash the column extensively in 20 mM sodium phosphate pH 7.4, then in 20 mM sodium phosphate pH 7.4, 100 mM NaCl.
4. Elute with 20 mM edrophonium chloride in the latter buffer.
5. Monitor all fractions for AChE activity by the Ellman assay as described above.
6. Pool the peak fractions from the eluate and dialyse in phosphate buffer to remove edrophonium chloride prior to further analysis.

*Procainamide can be similarly conjugated to Sepharose and utilised in an analagous procedure. Decamethonium bromide, also at 20 mM, provides an alternative ligand for elution of AChE.*

**Expression in *Pichia pastoris* and purification of recombinant enzyme**

Production of active secreted enzymes often requires expression in suitable eukaryotic systems. Although we have successfully used baculovirus vectors in insect cells for this purpose, the methylotrophic yeast *Pichia pastoris* has proved to be particularly useful for high level expression of secreted AChEs (Hussein et al., 1999a). This has the advantage of being commercially available in kit form (Invitrogen), with expression plasmids (pPICZαA,B,C) in all reading frames. The general procedure is to amplify appropriate cDNAs with appropriate restriction sites for ligation into the expression plasmid, deleting the native signal peptide (the plasmids encode an N-terminal signal peptide provided by the pre-pro sequence of the α-factor of *Saccharomyces cerevisiae*). The protein may be expressed with an authentic C-terminus by inclusion of an appropriate stop codon, or with a *C*-terminal tag containing a myc epitope (useful for antibody-based screening of expression) and a polyhistidine tag (useful for purification by nickel-chelating chromatography). Expression is driven by the alcohol oxidase 1 promoter (*AOX1*), is tightly regulated and induced by methanol to typically high levels. After cloning, the reading frame should be verified by sequencing of the recombinant plasmid, which is then linearised and integrated into competent *Pichia pastoris* via homologous recombination according to the manufacturer's instructions (Invitrogen). Transformants should be confirmed by PCR with oligonucleotide primers homologous to pPICZα sequences flanking the insertion site.

Reagents:
All components of media necessary for growth of *P. pastoris* can be obtained from Sigma. Zeocin, required for initial cloning steps, is purchased from Invitrogen. Nickel-agarose columns can be purchased from Qiagen, or made by charging Chelating Sepharose (Pharmacia) with nickel sulphate.

Procedures:

**Expression**

1. Using a single colony, inoculate 5 ml of BMGY medium (1% yeast extract, 2% peptone, 0.1M potassium phosphate pH 6.0, 1.34% yeast nitrogen base, 400 ng $ml^{-1}$ biotin, 1% glycerol) in a 50 ml Falcon tube. Grow at 28 to 30°C in a shaking incubator (230 to 300 rpm) until the culture reaches $OD_{600}$ of 2 to 6 (approximately 18 hr).
2. Use this 5 ml culture to inoculate 50 ml of BMGY medium in a 250 ml culture flask and grow at 28 to 30°C in a shaking incubator until culture reaches log phase ($OD_{600}$ of 2 to 6).
3. Harvest cells in sterile Falcon tubes by centrifugation at 1,500 to 3,000 x g for 5 min at room temperature.
4. To induce expression, decant the supernatant and resuspend the pellet in 50 ml of BMMY medium (as BMGY, but with glycerol replaced by 0.5% methanol). Grow at 28 to 30°C in a shaking incubator for up to 5 days.
5. Collect aliquots of 1 ml at time points 0, 1, 2, 4, 8, 24, 36, 48, 72 and 96 hr. Add fresh methanol daily to a final concentration of 1%.
6. Assay for AChE activity using ASCh under standard conditions. Determine protein content from each time point.
7. Concentrate the supernatant using Centricon 50 concentrators (Amicon) to 5 ml.

**Purification by nickel-chelating sepharose chromatography**

1. Dialyse the yeast culture supernatant against 500 ml 50 mM $NaPO_4$ pH 8.0, 0.3 M NaCl (buffer A) at 4°C, with 2 changes of buffer.
2. Wash 7.5 ml nickel-chelating Sepharose beads in 30 ml buffer A. Change buffer twice.
3. Prepare column and pack at 50 ml/hr with buffer A. Leave to equilibrate for >5hr.
4. Apply the dialysate onto the column at flow rate $\leq$ 50 ml/h. Wash with at least 5 column bed volumes buffer A (i.e. >40 ml).
5. Elute with a gradient of 0 to 200 mM of imidazole in buffer A. Alternatively, wash the column with 10 mM imidazole and then elute

with 200 mM imidazole.
6. Assay AChE activity in different fractions under standard conditions.
7. Pool fractions with peak activities and resolve 10 µl by 12% SDS-PAGE.

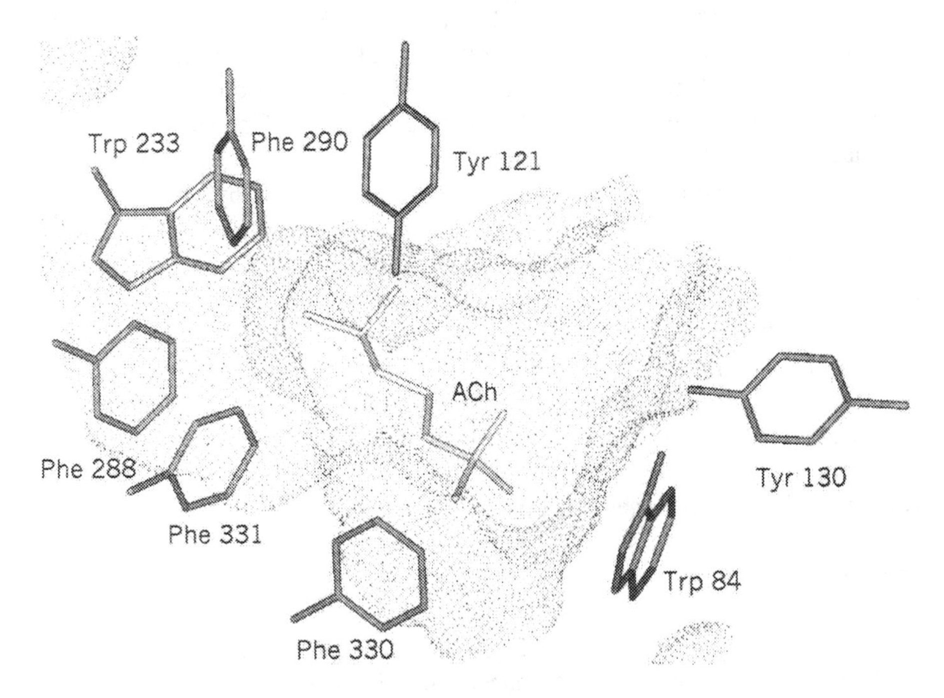


**Figure 3.** Active site of *Torpedo californica* AChE. Acetylcholine is shown docked into the active site. The solvent-accessible surface of the active site gorge is depicted along with side chains of aromatic residues. Data were obtained using coordinates from PDB accession code 2ACE.

**Enzymatic properties of nematode AChEs**

In addition to somatic enzyme(s), adult *Nippostrongylus brasiliensis* secrete three monomeric non-amphiphilic ($G_1^{na}$) variants of AChE designated A, B and C, with apparent masses of 74, 69 and 71 kDa respectively (Grigg et al., 1997). We have isolated cDNAs for two of these variants, AChE B and AChE C (Hussein, 1999a; Hussein et al., submitted). Both recombinant enzymes behave as true AChEs, with minimal activity towards BuSCh. Mutagenesis studies on *Torpedo* and human AChE have shown that Phe-288 and Phe-290 in the active site gorge dictate substrate specificity, most probably via steric occlusion, but also possibly by stabilising the substrate in an optimal position for catalysis (Harel et al., 1992; Vellom et al., 1993; Ordentlich et al., 1993).

It has been suggested that the intermediate substrate specificity of certain invertebrate enzymes such as *C. elegans* ACE-1 and *Drosophila melanogaster* AChE (both enzymes hydrolyse BuSCh at approximately 50% the rate of ASCh) could be explained by the substitution of Phe-288 by glycine and leucine respectively (Arpagaus et al., 1994; Gnagey et al., 1987). Replacement of Phe-288 in *Torpedo* and human AChE by non-aromatic residues greatly enhanced the ability of these enzymes to hydrolyse BuSCh, in addition to conferring sensitivity to inhibition by iso-OMPA (Harel et al., 1992; Vellom et al., 1993; Ordentlich et al., 1993). It was therefore surprising that Phe-288 was replaced by a methionine residue in both *N. brasiliensis* secreted AChEs sequenced to date, as they display little activity against BuSCh and no inhibition by iso-OMPA even at very high concentrations (Hussein et al., 1999a; Hussein et al., submitted). We suggested that the replacement of Phe-290 and Phe-331 by the bulkier residue Trp in the nematode enzymes might restrict the size of the acyl pocket and access of BuSCh, and have addressed this possibility by site-directed mutagenesis.

Procedures:

**Mutagenesis**

Mutant enzymes can be generated using a commercial kit (the Stratagene Quikchange™ site-directed mutagenesis kit) which accelerates and simplifies the procedure. The basic protocol utilises a double stranded DNA vector with the insert of interest and two synthetic oligonucleotide primers of about 30 bases, each of which contains the desired mutation in the middle, and anneal to the same sequence on opposite strands of the plasmid. The primers catalyse extension by PCR using the high fidelity Pfu DNA polymerase, and a mutated plasmid containing a staggered nick is therefore generated. Following PCR amplification, the products are treated with the endonuclease Dpn I. Dpn I is specific for methylated and hemimethylated DNA and digests the parental DNA template. The unmethylated PCR product (the mutated plasmid containing the staggered nick) is therefore selected for following transformation of reaction products into *E. coli*.

**Kinetic properties**

An accompanying chapter of this book (Kinch et al., page 93) gives a good coverage of enzyme kinetic analysis, which has been extensively investigated for vertebrate cholinesterases. Our specific interest in mutating residues lining the active site gorge of *N. brasiliensis* AChEs is to address the question of substrate specificity. The standard Ellman assay is therefore carried out with ASCh and BuSCh at final concentrations between 0.05 and 2.0 mM, and the velocity of the reaction (nmoles/min) plotted against substrate concentration. The $K_m$ and $V_{max}$ can be calculated from Lineweaver-Burk plots, using substrate concentrations between 0.05 to 0.75 mM (i.e. below excess substrate inhibition). These use the reciprocal of the Michaelis-Menten equation, i.e.:

$$1/v = (K_m / V_{max})\, 1/[S] + 1/V_{max}$$

This is a linear equation in $1/[S]$ and $1/v$. Thus, by plotting these values, the slope of the line is $K_m / V_{max}$, the $1/v$ intercept is $1/V_{max}$, and the extrapolated $1/[S]$ intercept is $-1/K_m$. The catalytic constant or turnover number $k_{cat}$ is determined from the $V_{max}$ and the enzyme concentration acccording to $k_{cat} = V_{max}/[E]_T$.

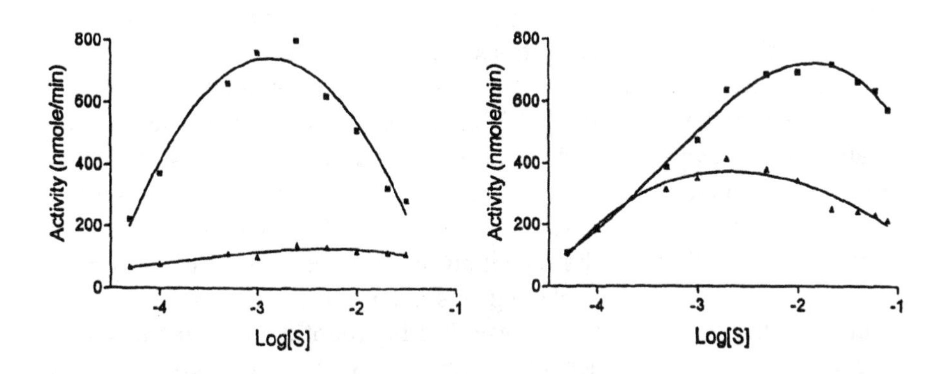


**Figure 4.** Substrate specificity of *N. brasiliensis* secreted AChE C. Hydrolysis of ASCh (squares) and BuSCh (triangles) is illustrated for the wild-type enzyme (left panel) and the triple mutant M300G/W302F/W345F (right panel).

The precise mechanism by which excess substrate inhibits AChE is unclear, but is one of the features that distinguishes it from BuChE. It has been proposed that excess substrate inhibition results from binding of a second ACh molecule to a low affinity site, blocking the reaction mechanism or entrance/exit of substrate/reaction products. The equilibrium constant for this binding event is represented by $K_{ss}$, and can be determined from the Haldane equation:

$$v = V_{max}/ (1 + K_m/[S] + [S]/K_{ss})$$

Alternatively, $K_{ss}$ may be determined by plotting $(1/v)$ against $1/[S]$, using substrate concentrations where excess substrate inhibition is clear. This would generally be in the range 5-40 mM, but may have to be extended to 80 mM ASCh for mutant enzymes (Fig. 4).

## DISCUSSION AND CONCLUSION

The secreted AChEs of *N. brasiliensis* differ from 'classical' AChEs in that they are truncated at the carboxyl terminus, and possess substitutions in the aromatic residues which line the active site gorge, corresponding to Tyr-70 (Ser/Thr), Trp-279 (Asn) and Phe-288 (Met). The enzymes are monomeric and hydrophilic, and although of invertebrate origin and possessing the above substitutions in the active site gorge residues, efficiently hydrolyse ASCh and show minimal activity against BuSCh. They display excess substrate inhibition with ASCh at concentrations over 2.5 mM, and are highly sensitive to classical active site and 'peripheral' site inhibitors. A double mutation to one of these enzymes (AChE C) W302F/W345F, corresponding to positions 290 and 331 in *Torpedo*, renders the enzyme 10x less sensitive to excess substrate inhibition, but does not radically affect substrate specificity. In contrast, a triple mutant (M300G/W302F/W345F) shows considerable activity against BuSCh, whilst remaining insensitive to the BuChE-specific inhibitor iso-OMPA, and displaying a similar profile of excess substrate inhibition as the double mutant. These data highlight a conserved pattern of active site architecture for nematode secreted AChEs characterised to date, and provide an explanation for the substrate specificity which might otherwise appear inconsistent with the primary structure of these enzymes. In addition, the parasite contains at least one non-secreted AChE, a tetrameric amphiphilic form which shows properties closer to other invertebrate AChEs, although we do not yet have information on the primary structure of this enzyme (Hussein et al., 1999b).

In vertebrate AChEs, peripheral site inhibitors interact with residues located at the mouth of the active site gorge. A model has been proposed in which binding of ligands to Trp-279 and surrounding amino acids induces a conformational change relayed via residues along the gorge (including Tyr-

70) resulting in reorientation of Trp-84 which is crucial for positioning substrate in the active site (Ordentlich et al., 1993). The nematode secreted AChEs are highly sensitive to both active site and peripheral site inhibitors, although Trp-279 is substituted by Asn, and Tyr-70 is replaced by threonine/serine. There thus appears to be a discrepancy in the basis for inhibition by peripheral site ligands in the nematode enzymes. We therefore aim to obtain structural data which will enable us to assess the potential for selective inhibition of these enzymes as a basis for the rational design of new drugs for intestinal nematodes, in addition to providing reagents with which to investigate the physiological function of the unusual secreted variants.

## REFERENCES

Arpagaus, M., Fedon, Y., Cousin, X., Chatonnet, A., Bergé, J.-B., Fournier, D., and Toutant, J.-P. (1994) cDNA sequence, gene structure, and in vitro expression of *ace-1*, the gene encoding acetylcholinesterase of class A in the nematode *Caenorhabditis elegans*. J. Biol. Chem. *269*, 9957-9965.

Arpagaus, M., Richier, P., Bergé, J.-B., and Toutant, J.-P. (1992) Acetylcholinesterases of the nematode *Steinernema carpocapsae*. Characterization of two types of amphiphilic forms differering in their mode of membrane association. Eur. J. Biochem. *207*, 1101-1108.

Austin, L., and Berry, W.K. (1953) Two selective inhibitors of cholinesterase. Biochem. J. *54*, 695-700.

Bazelyansky, M., Robey, E., and Kirsch, J. F. (1986) Fractional diffusion-limited component of reactions catalyzed by acetylcholinesterase. Biochemistry *25*, 125-30.

Bourne, Y., Taylor, P., and Marchot, P. (1995) Acetylcholinesterase inhibition by fasciculin: crystal structure of the complex. Cell *83*, 503-12.

Culetto, E., Combes, D., Fedon, Y., Roig, A., Toutant, J.-P., and Arpagaus, M. (1999) Structure and promoter activity of the 5' flanking region of *ace-1*, the gene encoding acetylcholinesterase of class A in *Caenorhabditis elegans*. J. Mol. Biol. *290*, 951-66.

Cygler, M., Schrag, J.D., Sussman, J.L., Harel, M., Silman, I., Gentry, M.K., and Doctor, B.P. (1993) Relationship between sequence conservation and three-dimensional structure in a large family of esterases, lipases, and related proteins. Prot. Sci. *2*, 366-382.

Dudai, Y., and Silman, I. (1974) Acetylcholinesterase. Meth. Enzymol. *34*, 571-580.

Ellman, G.L., Courtney, K.D., Andres, V., and Featherstone, R.M. (1961) A new and rapid colorimetric determination of acetylcholinesterase activity. Biochem. Pharmacol. *7*, 88-95.

Giacobini, E. (1998) Cholinesterase inhibitors for Alzheimer's disease therapy: from tacrine to future applications. Neurochem. Int. *32*, 413-9.

Giles, K. (1997) Interactions underlying subunit associations in cholinesterases. Prot. Eng. *10*, 677-685.

Gnagey, A.L., Forte, M., and Rosenberry, T.L. (1987) Isolation and characterization of acetylcholinesterase from *Drosophila*. J. Biol. Chem. *262*, 13290-13298.

Grauso, M., Culetto, E., Combes, D., Fedon, Y., Toutant, J.-P., and Arpagaus, M. (1998) Existence of four acetylcholinesterase genes in the nematodes *Caenorhabditis elegans* and *Caenorhabditis briggsae*. FEBS Lett. *424*, 279-284.

Greenfield, S.A. (1991) A non-cholineric action of acetylcholinesterase (AChE) in the brain: from neuronal secretion to the generation of movement. Cell. Mol. Neurobiol. *11*, 55-77.

Griffiths, G., and Pritchard, D.I. (1994) Purification and biochemical characterisation of acetylcholinesterase (AChE) from the excretory/secretory products of *Trichostrongylus colubriformis*. Parasitology *108*, 579-586.

Grigg, M.E., Tang, L., Hussein, A.S., and Selkirk, M.E. (1997) Purification and properties of monomeric ($G_1$) forms of acetylcholinesterase secreted by *Nippostrongylus brasiliensis*. Mol. Biochem. Parasitol. 90, 513-524.

Harel, M., Sussman, J.L., Krejci, E., Bon, S., Chanal, P., Massoulié, J., and Silman, I. (1992) Conversion of acetylcholinesterase to butyrylcholinesterase: modelling and mutagenesis. Proc. Natl. Acad. Sci. USA *89*, 10827-10831.

Helenius, A., and Simons, K. (1977) Charge shift electrophoresis: simple method for distinguishing between amphiphilic and hydrophilic proteins in detergent solution. Proc. Natl. Acad. Sci. USA *74*, 529-532.

Hussein, A.S., Chacón, M.R., Smith, A.M., Tosado-Acevedo, R., and Selkirk, M.E. (1999) Cloning, expression and properties of a non-neuronal secreted acetylcholinesterase from the parasitic nematode *Nippostrongylus brasiliensis*. J. Biol. Chem. *274*, 9312-9319.

Hussein, A.S., Grigg, M.E., and Selkirk, M.E. (1999) *Nippostrongylus brasiliensis*: Characterisation of a somatic amphiphilic acetylcholinesterase with properties distinct from the secreted enzymes. Exp. Parasitol. *91*, 144-150.

Hussein, A.S., Smith, A.M., Chacón, M.R., and Selkirk, M.E. (2000) A second non-neuronal secreted acetylcholinesterase from the parasitic nematode *Nippostrongylus brasiliensis*: determinants of substrate specificity. (submitted for publication)

Johnson, C.D., Rand, J.R., Herman, R.K., Stern, B.D., and Russell, R.L. (1988) The acetylcholinesterase genes of *C. elegans*: identification of a third gene (*ace-3*) and mosaic mapping of a synthetic lethal phenotype. Neuron *1*, 165-173.

Johnson, C.D., and Russell, R.L. (1975) A rapid, simple radiometric assay for cholinesterase, suitable for multiple determinations. Anal. Biochem. *64*, 229-38.

Johnson, C.D., and Russell, R.L. (1983) Multiple molecular forms of acetylcholinesterase in the nematode *Caenorhabditis elegans*. J. Neurochem. *41*, 30-46.

Karnovsky, M. J., and Roots, L. (1964) A 'direct-coloring' method for cholinesterases. J. Histochem. Cytochem. *12*, 219-221.

Kolson, D.L., and Russell, R.L. (1985) A novel class of acetylcholinesterase, revealed by mutations, in the nematode *Caenorhabditis elegans*. J. Neurogenet. *2*, 93-110.

Lee, D.L. (1970) The fine structure of the excretory system in adult *Nippostrongylus brasiliensis* (Nematoda) and a suggested function for the "excretory glands". Tissue Cell *2*, 225-231.

Lee, D.L. (1996) Why do some nematode parasites of the alimentary tract secrete acetylcholinesterase? Int. J. Parasitol. *26*, 499-508.

Martin, R.J. (1997) Modes of action of anthelmintic drugs. Vet. J. *154*, 11-34.

Massoulié, J., Sussman, J.L., Bon, S., and Silman, I. (1993) Structure and functions of acetylcholinesterase and butyrylcholinesterase. Prog. Brain Res. *98*, 139-146.

McLaren, D., Burt, J.S., and Ogilvie, B.M. (1974) The anterior glands of adult *Necator americanus* (nematoda: strongyloidea)-II. Cytochemical and functional studies. Int. J. Parasitol. *4*, 39-46.

Ogilvie, B.M., Rothwell, T.L.W., Bremner, K.C., Schitzerling, H.J., Nolan, J., and Keith, R. K. (1973) Acetylcholinesterase secretion by parasitic nematodes, 1. Evidence for secretion of the enzyme by a number of species. Int. J. Parasitol. *3*, 589-597.

Ordentlich, A., Barak, D., Kronman, C., Flashner, Y., Leitner, M., Segall, Y., Ariel, N., Cohen, S., Velan, B., and Shafferman, A. (1993) Dissection of the human acetylcholinesterase active center determinants of substrate specificity. J. Biol. Chem. *268*, 17083-17095.

Pritchard, D.I., Leggett, K.V., Rogan, M.T., McKean, P.G., and Brown, A. (1991) *Necator americanus* secretory acetylcholinesterase and its purification from excretory-secretory products by affinity chromatography. Parasite Immunol. *113*, 187-199.

Radic, Z., Pickering, N.A., Vellom, D.C., Camp, S., and Taylor, P. (1993) Three distinct domains in the cholinesterase molecule confer selectivity for acetyl-and butyrylcholinesterase inhibitors. Biochemistry *32*, 12074-12084.

Rand, J.B., and Nonet, M.L. (1997) Synaptic transmission. *C. elegans* II. Cold Spring Harbor, New York: Cold Spring Harbor Press, 611-643.

Rhoads, M.L. (1984) Secretory cholinesterases of nematodes: possible functions in the host-parasite relationship. Trop. Vet. *2*, 3-10.

Ripoll, D.R., Faerman, C.H., Axelson, P.H., Silman, I., and Sussman, J.L. (1993) An electrostatic mechanism for substrate guidance down the aromatic gorge of acetylcholinesterase. Proc. Natl. Acad. Sci. USA *90*, 5128-5132.

Sanders, M., Mathews, B., Sutherland, D., Soong, W., Giles, H., and Pezzementi, L. (1996) Biochemical and molecular characterization of acetylcholinesterase from the hagfish *Myxine glutinosa*. Comp. Biochem. Physiol. *115B*, 97-109.

Selkirk, M.E., Hussein, A.S., Russell, W.S., Grigg, M.E., Chacón, M.R., Smith, A.M., Henson, S., and Tippins, J.R. (1998) Secretory acetylcholinesterases of

*Nippostrongylus brasiliensis*: properties and implications for mucosal immunity. Structure and Function of Cholinesterases and Related Proteins. New York: Plenum Press, 515-522.

Silver, A. (1974) The Biology of Cholinesterases. Amsterdam: North-Holland Publishing Company.

Soreq, H., Patinkin, D., Lev-Lehman, E., Grifman, M., Ginzberg, D., Eckstein, F., and Zakut, H. (1994) Antisense oligonucleotide inhibition of acetylcholinesterase gene expression induces progenitor cell expansion and suppresses hematopoietic apoptosis ex vivo. Proc. Natl. Acad. Sci. USA *91*, 7907-11.

Sussman, J.L., Harel, M., Frolow, F., Oefner, C., Goldman, C., Toker, L., and Silman, I. (1991) Atomic structure of acetylcholinesterase from *Torpedo californica*: a prototypic acetylcholine-binding protein. Science *253*, 872-879.

Taylor, P. (1991) The cholinesterases. J. Biol. Chem. *266*, 4025-4028.

Taylor, P., and Radic, Z. (1994) The cholinesterases: from genes to proteins. Annu. Rev. Pharmacol. Toxicol. *34*, 281-320.

Toutant, J.-P. (1986) An evaluation of the hydrophobic interactions of chick muscle acetylcholinesterase by charge shift electrophoresis and gradient centrifugation. Neurochem. Int. *9*, 111-119.

Toutant, J.-P. (1989) Insect acetylcholinesterase: catalytic properties, tissue distribution and molecular forms. Prog. Neurobiol. *32*, 423-446.

Vellom, D.C., Radic, Z., Li, Y., Pickering, N.A., Camp, S., and Taylor, P. (1993) Amino acid residues controlling acetylcholinesterase and butyrylcholinesterase specificity. Biochemistry *32*, 12-17.

Wilson, I.B., Bergmann, F., and Nachmansohn, D. (1950) Acetylcholinesterase X. Mechanism of the catalysis of acylation reactions. J. Biol. Chem. *186*, 781-790.

# 7

# *TOXOPLASMA* AS A MODEL APICOMPLEXAN PARASITE: BIOCHEMISTRY, CELL BIOLOGY, MOLECULAR GENETICS, GENOMICS AND BEYOND

David S. Roos, John A. Darling, Mary G. Reynolds, Kristin M. Hager, Boris Striepen and Jessica C. Kissinger
*Department of Biology, University of Pennsylvania, Philadelphia PA 19104-6018*

## INTRODUCTION

More than twenty years ago, virologist Elmer Pfefferkorn recognised that the plaques formed following infection of mammalian host cell monolayers with the protozoan parasite *Toxoplasma gondii* should permit clonal isolation. Seeking to develop a genetic system suitable for the analysis of this intracellular pathogen, Pfefferkorn and his colleagues produced various mutant parasite lines. Carrying out classical genetic crosses (in cats, the definitive host for *T. gondii*), they demonstrated that tachyzoites are haploid, and that Mendelian segregation occurs through meiotic production of sporozoites within the oocyst (Pfefferkorn and Pfefferkorn, 1980). These investigators also exploited the ability of *T. gondii* to infect virtually any nucleated cell to develop somatic cell genetic approaches, infecting defined mammalian cell mutants to determine what wild-type and mutant parasites can do for themselves, and what they require from their host (Pfefferkorn et al., 1983).

Nucleoside metabolic pathways have traditionally supplied the most successful targets for parasiticidal drug design, and mapping these pathways provided a primary focus for early studies, as described in the first volume of articles to emerge from the Biology of Parasitism course (Pfefferkorn, 1988). In recent years, the development of molecular genetic tools has made it possible to test these early models of parasite metabolism. On the occasion of Professor Pfefferkorn's retirement, it seems appropriate to review these studies.

*T. gondii* is remarkably accessible to examination in other ways as well, permitting recombinant protein expression, ultrastructural studies, pharmacological analysis, and the validation of data gleaned from genome databases. Recapitulating the Biochemistry, Cell Biology and Molecular Genetics components of the Biology of Parasitism course, this chapter provides several example of how and why *T. gondii* has proved such a useful experimental organism and a model for other apicomplexan parasites, including *Plasmodium* species. Throughout the text, open questions deserving further study have been highlighted in brackets.

## MOLECULAR GENETIC DISSECTION OF PURINE SALVAGE PATHWAYS

Complementing the classical genetic approaches pioneered by Pfefferkorn et al., a wide variety of molecular tools are now available for the genetic manipulation of *T. gondii* (Boothroyd et al., 1994; Roos et al., 1994 and 1997; Black and Boothroyd, 1998; Donald and Roos, 1998; Knoll and Boothroyd 1998; Nakaar and Joiner, 1999). These include:

- Vectors for transient and stable expression (or overexpression) of recombinant transgenes,
- Integrating and episomal vectors, employing various positive and negative selectable markers,
- Strategies for nonhomologous recombination, permitting saturation mutagenesis of the entire parasite genome (including gene/promoter trap experiments),
- Strategies for homologous recombination, permitting targeted gene knock-outs and allelic replacements, and
- Antisense expression vectors, permitting down-regulation of steady-state transcript levels.

*[Note: while the development of molecular genetic technologies for Toxoplasma has proceeded rapidly in recent years, additional tools are still*

*needed, including more efficient strategies for complementation cloning, and improved techniques for regulating gene expression.]*

Perhaps the greatest surprise to emerge from the development of *T. gondii* molecular genetics has been the remarkably high frequency of molecular transformation observed in this parasite: using suitable vectors, >50% of viable tachyzoites transiently express recombinant reporters, and up to 5% stably integrate these plasmids (Donald and Roos, 1994). Nonhomologous integrating vectors have been particularly useful, permitting saturation mutagenesis of the entire parasite genome in a single electroporation cuvette (Roos et al., 1997). This process is conceptually analogous to transposon mutagenesis in *Drosophila, Saccharomyces, Arabidopsis,* and other experimentally accessible genetic systems, but with several additional advantages. Introduced transgenes can be engineered to provide single-copy tags that are distinct from other chromosomal sequences, readily recognisable, and easily rescued. Because *T. gondii* tachyzoites are haploid, it is a relatively straightforward matter to clone any gene whose function is not essential for tachyzoite survival, provided that the knock-out generated by insertional tagging yields an identifiable phenotype.

In his pioneering genetic studies, Pfefferkorn mapped out the nucleotide metabolic pathways of *T. gondii* in some detail, noting that: (i) adenine arabinoside (araA) acts as a subversive substrate of the parasite's adenosine kinase (AK) (Pfefferkorn and Pfefferkorn, 1976 and 1978), (ii) fluorouracil acts as a subversive substrate of uracil phosphoribosyl transferase (UPRT) (Pfefferkorn, 1977; Pfefferkorn and Pfefferkorn, 1977), and (iii) thioxanthine is a subversive substrate of hypoxanthine-xanthine-guanine phosphoribosyl transferase (HXGPRT) (Pfefferkorn and Borotz, 1994).

While *T. gondii* is unable to synthesise purines de novo (Schwartzman and Pfefferkorn, 1982; Krug et al., 1989), redundant salvage pathways are provided through the ability to interconvert AMP and IMP, as illustrated in Figure 1. These pathways provide an ideal test case for the exploitation of insertional mutagenesis: inactivation of AK or HXGPRT is not lethal to the parasite, due to the functional redundancy of these pathways, but mutation of either enzyme yields a phenotype that can readily be identified by virtue of resistance to the relevant subversive substrate (araA or thioxanthine). Accordingly, *T. gondii* tachyzoites were subjected to insertional mutagenesis, selection of transformants harbouring the pyrimethamine-resistant transgene, and further selection for resistance to either araA or thioxanthine. These studies led to the successful cloning of the AK and HXGPRT genes (Donald et al., 1996; Sullivan et al., 1999). Kinetic and crystallographic characterisation of these enzymes (Schumacher et al., 1996; Darling et al., 1999) should facilitate structure-based approaches to drug design.

In addition to permitting the molecular cloning of these enzymes, the insertional mutagenesis strategy serves to illustrate the potential of genetic approaches (Sullivan et al., 1999). Among the 13 araA-resistant clones that were obtained in this study, only five were tagged at the AK locus. An additional four clones were tagged at a second locus, which proved to encode the parasite's major adenosine transporter (Chiang et al., 1999). A third, as yet uncharacterised, group of araA-resistant clones exhibit no AK activity, but appear to harbour no mutations in the AK locus. Thus, a single screen may identify multiple genetic loci of interest.

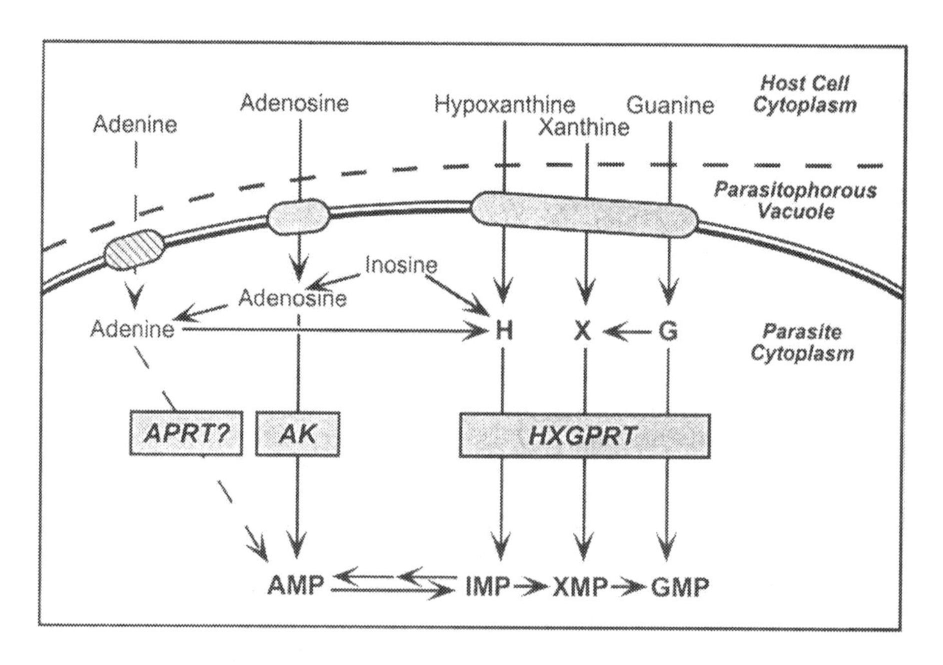


**Figure 1.** Purine metabolism in *T. gondii.* Like all intracellular protozoan parasites, *Toxoplasma* relies on salvage for all of its purines (Ullman and Carter, 1995). Host cell nucleotides and/or nucleosides pass through pores in the parasitophorous vacuole membrane, which permit passage of small molecules up to about 1,300 Da (Schwab et al., 1994); transport across the parasite's plasma membrane is mediated by specific transporters (Schwab et al., 1995; Chiang et al., 1999). Within the

parasite cytoplasm, adenosine is phosphorylated by AK to yield AMP, while hypoxanthine (or xanthine, or guanine) is phosphoribosylated by HXGPRT to yield IMP (or XMP, or GMP). Subversive substrates such as araA (a toxic adenosine analogue) or thioxanthine are processed to toxic nucleotides, killing the parasite. Insertional mutagenesis of the parasite genome and selection of drug-resistant mutants has permitted cloning of AK, HXGPRT, the adenosine transporter, and other genes of interest (Donald and Roos, 1995; Donald et al., 1996; Sullivan et al., 1999). Further genetic studies (see text) demonstrate that *T. gondii* possesses no biologically-relevant APRT activity.

No genetic study is complete without regenerating the observed phenotype through a targeted gene knock-out (or allelic replacement), and reversing the phenotype by complementation with the wild-type gene, fulfilling the molecular equivalent of Koch's principles (Falkow, 1988). Unlike small cDNA-derived vectors, which integrate at random throughout the *T. gondii* genome by nonhomologous recombination (permitting insertional mutagenesis), vectors containing large of contiguous genomic DNA often integrate by homologous recombination. *[Note: the length of genomic sequence required for homologous recombination appears to differ at different loci. Determining the requirements and constraints for homologous vs. nonhomologous recombination is an important open problem in T. gondii molecular biology.]* Where negative selection is available (e.g. resistance to a subversive substrate), knock-outs can be created by direct recombination. For loci where no direct selection is feasible, a range of selectable markers can be introduced into the transfection vector disrupting the target gene of interest. Where allelic replacements are required (e.g. to modify enzyme substrate specificity rather than inactivating the gene), hit-and-run mutagenesis strategies have been devised to permit more subtle mutations at the endogenous genomic locus in otherwise wild-type parasites (Donald and Roos, 1998).

For purine salvage genes, the drug-resistance phenotype of AK- or HXGPRT-deficient parasites permits direct selection of knock-out mutants. Knock-out vectors targeted to the AK or HXGPRT loci also permit the genetic dissection of functional redundancy in purine salvage pathways: while either locus can be deleted with high efficiency, it has not been possible to generate double knock-outs unless a complementing activity is provided in trans (unpublished). Thus, while the ability to interconvert AMP and IMP means that neither AK nor HXGPRT activity is essential for parasite survival, one of these two pathways is required. It thus appears that

APRT activity is not present in *T. gondii,* despite previous reports to the contrary (Krug et al., 1989).

## ANALYSIS OF 'MALARIA-IZED' *T. GONDII* DHFR-TS: THE EVOLUTION OF DRUG-RESISTANCE

Turning to consider pyrimidine metabolism (Schwartzman and Pfefferkorn, 1981), *T. gondii* is capable of salvaging uracil, and the parasite's UPRT gene has been identified by insertional mutagenesis, isolation of fluorouracil-resistant mutants, and cloning of the tagged gene by plasmid rescue (Donald and Roos, 1995). Cloned cDNA was expressed as recombinant enzyme for kinetic and crystallographic characterisation (Carter et al., 1997; Schumacher et al., 1998). UMP can also be synthesised de novo, but lacking thymidine kinase, the parasite is absolutely dependent on thymidylate synthase (TS) activity for converting either salvaged or newly-synthesised UMP to TMP. TS, in turn, depend on a functional folate cycle to provide the necessary methyl donor. Unlike animals and fungi, all protozoa examined to date express a bifunctional enzyme in which TS is fused to dihydrofolate reductase (DHFR). DHFR has been a prominent target for parasiticidal drugs, including trimethoprim, pyrimethamine, and cycloguanil. These compounds are usually administered in combination with sulfonamides (Hitchings, 1983), which target the prior step in folate biosynthesis, catalysed by dihydropteroate reductase (DHPS). The common antimalarial Fansidar, for example, is a formulation containing pyrimethamine and sulfadoxine.

Widespread use of antifolate chemotherapy against malaria has led to the emergence of drug-resistant strains of *Plasmodium falciparum,* and these mutants frequently harbour unusual DHFR and/or DHPS alleles (Hyde et al., 1990; Triglia et al., 1997; Plowe et al., 1998). Unfortunately, the extreme A/T bias of the *P. falciparum* genome (and consequent codon bias) complicates protein expression in recombinant systems, and it has been difficult to express full-length *P. falciparum* DHFR-TS at high specific activity (Sirawaraporn et al., 1997). Prior to the development of transfection systems for *Plasmodium,* it was therefore difficult to determine the precise role of individual DHFR point mutations in drug resistance. *[Note: anecdotal evidence suggests that the difficulty in expressing functional P. falciparum proteins may not be entirely attributable to codon bias; comparisons of Plasmodium protein folding in vivo and in recombinant systems is a problem worthy of further study.]*

Unlike *Plasmodium,* the *Toxoplasma* genome exhibits no unusual sequence biases, and *T. gondii* proteins are readily expressed in recombinant systems. In order to assess the role of DHFR polymorphisms in parasite drug resistance, various mutations observed in *P. falciparum* were engineered into the *T. gondii* DHFR-TS homologue, for expression as recombinant enzyme and in transgenic parasites (Reynolds and Roos, 1998). Results of these studies are summarised in Figure 2. Wild-type (drug-sensitive) *P. falciparum* DHFR-TS contains either alanine or valine at position 16, coupled with serine or threonine (respectively) at position 108. Studies in *T. gondii* and *E. coli* confirm that any combination of Ala/Val-16 + Ser/Thr-108 yields active enzyme. $K_m$ values are relatively high for the combinations observed in wild-type *P. falciparum,* but the TS domain of this bifunctional enzyme probably maintains a high local concentration of dihydrofolate (Trujillo et al., 1996), making $K_m$ values relatively unimportant for overall efficiency.

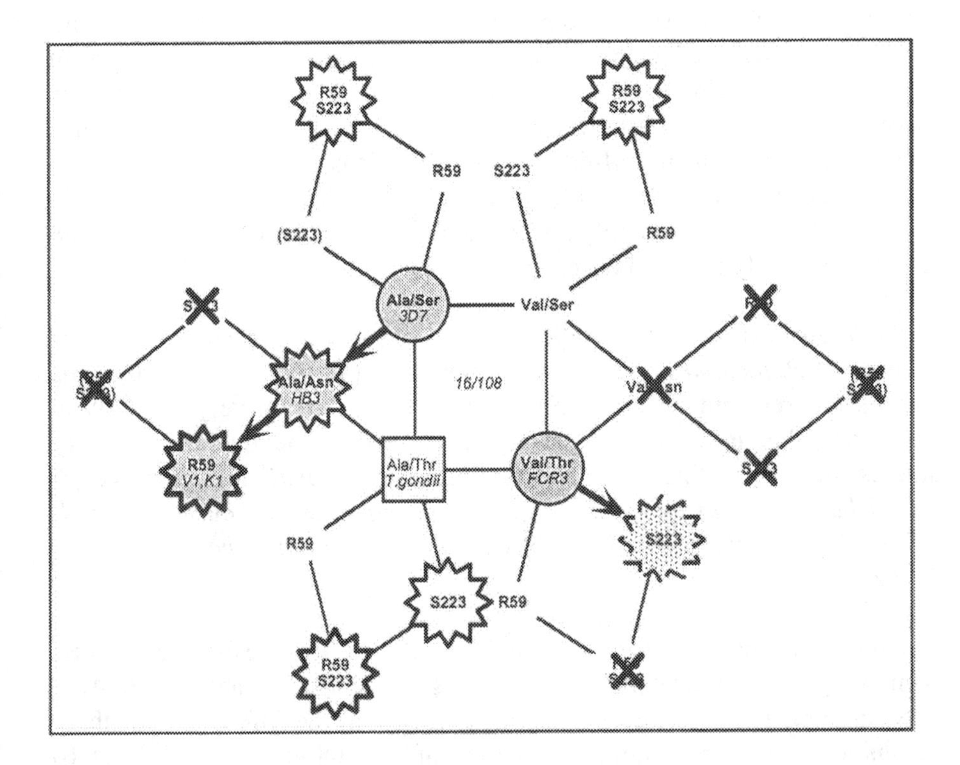


**Figure 2**. The evolution of pyrimethamine-resistant DHFR. Each node represents a different *T. gondii* DHFR-TS allele harbouring mutations based on polymorphisms observed in drug-resistant *P. falciparum*. Enzyme activity, kinetics and drug sensitiv-

ity was assessed for all alleles (except those shown in parentheses), through studies on recombinant enzyme, *E. coli* complementation, and transgenic *T. gondii* parasites. The center-most six nodes indicate amino acids at positions 16 and 108 (*P. falciparum* numbering). Wild-type *P. falciparum* sequences are denoted by circles; wild-type *T. gondii* by a square. Alleles distal to each of these central six represent the addition of mutations at positions 59 and/or 223. Representative field isolates of *P. falciparum* are shown in italics. Inactive enzyme is indicated by 'X', while increasing resistance to pyrimethamine is denoted by stars of increasing line thickness; arrows track the evolution of drug-resistant parasites. Broken lines indicate a drug-resistant allele that exhibits very poor kinetics, and has only been observed in the laboratory (Tanaka et al., 1990). See text and Reynolds and Roos (1998) for further details.

Both Ala-16/Ser-108 DHFR-TS and Val-16/Thr-108 are sensitive to pyrimethamine, although the latter allele is relatively resistant to cycloguanil. Introduction of Asn-108 in combination with Ala-16 confers significant resistance to pyrimethamine, and this is the predominant polymorphism observed in the field (Hyde, 1990; Plowe, 1998). The Asn-108 mutation is incompatible with Val-16, however: this allele fails to confer drug-resistance and cannot complement *folA*-deficient *E. coli.* Thus, failure to identify Val-16/Asn-108 alleles in the field reflects the fact that this is not a viable combination, rather than implying that drug resistance based on the Asn-108 mutation has only emerged once.

In the Val-16 background, mutation Ser-223 has been observed in laboratory mutants of *P. falciparum,* but never in the clinic (Tanaka et al., 1990). This enzyme is resistant to pyrimethamine, but exhibits a very poor $k_{cat}$, suggesting that it would be unlikely to survive in the field. Preliminary studies suggest that this allele renders transgenic *T. gondii* cold-sensitive. *[It would be interesting to see if P. falciparum parasites harbouring the Val-16/Ser-223 mutations are also cold-sensitive, and whether they can survive in the mosquito.]*

Mutations at other loci, such as Arg-59, confer no drug-resistance on their own, but yield high-level pyrimethamine resistance in combination with Asn-108. In sum, pyrimethamine-resistant malaria may be inferred from these results to have emerged through mutation of Ser-108 to Asn (in the Ala-16 background), followed by subsequent acquisition of Arg-59 (and other mutations; Hyde, 1990; Sirawaraporn, 1997; Plowe, 1998). Other alleles are either inactive (e.g. Val-16/Asn-108), unfavourable in kinetic terms (e.g.

Val-16/Ser-223), confer no resistance to pyrimethamine (e.g. Arg-59), or difficult to evolve from wild-type DHFR due to intervening steps that confer no drug resistance (e.g. Arg-59/Ser-223). Pyrimethamine-resistant toxoplasmosis is rarely observed in the clinic, because human *T. gondii* infections are unlikely to be transmitted (excluding cannibalism or vampire cats). For this reason, laboratory mutants pose no unusual biohazard outside the research environment itself.

In addition to their utility for understanding drug-resistant malaria, pyrimethamine-resistant *T. gondii* DHFR-TS alleles have also proved very useful in developing transformation schemes for both *Toxoplasma* and *Plasmodium* (Wu et al., 1996; Crabb et al., 1997; Waters et al., 1997). Successful transfection of *P. falciparum* has, in turn, permitted the role of DHPS point mutations in sulfonamide resistance to be studied in *Plasmodium* itself (Triglia et al., 1998).

## ULTRASTRUCTURE OF THE SECRETORY PATHWAY: *T. GONDII* AS A "MINIMAL EUKARYOTE" AND A GUIDE FOR UNDERSTANDING *PLASMODIUM*

The *T. gondii* tachyzoite is a highly-polarised secretory machine. In ultrastructural terms, this parasite can be viewed as a textbook eukaryote, in which most of the organelles familiar from studies on mammalian cells (nucleus, mitochondrion, endoplasmic reticulum (ER), Golgi apparatus, secretory organelles, cytoskeleton) are readily visualised, but occupy essentially fixed positions within an organism whose entire volume is only about 40 $\mu m^3$ (the volume of a typical fibroblast is about 10,000 $\mu m^3$). In this minute eukaryote, it is feasible to visualise the entire secretory pathway in a single electron micrograph (Hager et al., 1999).

At the light microscopic level, ultrastructural identification of subcellular organelles in living cells has been greatly aided by the use of the Green Fluorescent Protein (GFP) as a reporter molecule (Striepen et al., 1998). Stable transgenic parasites in which GFP has been targeted to the nucleus, ER, Golgi, rhoptries, micronemes, plasma membrane, etc define the distinctive morphology of these organelles, permitting rapid analysis of new antigens by co-localisation (Striepen et al., 1998; Hager et al., 1999; http://www.sas.upenn.edu/~striepen/gfp.html).

T. gondii

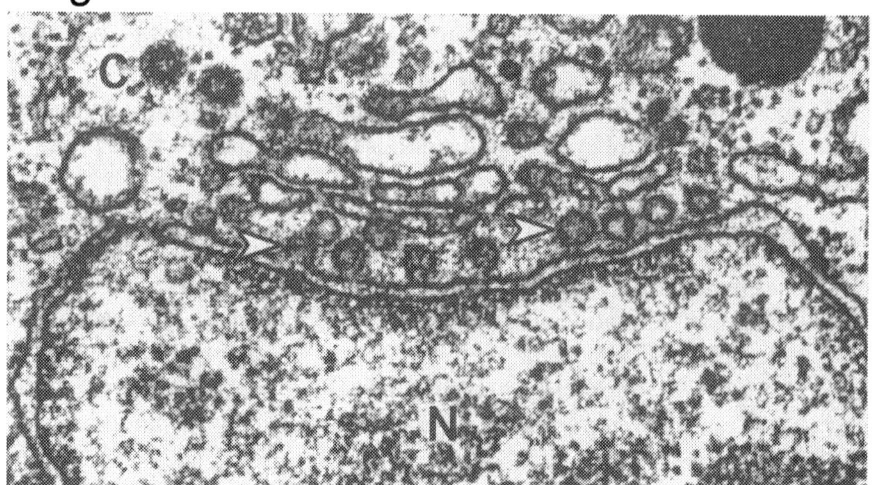


P. falciparum

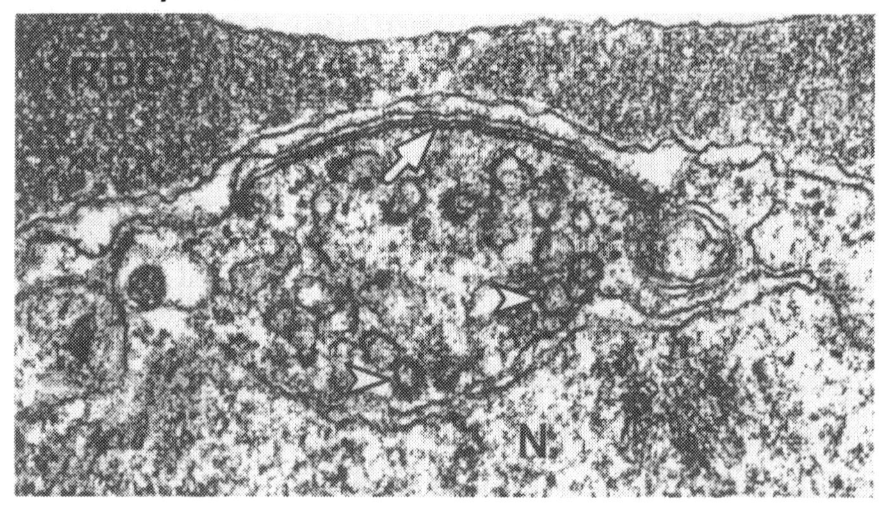


**Figure 3.** The Golgi apparatus of apicomplexan parasites. In *T. gondii* (top), vesicular trafficking between the ER and cis-Golgi takes place near the apical end of the nucleus. Coatomer-coated vesicles (arrowheads) bud from and/or fuse with the nuclear envelope; clathrin-coated vesicles (C) bud from the trans-Golgi. In *P. falciparum* (bottom), vesicular trafficking is most readily detected in schizonts, when new merozoites are being assembled (arrow indicates a developing merozoite). Using *T. gondii* as a guide for understanding *Plasmodium* permits identification of

putative coatomer-coated vesicles associated with the apical end of the nuclear envelope (arrowheads). N, parasite nucleus; RBC, red blood cell cytoplasm. Micrographs courtesy of Dr. L.G. Tilney.

Several investigators have attempted to exploit *T. gondii* as a model 'minimal eukaryote', dissecting basic processes such as protein secretion and organellar targeting in a system with excellent ultrastructural clarity and reasonably accessible genetics (Chaturvedi et al., 1998; Hager et al., 1999). Although these studies are still very much in their infancy, it is already clear that similarities between *T. gondii* and 'higher' eukaryotes extend to the biochemical level: vesicular trafficking involves coatomer proteins, rabs, SNAP/SNARE complexes, etc with very similar characteristics to those defined in mammalian cells (Chaturvedi et al., 1998). These studies have defined the apical end of the nuclear envelope as a critical nexus in the secretory pathway: all traffic between the ER and Golgi appears to take place in this region, which is filled with coatomer-coated vesicles (Hager et al., 1999).

The ultrastructural clarity of the Golgi in *Toxoplasma* contrasts strikingly with *Plasmodium* merozoites, where sub-cellular architecture is very difficult to discern (Elmendorf and Haldar, 1993). Despite the many proteins known to traffic through the secretory pathway in *P. falciparum* (Haldar, 1998), it has not been possible to unequivocally define the Golgi apparatus during the erythrocytic cycle. Fortunately, just as *T. gondii* has provided a useful model for studying the certain aspects of the molecular genetics and biochemistry of *Plasmodium,* the same is true for cell biological analysis.

*T. gondii* traverses the mitotic cell cycle rapidly, dividing every 6 to 9 hr, depending on the strain and culture conditions. As a consequence, the Golgi is very active, constantly assembling new secretory organelles and building the inner membrane complex (which serves as a scaffold for daughter parasite formation). In contrast, *P. falciparum* replicates its DNA many times before cytokinesis, assembling multiple daughters simultaneously during schizogony (Roos et al., 1999a). The *T. gondii* model suggests that the *Plasmodium* Golgi will also be closely associated with the apical end of the nuclear envelope, and should be particularly active during merozoite assembly. As seen in Figure 3, electron micrographs of *P. falciparum* schizonts show that the cytoplasmic space between the nucleus and the flattened alveolae of the inner membrane complex, where new daughter parasites will develop, is filled with numerous vesicles similar in size and appearance to coatomer-coated vesicles in *Toxoplasma.*

## IDENTIFICATION OF THE APICOPLAST: A NOVEL DRUG TARGET

The recent identification of a plastid in apicomplexan parasites has served to resolve several long-standing enigmas: the nature of the "spherical body", the function of a 35 kb episomal DNA, and the mechanism of action by which prokaryotic transpeptidase inhibitors act against *Toxoplasma, Plasmodium* and other apicomplexan parasites. Studies on the apicomplexan plastid, or apicoplast, have recently been reviewed in detail (McFadden and Roos, 1999; Roos et al., 1999b; Soldati, 1999), and will therefore be summarised only briefly.

In addition to organelles common to all eukaryotes, structures associated with replication by endodyogeny/schizogony, and secretory organelles required for host cell invasion, apicomplexan parasites harbour an unexpected organelle: a plastid closely related to the chloroplast (or rhodoplast) of eukaryotic algae/plants (Köhler et al., 1997; Roos et al., 1999b). Discovery of the apicoplast emerged from detailed analyses of an episomal DNA, originally thought to constitute these parasites' mitochondrial genome. Discovery of the true mitochondrial genome in *Plasmodium* and related fragmentary sequences in *Toxoplasma* left the origin of this 35 kb circle a mystery, until phylogenetic analysis revealed it to be similar to algal plastids (Feagin, 1994; Köhler et al., 1997; Wilson et al., 1997). In fact, this DNA is so closely related to the plastid genomes of algae and higher plants that it is unlikely to have derived from an independent cyanobacterial endosymbiont. If apicomplexans are not plants, as argued by abundant molecular and morphological evidence, then the apicoplast genome must have arisen by lateral genetic transfer. In other words, *Toxoplasma* and *Plasmodium* are not degenerate plants, but a common ancestor ate a plant and retained the plastid, as shown in Figure 4. *[The precise origin of the apicoplast is still an open question; some scientists argue for a red algal ancestry, while others argue for closer affinity to green algae (Köhler et al., 1997; McFadden et al., 1997; Blanchard and Hicks, 1999).]*

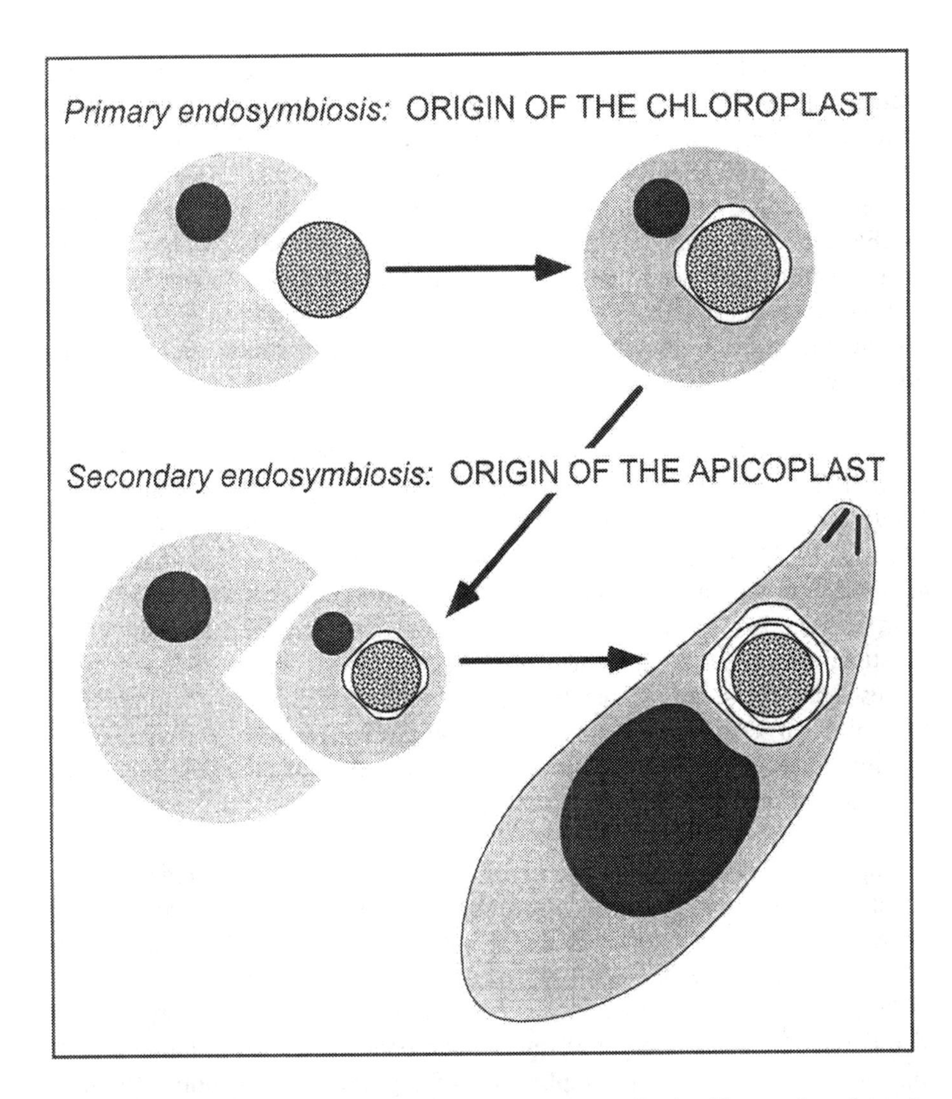


**Figure 4.** 'Pac-man' model for apicoplast evolution. All plastids are thought to be monophyletic in origin, having arisen when a heterotrophic eukaryote engulfed, or was invaded by, a cyanobacterium (top). The resulting alga might be expected to contain an organelle surrounded by a double membrane, as observed in modern chloroplasts. The apicoplast arose as the result of a *secondary* endosymbiotic event (bottom), in which the ancestor of all apicomplexans engulfed (or was invaded by) a eukaryotic alga. Subsequent loss of the algal nucleus, mitochondrion, and most photosystem genes gave rise to the apicoplast, surrounded by four membranes. See Roos et al (1999b) for further details.

Several examples of 'secondary' endosymbiosis are known among the protozoa (Palmer and Delwiche, 1996), the free-living flagellate *Euglena,* which is related to kinetoplastid parasites (Yasuhira and Simpson, 1997) but harbours a plastid (Schwartzbach et al., 1998). In contrast to the apicoplast, the euglenoplast retains photosynthetic function. In situ hybridisation (McFadden et al., 1996; Köhler et al., 1997) localises the apicoplast genome to a structure previously described as the Golgi adjunct, spherical body, Hohlzylinder, organelle plurimembranaire, etc (Siddall, 1992), but its association with a genome reveals this to be a distinct endosymbiotic organelle.

The apicoplast of *T. gondii* consists of four concentric bilayer membranes surrounding the plastid lumen (Köhler et al., 1997; McFadden and Roos, 1999), supporting its origin as a secondary endosymbiont (Fig. 4). Recent evidence (L.H. Bannister, personal communication) suggests that the *P. falciparum* apicoplast may be surrounded by three membranes rather than four, as found in the *Euglena* plastid. It will be interesting to resolve this question, but note that the conclusive evidence for secondary endosymbiosis comes from phylogeny rather than ultrastructure. Moreover, *P. falciparum* apicoplast targeting signals function in transgenic *T. gondii,* and vice-versa, arguing that any ultrastructural differences are of limited functional importance (unpublished observations).

The apicoplast provides the most plausible target for the parasiticidal activity of clindamycin and various macrolide antibiotics, classical prokaryotic protein synthesis inhibitors that might not have been expected to kill either *Toxoplasma* or *Plasmodium* (Pfefferkorn et al., 1992; Beckers et al., 1995; Fichera et al., 1995). Further studies have demonstrated that this organelle is also the most plausible target of fluoroquinolones, presumed to target type II topoisomerases required for replication of the apicoplast genome (Fichera and Roos, 1997; Weissig et al., 1997). *[Parasites treated with any of these drugs exhibit a peculiar 'delayed-death' phenotype, the basis of which remains unexplained (Pfefferkorn et al., 1992; Fichera et al., 1995).]* Thus, the apicoplast is essential for parasite survival, and hence a target for parasiticidal chemotherapy. Fluoroquinolones, rifamycins, and macrolides are thought to target DNA replication, RNA transcription, and protein synthesis in the apicoplast (McFadden and Roos, 1999), but the ultimate function of this organelle remains to be elucidated.

## MINING GENOME DATABASES TO DEFINE THE ORGANELLAR 'METABOLOME'

The apicoplast genome encodes a very limited number of proteins and RNAs, and all of these appear to be involved in 'housekeeping' functions (Wilson et al., 1997; http://www.sas.upenn.edu/~jkissing/toxomap.html). Any metabolic enzymes must be encoded in the nuclear genome and post-translationally imported into the apicoplast, as is commonly observed in other endosymbiotic organelles (mitochondria, chloroplasts, etc; Schatz and Dobberstein, 1996). Fortunately, extensive sequence information is now available from publicly accessible databases (Ajioka et al., 1998; Gardner et al., 1998; Bowman et al., 1999):

*T. gondii* expressed sequence tags (ESTs):
http://www.cbil.upenn.edu/ParaDBs/Toxoplasma/index.html

*P. falciparum* genomic sequence:
http://www.sanger.ac.uk/Projects/P_falciparum/
http://www.tigr.org/tdb/edb/pfdb/pfdb.html
http://sequence-www.stanford.edu/group/malaria/index.html
http://www.ncbi.nlm.nih.gov/Malaria/index.html

Searching *T. gondii* ESTs and *P. falciparum* genome sequences for proteins likely to be imported into the apicoplast identified several candidate genes, and the subcellular localisation of the proteins they encode has been confirmed by immunoelectron microscopy (Waller et al., 1998).

Examination of the predicted sequence for several nuclear-encoded apicoplast proteins reveals a long N-terminal domain, consisting of a hydrophobic region similar to secretory signal sequence, followed by a relatively basic region bearing some similarity to mitochondrial and plastid targeting sequences (Waller et al., 1998). This structure suggests that entry into the apicoplast might proceed via a two-step process: entry into to the secretory pathway, followed by cleavage of the signal sequence to reveal a plastid targeting domain capable of translocation into the apicoplast (Roos et al., 1999b). This model differs significantly from the post-translational import of proteins into mitochondria and chloroplasts (Schatz and Dobberstein, 1996), but derives support from studies on in vitro import into other secondary endosymbiotic organelles, where nuclear-encoded plastid proteins exhibit a similar structure (Bodyl, 1997; Schwartzbach et al., 1998;

McFadden, 1999). This mechanism also provides a solution to the problem of how nuclear-encoded proteins might traverse the four (or three) membranes surrounding the apicoplast.

The essential features of this model predicting a bipartite targeting signal have been confirmed experimentally by transfection of various recombinant constructs into *T. gondii.* The N-terminal domain is sufficient to target a GFP reporter into the apicoplast (Waller et al., 1998); deletion of the signal sequence yields a cytosolic protein; and internal deletion of the plastid-targeting domain results in secretion of GFP into the parasitophorous vacuolar space (Roos et al., 1999b and unpublished data). *[Many important questions regarding the nature of nuclear-encoded protein targeting to the apicoplast still remain to be addressed. How are proteins within the secretory pathway directed to the apicoplast? What proteins mediate import across the internal plastid membranes?]*

Combining the distinctive features that define nuclear-encoded apicoplast proteins (signal sequence, targeting domain, plastid protein), with the growing *P. falciparum* sequence database, and the experimental accessibility of *T. gondii* offers the prospect of a high throughput genomics-based scheme for identifying nuclear encoded apicoplast proteins, as outlined in Figure 5. To elucidate apicoplast function(s), several screens can be applied to filter information from the genome and EST databases of apicomplexan parasites. Similarity to known genes from plants or algal plastids (Martin et al., 1998) provides one source of possible nuclear-encoded apicoplast genes. Similarity to non-plastid genes of plants or algae may highlight other genes of interest, even if these are not normally associated with plastid function, as it is possible that novel metabolic functions may have taken up residence in the apicoplast. *[Plant-like genes may include other genes of evolutionary and pharmacological interest as well, including any vestiges of genes acquired from the algal nucleus during the secondary endosymbiotic event; see Fig. 4.]* Open reading frames predicted to encode a secretory signal sequence will include not only apicoplast proteins, but also other organellar and secreted proteins of interest. A database of long N-terminal extensions upstream of recognisable coding sequences will include nuclear-encoded mitochondrial proteins, as well as proteins destined for the apicoplast. Additional candidate apicoplast genes may be identified experimentally, through organelle purification and partial protein sequencing, or using genetic approaches such as insertional mutagenesis with a GFP reporter and screening for apicoplast localisation (Roos et al., 1997).

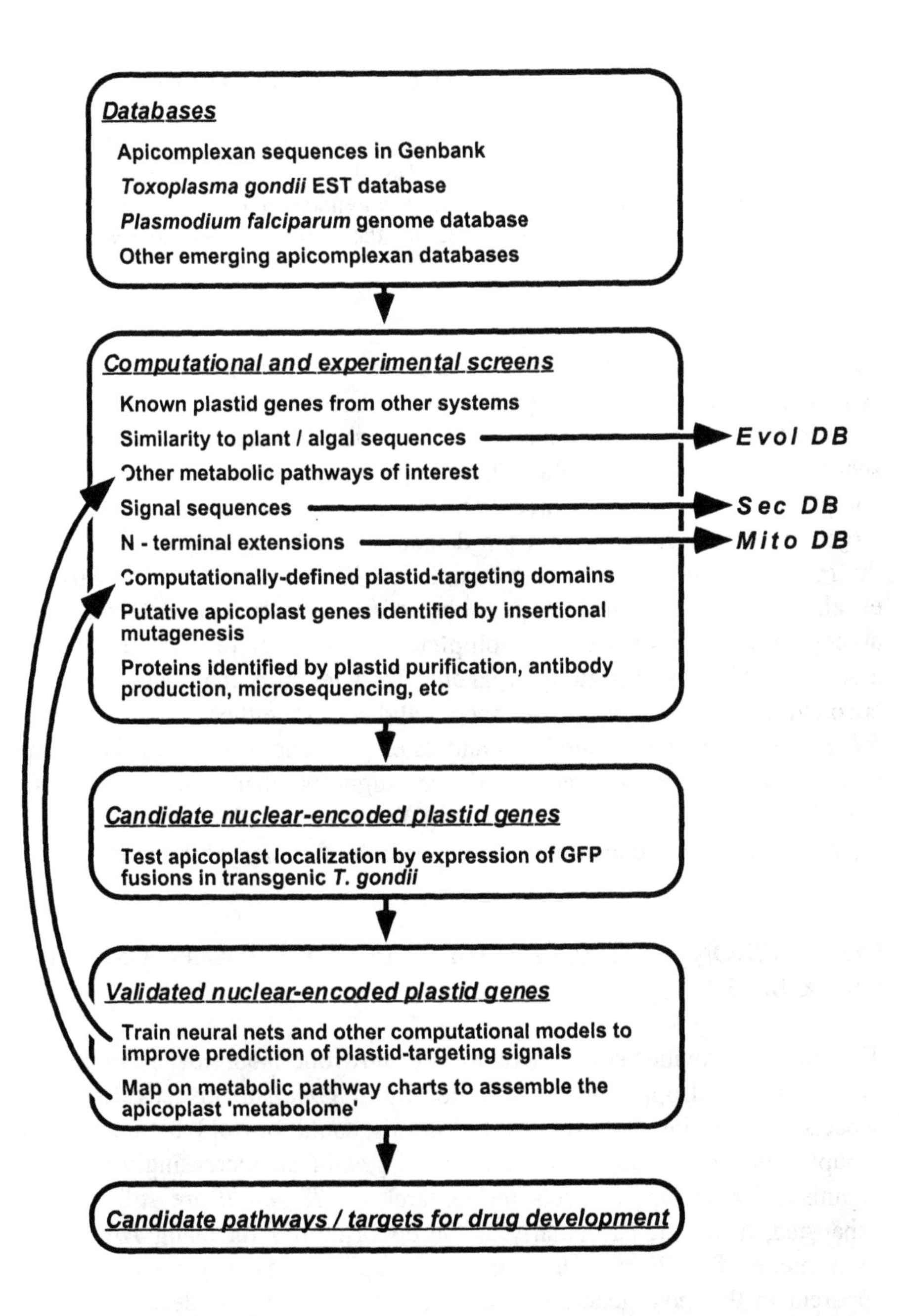


**Figure 5.** Flow-chart for computational database mining to identify nuclear-encoded apicoplast genes. Although no unique sequence motif defines apicoplast genes, various filters can be applied to apicomplexan genome and EST databases to identify

candidate plastid genes (see text). Introduction of putative targeting signals into transgenic *T. gondii* (as GFP fusions) provides a rapid assay for apicoplast localisation, yielding a validated database for refining computational/statistical models. These computational screens are also likely to identify other proteins of interest (nuclear proteins acquired from the algal endosymbiont, secretory proteins, mitochondrial proteins, etc), as indicated by the various sub-databases (outlined text).

Putative apicoplast genes can be assayed by fusion of PCR-amplified targeting domains to a GFP reporter, and transient transfection into *T. gondii* tachyzoites, yielding a database of validated nuclear-encoded plastid sequences (Waller et al., 1998). These sequences, in turn, can be used to train computational or statistical models to improve the recognition of apicoplast targeting signals directly from the database. As the reference sequence for the *P. falciparum* genome nears completion (Gardner et al., 1998; Bowman et al., 1999), it should be possible to identify every nuclear-encoded apicoplast protein, providing a complete metabolic picture of all pathways associated with this intriguing organelle. To date, more than twenty nuclear-encoded plastid proteins have been validated in either *T. gondii* or *P. falciparum,* and several hundred candidates pass one or more of the screens outlined above. Preliminary evidence suggests that lipid metabolism, including synthesis of both fatty acids (Waller et al., 1998) and isoprenoids (Jomaa et al., 1999), constitutes a key metabolic function of the apicoplast.

## CONCLUSION: *TOXOPLASMA* AS AN EXPERIMENTAL ORGANISM

This chapter provides several examples of how one laboratory has exploited the experimental opportunities afforded by *T. gondii* to investigate various aspects of parasite biology; similar stories could be told by many other groups. Such opportunities have made *T. gondii* an increasingly attractive organism for study. Prospects for research on *T. gondii* are still far from exhausted, however, particularly as the opportunities for using *Toxoplasma* as a model for *Plasmodium* and other apicomplexan parasites become apparent in the 'post-genomic' era. Progress over the past decade serves to vindicate Pfefferkorn's early vision in focusing on the development of *T. gondii* as a genetic system. In the future, it is likely that continued success will come from taking advantage of the opportunities afforded by many systems to focus on problems of general interest, integrating, for example, the molecular genetic and cell biological potential of *Toxoplasma,* genomic approaches available in *Plasmodium,* biochemical opportunities afforded by free-living parasites such as *Perkinsus,* and the accessibility of different life-cycle stages in *Eimeria.*

**ACKNOWLEDGEMENTS**

We thank past and present members of the Roos laboratory for developing the molecular genetic and cell biological tools that made this work possible. Thanks also to the broader *T. gondii* research community for following the tradition established by Dr. Pfefferkorn of freely discussing ideas and exchanging reagents, even in advance of publication. We are grateful to the *T. gondii* EST project and *P. falciparum* genome project for making sequence information available. J.A.D. and M.G.R. were supported by training grant in Cell and Molecular Biology from the National Institutes of Health, B.S. by a fellowship from the Deutsche Forschungsgemeinschaft, and J.C.K. by a training grant in Computational Biology from the National Science Foundation. D.S.R. is a Burroughs Wellcome Scholar in Molecular Parasitology.

## REFERENCES


Ajioka, J., Boothroyd, J.C., Brunk, B.P., Hehl, A., Hillier, L., Manger, I.D., Overton, G.C., Marra, M., Roos, D.S., Wan, K.L., Waterston, R., & Sibley, L.D. (1998). Gene discovery by EST sequencing in *Toxoplasma gondii* reveals sequences restricted to the apicomplexa. Genome Res. *8,* 18-28.

Beckers, C.J.M., Roos, D.S., Donald, R.G.K., Luft, B.J., Schwab, J.C., Cao, Y., & Joiner, K.A. (1995). Inhibition of cytoplasmic and organellar protein synthesis in *Toxoplasma gondii:* Implications for the target of macrolide antibiotics. J. Clin. Invest. *95,* 367-376.

Black, M.W., & Boothroyd. J.C. (1998). Development of a stable episomal shuttle vector for *Toxoplasma gondii.* J. Biol. Chem. *273,* 3972–3979.

Blanchard, J.L., & Hicks, J.S. (1999). The non-photosynthetic plastid in malaria parasites and other apicomplexans is derived from outside the green plastid lineage. J. Eukaryot. Microbiol. *46,* 367-375.

Bodyl, A. (1997). Mechanism of protein targeting to the chlorarachniophyte plastids and the evolution of complex plastids with four membranes: A hypothesis. Botan. Acta *110,* 395-400.

Boothroyd, J.C., Kim, K., Pfefferkorn, E.R., Sibley, L.D., & Soldati. D. (1994). Forward and reverse genetics in the study of the obligate intracellular parasite *Toxoplasma gondii.* Methods Mol. Genet. *3,* 1–29.

Bowman, S., Lawson, D., Basham, D., Brown, D., Chillingworth, T., Churcher, C.M., Craig, A., Davies, R.M., Devlin, K., Feltwell, T., Gentles, S., Gwilliam, R., Hamlin, N., Harris, D., Holroyd, S., Hornsby, T., Horrocks, P., Jagels, K., Jassal, B., Kyes, S., McLean, J., Moule, S., Mungall, K., Murphy, L., Oliver, K., Quail, M.A., Rajendream, M.-A., Rutter, S., Skelton, J., Squares, R., Squares, S., Sulston, J.E., Whitehead, S., Woodward, J.R., Newbold, C., & Barrell, B.G. (1999). The complete nucleotide sequence of chromosome 3 of *Plasmodium falciparum.* Nature *400,* 532-538.

Carter, D., Donald, R.G.K., Roos, D.S., & Ullman, B. (1997). Expression, purification, and characterization of uracil phosphoribosyltransferase from *Toxoplasma gondii* Mol. Biochem. Parasitol. *87,* 137-144.

Chaturvedi, S., Qi, H., Coleman, D., Rodriguez, A., Hanson, P.S., Striepen, B., Roos, D.S., & Joiner, K.A. (1999). Constitutive calcium independent release of *Toxoplasma gondii* dense granules occurs through the NSF/SNAP/SNARE/Rab machinery. J. Biol. Chem. *274,* 2424-2431.

Chiang, C.-W., Carter, N., Sullivan, W.J., Jr., Donald, R.G.K., Roos, D.S., Naguib, F.N.M., el Kouni, M.H., Ullman, B., & Wilson, C.M. (1999). The adenosine transporter of *Toxoplasma gondii:* Identification by insertional mutagenesis, cloning and recombinant expression. J. Biol. Chem., in press.

Crabb, B.S., Triglia. T., Waterkeyn, J.G., & Cowman, A.F. (1997). Stable transgene expression in *Plasmodium falciparum.* Mol. Biochem. Parasitol. *90,* 131-144.

Darling, J., Sullivan, W.J., Jr., Carter, D., Ullman, B., & Roos, D.S. (1999). Recombinant expression, purification, and characterization of *Toxoplasma gondii* adenosine kinase. Mol. Biochem. Parasitol. *103,* 15-23.

Donald, R.G.K., & Roos, D.S. (1994). Homologous recombination and gene replacement at the dihydrofolate reductase/thymidylate synthase locus in *Toxoplasma gondii.* Mol. Biochem. Parasitol. *63,* 243-253.

Donald, R.G.K., & Roos, D.S. (1995). Insertional mutagenesis in a protozoan parasite: Direct cloning of the uracil phosphoribosyl transferase gene from *Toxoplasma gondii.* Proc. Nat'l Acad. Sci. U.S.A. *92,* 5749-5753.

Donald, R.G.K., & Roos, D.S. (1998). Gene knock-outs and allelic replacements in *Toxoplasma gondii*: HXGPRT as a selectable marker for hit-and-run mutagenesis. Mol. Biochem. Parasitol. *91,* 295–305.

Donald, R.G.K., Carter, D., Ullman, B., & Roos, D.S. (1996). Insertional tagging, cloning and expression of the *Toxoplasma gondii* hypoxanthine-xanthine-guanine phosphoribosyl transferase gene: Use as a selectable marker for stable transformation. J. Biol. Chem. *271,* 14010-14019.

Elmendorf, H.G., & Haldar, K. (1993). Identification and localization of ERD2 in the malaria parasite Plasmodium falciparum: separation from sites of sphingomyelin synthesis and implications for organization of the Golgi. EMBO Journal. *12,* 4763-73.

Falkow, S. (1988) Molecular Koch's postulates applied to microbial pathogenicity. Rev. Infect. Dis. 10 suppl. *2,* S274-276.

Feagin, J.E. (1994). The extrachromosomal DNAs of apicomplexan parasites. Annu. Rev. Microbiol. *48,* 81-104.

Fichera, M.E., & Roos, D.S. (1997). A plastid organelle as a drug target in apicomplexan parasites. Nature *389,* 407–409.

Fichera, M.E., Bhopale, M.K., & Roos, D.S. (1995). In vitro assays elucidate the peculiar mechanism of macrolide/lincosamide action against *Toxoplasma gondii.* Antimicr. Agents Chemother. *39,* 1530–1537.

Gardner, M.J., Tettelin, H., Carucci, D.J., Cummings, L.M., Aravind, L., Koonin, E.V., Shallom, S., Mason, T., Yu, K., Fujii, C., Pederson, J., Shen, K., Jing, J., Aston, C., Lai, Z., Schwartz, D.C., Pertea, M., Salzberg, S., Zhou, L., Sutton, G.G., Clayton, R., White, O., Smith, H.O., Fraser, C.M., Adams, M.D., Ventner, J.C., Hoffman, S.L. (1998). Chromosome 2 sequence of the human malaria parasite *Plasmodium falciparum.* Science *282,* 1126-1132.

Hager, K.M., Striepen, B., Tilney, L.G., & Roos, D.S. (1999). The nuclear envelope serves as an intermediary between the ER and Golgi complex in the intracellular parasite *Toxoplasma gondii.* J. Cell Sci. *112,* 2631-2638.

Haldar, K. (1998). Intracellular trafficking in Plasmodium-infected erythrocytes. Curr. Opin. Microbiol. *1,* 466-71.

Hitchings, G.H., ed. (1983). Inhibition of Folate Metabolism in Chemotherapy. Springer-Verlag, Berlin.

Hyde, J.E. (1990). The dihydrofolate reductase-thymidylate synthase gene in the drug resistance of malaria parasites. Pharmacol. Ther. *48,* 45-59.

Jomaa, H., Wiesner, J., Sanderbrand, S., Altincicek, B., Weidemeyer, C., Hintz, M., Türbachova, I., Eberl, M., Zeidler, J., Lichtenthaler, H.K., Soldati, D., and Beck, E. (1999). Inhibitors of the Non-mevalonate Pathway of Isoprenoid Biosynthesis as Antimalarial Drugs. Science *285,* 1573-1576.

Köhler, S., Delwiche, C.F., Denny, P.W., Tilney, L.G., Webster, P., Wilson, R.J.M., Palmer, J.D., & Roos, D.S. (1997). A plastid of probable green algal origin in apicomplexan parasites. Science *275,* 1485-1488.

Knoll, L.J., & Boothroyd, J.C. (1998). Isolation of developmentally regulated genes from *Toxoplasma gondii* by a gene trap with the positive and negative selectable marker hypoxanthine-xanthine-guanine phosphoribosyltransferase. Mol. Cell. Biol. *18,* 807-814.

Krug, E.C., Marr, J.J., & Berens, R.L. (1989) Purine metabolism in *Toxoplasma gondii.* J. Biol. Chem. *264,* 10601-10607.

Martin, W., Stoebe, B., Goremykin, V., Hansmann, S., Hasegawa, M., & Kowallik, K.V. (1998). Gene transfer to the nucleus and the evolution of chloroplasts. Nature *393,* 162-165.

McFadden, G.I. (1999). Plastids and protein targeting. J. Eukaryot. Microbiol. *46,* 339-346.

McFadden, G.I., & Roos, D.S. (1999). Apicomplexan plastids as drug targets. Trends Microbiol. *7,* 328-333.

McFadden, G.I., Reith, M.E., Mulholland, J., & Lang-Unnasch, N. (1996). Plastid in human parasites. Nature *381,* 482.

McFadden, G.I., Waller, R.F., Reith, M.E., & Lang-Unnasch, N. (1997). Plastids in apicomplexan parasites. In: Origins of Algae and their Plastids, Bhattacharya, D., ed. Springer-Verlag, Vienna, New York; pp. 261-287.

Nakaar, V., Samuel, B.U., Ngo, E.O., & Joiner K.A. (1999). Targeted reduction of nucleoside triphosphate hydrolase by antisense RNA inhibits *Toxoplasma gondii* proliferation. J. Biol. Chem. *274,* 5083-5087.

Palmer, J.D., & Delwiche, C.F. (1996). Second-hand chloroplasts and the case of the disappearing nucleus. Proc. Nat'l Acad. Sci. U.S.A. *93,* 7432-7435.

Pfefferkorn, E.R. (1977). *Toxoplasma gondii:* The enzymic defect of a mutant resistant to 5-fluorodeoxyuridine. Exper. Parasitol. *44,* 26-35.

Pfefferkorn, E.R. (1988). *Toxoplasma gondii* as viewed from a virological perspective. In: The Biology of Parasitism, Englund, P.T., & Sher, A., eds. Alan R. Liss, New York. MBL Lect. Biol. *9,* 479-501.

Pfefferkorn, E.R., & Borotz, S.E. (1994). *Toxoplasma gondii:* characterization of a mutant resistant to 6-thioxanthine. Exper. Parasitol. *79,* 374-382.

Pfefferkorn, E.R., & Pfefferkorn, L.C. (1976). Arabinosyl nucleosides inhibit *Toxoplasma gondii* and allow the selection of resistant mutants. J. Parasitol. *62,* 993-999.

Pfefferkorn, E.R., & Pfefferkorn, L.C. (1977). *Toxoplasma gondii:* Characterization of a mutant resistant to 5-fluorodeoxyuridine. Exper. Parasitol. *42,* 44-55.

Pfefferkorn, E.R., & Pfefferkorn, L.C. (1978). The biochemical basis for resistance to adenine arabinoside in a mutant of *Toxoplasma gondii.* J. Parasitol. *64,* 486-492.

Pfefferkorn, E.R., & Pfefferkorn, L.C. (1980). *Toxoplasma gondii:* genetic recombination between drug resistant mutants. Exper. Parasitol. *50,* 305-316.

Pfefferkorn, E.R., Schwartzman, J.D., & Kasper, L.H. (1983). *Toxoplasma gondii*: use of mutants to study the host-parasite relationship. Ciba Fdn. Symp. *99,* 74-91.

Pfefferkorn, E.R., Nothnagel, R.F., & Borotz, S.E. (1992). Parasiticidal effect of clindamycin on *Toxoplasma gondii* grown in cultured cells and selection of a drug-resistant mutant. Antimicr. Agents Chemother. *31,* 1091-1096.

Plowe, C., Kublin, J., & Doumbo, O. (1998). *P. falciparum* DHFR and DHPS mutations: epidemiology and role in clinical resistance to antifolates. Drug Resistance Updates *1,* 389–396.

Reynolds, M.G., & Roos, D.S. (1998). A biochemical and genetic model for parasite resistance to antifolates. *Toxoplasma gondii* provides insights into pyrimethamine and cycloguanil resistance in *Plasmodium falciparum.* J. Biol. Chem. *273,* 3461-3469.

Roos, D.S., Donald, R.G.K., Morrissette, N.S., & Moulton, A.L.C. (1994). Molecular tools for genetic dissection of the protozoan parasite *Toxoplasma gondii.* Methods Cell Biol. *45,* 27–63.

Roos, D.S., Sullivan, W.J., Jr., Striepen, B., Bohne, W., & Donald, R.G.K. (1997). Tagging genes and trapping promoters in *Toxoplasma gondii* by insertional mutagenesis. Methods *13,* 112-122.

Roos, D.S., Crawford, M.J., Donald, R.G.K., Fohl, L.M., Hager, K.M., Kissinger, J.C., Reynolds, M.G., Striepen, B., & Sullivan, W.J., Jr. (1999a). Transport and trafficking: *Toxoplasma* as a model for *Plasmodium.* Novartis Fdn. Symp. *226,* 176-198.

Roos, D.S., Crawford, M.J., Donald, R.G.K., Kissinger, J.C., Klimczak, L.J., & Striepen, B. (1999b). Origins, targeting, and function of the apicomplexan plastid. Curr. Opin. Microbiol. *2,* 426-432.

Schatz, G., & Dobberstein, B. (1996). Common princiles of protein translocation across membranes. Science *278,* 1467-1470.

Schumacher, M.A., Carter, D., Roos, D.S., Ullman, B., & Brennan, R.G. (1996). Crystal structures of *Toxoplasma gondii* HXGPRTase reveal the catalytic role of a long flexible loop. Nature (Struct. Biol.). *3,* 881-887.

Schumacher, M.A., Carter, D., Scott, D.M., Roos, D.S., Ullman, B., & Brennan, R.G. (1998). Crystal structures of *Toxoplasma gondii* uracil phosphoribosyltransferase reveal the atomic structure of pyrimidine discrimination and prodrug binding. EMBO J. *17,* 3219-3232.

Schwab, J.C., Beckers, C.J.M., & Joiner, K.A. (1994). The parasitophorous vacuole membrane surrounding intracellular *Toxoplasma gondii* functions as a molecular sieve. Proc. Nat'l Acad. Sci. U.S.A. *91,* 509-513.

Schwab, J.C., Afifi, M.A., Pizzorno, G., Handschumacher, R.E., & Joiner, K.A. (1995). *Toxoplasma gondii* tachyzoites possess an unusual plasma membrane adenosine transporter. Mol. Biochem. Parasitol. *70,* 59-69.

Schwartzbach, S.D., Osafune, T., & Löffelhardt, W. (1998). Protein import into cyanelles and complex chloroplasts. Plant Mol. Biol. *38,* 247-263.

Schwartzman, J.D., & E.R. Pfefferkorn. 1981. Pyrimidine synthesis by intracellular *Toxoplasma gondii.* J. Parasitol. *67,* 150-158.

Schwartzman, J.D., & E.R. Pfefferkorn. 1982. *Toxoplasma gondii:* Purine synthesis and salvage in mutant host cells and parasites. Exper. Parasitol. *53,* 77-86.

Siddall, M.E. (1992). Hohlzylinders. Parasitol. Today *8,* 90-91.

Sirawaraporn, W., Sathitkul, T., Sirawaraporn, R., Yuthavong, Y., & Santi, D.V. (1997). Antifolate-resistant mutants of *Plasmodium falciparum* dihydrofolate reductase. Proc. Nat'l Acad. Sci. U.S.A. *94,* 1124-1129.

Soldati, D. (1999). The apicoplast as a potential therapeutic target in *Toxoplasma* and other Apicomplexan parasites. Parasitol. Today *15,* 5-7.

Striepen, B., He, C.Y., Matrajt, M., Soldati, D., & Roos, D.S. (1998). Expression, selection, and organellar targeting of the Green Fluorescent Protein in *Toxoplasma gondii.* Mol. Biochem. Parasitol. *92,* 328-338.

Sullivan, W.J., Jr., Chiang, C.-W., Wilson, C.M., Naguib, F.N.M., el Kouni, M.H., Donald, R.G.K., & Roos, D.S. (1999). Insertional tagging and cloning of at least two loci associated with resistance to adenine arabinoside in *Toxoplasma gondii,* and cloning of the adenosine kinase locus. Mol. Biochem. Parasitol. *103,* 1-14.

Tanaka, M., Gu, H., Bzik, D.J., Li, W., & Inselburg, J.W. (1990). Dihydrofolate reductase mutations and chromosomal changes associated with pyrimethamine resistance of *Plasmodium falciparum.* Mol. Biochem. Parasit. *39,* 127-134.

Triglia, T., Menting, J.G.T., Wilson, C., & Cowman, A.F. (1997). Mutations in dihydropteroate synthase are responsible for sulfone and sulfonamide resistance in *Plasmodium falciparum.* Proc. Nat'l Acad. Sci. U.S.A. *94,* 13944-13949.

Triglia, T., Wang, P., Sims, P.F.G., Hyde, J.E., & Cowman, A.F. (1998). Allelic exchange at the endogenous genomic locus in *Plasmodium falciparum* proves the role of dihydropteroate synthase in sulfadoxine-resistant malaria. EMBO J. *17,* 3807-3815.

Trujillo, M., Donald, R.G.K., Roos, D.S., Greene, P.J., & Santi, D.V. (1996). Heterologous expression and characterization of the bifunctional dihydrofolate reductase-thymidylate synthase enzyme of *Toxoplasma gondii.* Biochemistry *35,* 6366-6374.

Ullman, B., & Carter, D. (1995) Hypoxanthine-guanine phosphoribosyltransferase as a therapeutic target in protozoal infections. Infect. Agents Dis. *4,* 29-40.

Waller, R.F., Keeling, P.J., Donald, R.G.K., Striepen, B., Handman, E., Lang-Unnasch, N., Cowman, A.F., Besra, G.S., Roos, D.S., & McFadden, G.I. (1998). Nuclear-encoded proteins target to the plastid in *Toxoplasma gondii* and *Plasmodium falciparum*. Proc. Nat'l Acad. Sci. U.S.A. *95,* 12352-12357.

Waters, A.P., Thomas, A.W., van Dijk, M.R., & Janse, C.J. (1997). Transfection of malaria parasites. Methods *13,* 134-147.

Weissig, V., Vetro-Widenhouse, T.S., & Rowe, T.C. (1997). Topoisomerase II inhibitors induce cleavage of nuclear and 35 kb plastid DNAs in the malaria parasite *Plasmodium falciparum*. DNA Cell Biol. *16,* 1483- 1492.

Wilson, R.J.M., Denny, P.W., Preiser, P.R., Roberts, K., Roy, A., Whyte, A., Strath, M., Moore, D.J., & Williamson, D.H. (1997). Complete gene map of the plastid-like DNA of the malaria parasite *Plasmodium falciparum*. J. Mol. Biol. *261,* 155-172.

Wu, Y., Kirkman, L.A., & Wellems, T.E. (1996). Transformation of *Plasmodium falciparum* malaria parasites by homologous integration of plasmids that confer resistance to pyrimethamine. Proc. Nat'l Acad. Sci. U.S.A. *93,* 1130-1134.

Yasuhira, S., & Simpson, L. (1997). Phylogenetic affinity of mitochondria of *Euglena gracilis* and kinetoplastids using cytochrome oxidase I and hsp60. J. Mol. Evol. *44,* 341-347.

# 8

# THE TRICHOMONAD HYDROGENOSOME

Sabrina D. Dyall and Patricia J. Johnson
*Department of Microbiology and Immunology, University of California, Los Angeles, CA 90095*

## INTRODUCTION

Trichomonads are deep-branching protists that are thought to be early-diverging eukaryotes (Sogin, 1991). These organisms belong to the phylum Parabasalia which encompasses both non-parasitic and parasitic trichomonads. The two best studied parasitic trichomonads are the cattle-infective parasite, *Tritrichomonas foetus* and the human-infective parasite, *Trichomonas vaginalis*. These parasites are flagellated, extracellular organisms that are sexually transmitted and reside in the urogenital tracts of their hosts. Over 150 million cases of human trichomoniasis are reported each year and significant financial losses are frequently suffered due to trichomoniasis in cattle, making these parasites important in both the medical and agricultural communities. Aside from their medical and agricultural importance, a number of unusual biochemical properties of *Trichomonas* have captured the attention of scientists. The appeal of trichomonads from a biological viewpoint stems, in large part, from properties that reflect both their primitive nature and parasitic lifestyle. For example, trichomonads lack two organelles typically found in eukaryotes, the mitochondrion and the peroxisome, but instead contain an organelle involved in carbohydrate metabolism called the hydrogenosome.

## SCOPE OF HYDROGENOSOMAL RESEARCH

Hydrogenosomes or hydrogenosome-like organelles exist in a variety of eukaryotes; however, these organelles have been far better studied in *T. vaginalis* than in any other organism. Other eukaryotes that possess hydrogenosomes or hydrogenosome-like organelles are distributed throughout the phylogenetic tree and include free-living amoebaflagellates, rumen ciliates, and certain anaerobic fungi (Johnson et al., 1995; Müller, 1993). Although phylogenetically diverse, these organisms have in common life within anaerobic niches and the absence of mitochondria. Interestingly, hydrogenosomes, like mitochondria are involved in carbohydrate metabolism and ATP production. The mutually exclusive presence of hydrogenosomes and mitochondria in eukaryotic cells and the fact that all organisms known to contain hydrogenosomes are microaerophilic have led to the notion that hydrogenosomes are the anaeorobic functional equivalent of mitochondria.

In recent years, molecular analyses and phylogenetic reconstruction of proteins that constitute the hydrogenosome, as well as studies on the biogenesis of *T. vaginalis* hydrogenosomes, have provided insight into the relationship of this organelle and mitochondria. Strongly convincing evidence that mitochondria arose via a symbiosis of an α-proteobacteria with an ancestor of the eukaryotic cell (Gray et al., 1999) has been obtained by analysing mitochondrial DNA and comparing these with eubacterial DNA. As no DNA exists in trichomonad hydrogenosomes, it has been necessary to turn to analyses of other features of the organelle to discern its evolutionary history.

In this chapter, we will review what is known about hydrogenosomes in terms of their metabolism, ultrastructure, protein constituents and biogenesis. The similarities and differences with mitochondria will be assessed and current theories on the origin of hydrogenosomes will be discussed.

## METABOLIC PATHWAYS IN THE HYDROGENOSOME

The hydrogenosome participates in carbohydrate metabolism in the cell by the conversion of pyruvate to acetyl CoA, producing ATP from ADP by substrate-level phosphorylation. In this respect, hydrogenosomes and mitochondria can be regarded as similar organelles. However, the hydrogenosome uses a fermentative pathway for metabolism whereas in the mitochondrion, pyruvate metabolism proceeds via the Krebs cycle. No cytochromes or enzymes of the respiratory chain have been detected in hydrogenosomes (Müller, 1993). The yield of ATP from the two organelles shows their relative efficiencies: in mitochondria, metabolism of pyruvate produced by glycolysis in the cytosol yields 34 to 36 mol ATP per mol glucose whereas in hydrogenosomes, 2 mol of ATP are produced per mol glucose (see Martin and Müller, 1998 for a comparison of metabolism in mitochondriate and amitochondriate eukaryotes).

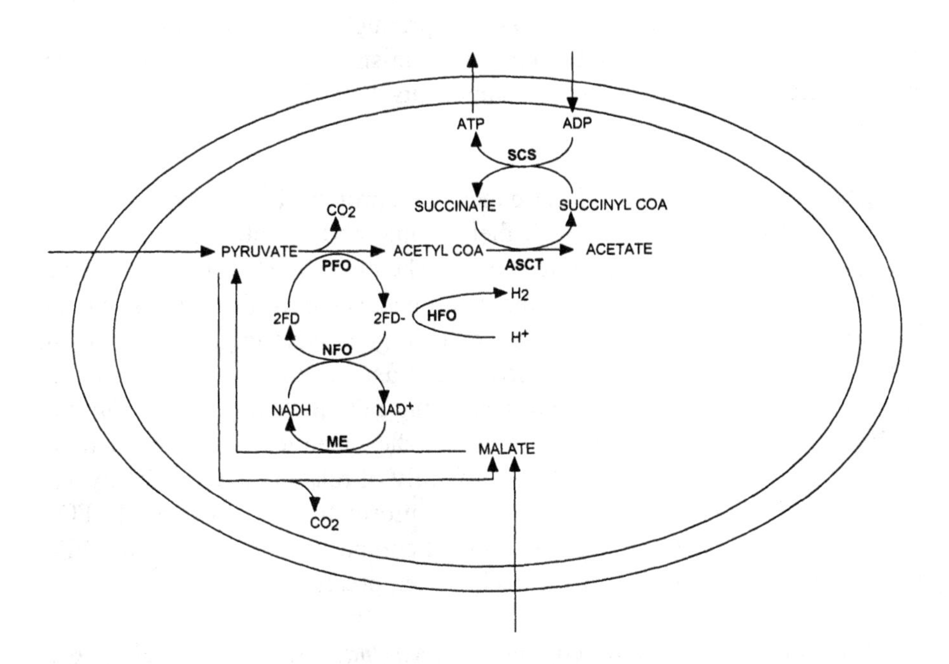


**Figure 1**. Schematic representation of pyruvate metabolism in the hydrogenosome. PFO = pyruvate:ferredoxin oxidoreductase, Fd = ferredoxin, NFO = NAD:ferredoxin oxidoreductase, HFO = $H_2$:ferredoxin oxidoreductase, ASCT = acetate:succinate CoA transferase, SCS = succinyl CoA synthetase, ME = malic enzyme (decarboxylating).

Detailed descriptions of the pyruvate metabolic pathway in the hydrogenosome can be found in comprehensive reviews by Biagini et al. (1997), Johnson et al. (1995) and Müller (1993). Briefly, the pathway consists of a short electron transfer chain (Fig. 1) comprising pyruvate:ferredoxin oxidoreductase, ferredoxin and ferredoxin:$H_2$

oxidoreductase (hydrogenase). Pyruvate, produced by glycolysis or by oxidative decarboxylation of malate by malate dehydrogenase (decarboxylating), is oxidatively decarboxylated by the enzyme pyruvate:ferredoxin oxidoreductase (PFO) to form carbon dioxide, acetyl CoA and reduced ferredoxin. Ferredoxin is reoxidised by transfer of its electrons to protons via the action of hydrogenase, resulting in the formation of molecular hydrogen. Acetyl CoA is converted to acetate by the action of acetate-succinyl CoA transferase, which results in the formation of succinyl CoA from succinate. Finally, an ADP-specific succinyl CoA synthetase (SCS) enzyme uses succinyl CoA to produce ATP by substrate-level phosphorylation. There may be other mechanisms involved in this pathway as other hydrogenosomal redox components have been detected (Müller, 1993).

The genes encoding several of these metabolic proteins have been cloned and have permitted a comparison with their counterparts in other systems in an attempt to establish their possible origins. The enzyme PFO, involved in the decarboxylation of pyruvate in the hydrogenosome is different from the functionally equivalent pyruvate dehydrogenase complex found in mitochondria and in certain eubacteria. PFO and related enzymes are found in eubacteria and archaea. The genes encoding PFO in *T. vaginalis*, PfoA and PfoB, have been found to be 40% identical to eubacterial pyruvate:flavodoxin oxidoreductases, a family of related enzymes (Hrdy and Müller, 1995). Given the limited sequence information on eubacterial PFOs, a wide phylogenetic sampling and detailed comparison of *T. vaginalis* PFO and its counterpart in other organisms has been precluded.

The iron-sulphur protein ferredoxin in *T. vaginalis* (Johnson et al., 1990) belongs to a family of [2Fe-2S] ferredoxins that occur in certain aerobic bacteria and in mitochondria. The hydrogenosomal ferredoxin was found to be distinct from the usual electron carrier in hydrogen-producing bacteria that are of the [4Fe-4S] type. Another hydrogenosomal ferredoxin of the [2Fe-2S] type was characterised from the free living anaerobic protist *Psalteriomonas lanterna*: this protein showed 47% similarity to the *T. vaginalis* protein and was also homologous to mammalian mitochondrial ferredoxins (Brul et al., 1994). Thus hydrogenosomal ferredoxins appear to be more related to aerobic ferredoxins than to anaerobic types.

Analyses of the two genes encoding *T. vaginalis* hydrogenase, hydA and hydB, revealed that they were of the [Fe] hydrogenase type, with 57 to 64%

sequence similarity to the hydrogenases of the strict bacterial anaerobes *Clostridium* and *Desulfovibrio* (Bui and Johnson, 1996). These proteins have been immunolocalised to the hydrogenosome. Unfortunately, in this case as well, phylogenetic analysis of this key enzyme is difficult in the absence of sequence data from a diversity of organisms. Recently, a putative [Fe] hydrogenase gene has been isolated from the hydrogenosome-bearing chytrid ciliate *Nyctotherus ovalis* (Akhmanova et al., 1998) which shows 41% identity to its *T. vaginalis* homologue and 35 to 41% identity to [Fe] hydrogenases from *Clostridium and Desulfovibrio* bacteria.

Malate dehydrogenase(decarboxylating), or malic enzyme, is a ubiquitous enzyme involved in the oxidative decarboxylation of malate to pyruvate and carbon dioxide. Phylogenetic analysis of this enzyme from *T. vaginalis* did not produce a conclusive relationship with any specific group of eukaryotic malic enzymes, but it formed a long separate branch within the eukaryotic group suggesting a long independent evolutionary history (Hrdy and Müller, 1995). An equally ambiguous result was obtained with similar analyses of hydrogenosomal malic enzyme from the anaerobic fungus *Neocallimastix frontalis* (van der Giezen et al., 1997) which did not even group with its *T. vaginalis* homologue. It would appear from these studies that malic enzyme is not an appropriate protein for phylogenetic analyses.

Succinyl CoA synthetase (SCS) is typically located in mitochondria. The β-subunit of *T. vaginalis* SCS (Lahti et al., 1992) revealed 65% similarity with *Escherichia coli* β-SCS. The α–SCS subunit from *T. vaginalis* showed 72% to 74% similarity to its homologues from rat mitochondria, *E. coli* and *Thermus flavus* (Lahti et al., 1994). The enzyme in *E. coli* is an $\alpha_2\beta_2$ tetramer of mass 140 kDa, consisting of two dimers of the α- and β- subunits respectively whereas in mitochondria and gram positive bacteria, the holoenzyme is an αβ dimer of 70 kDa. In *T. vaginalis*, it has been reported that the holoenzyme has a mass of about 140 kDa, suggesting an *E. coli* type tetrameric structure. However, in *T. foetus*, the holoenzyme is 70 kDa, suggesting a dimeric composition (Jenkins et al., 1991). Hydrogenosomal β-SCS in the anaerobic fungus *N. frontalis* grouped with the pig mitochondrial homologue and not with the *T. vaginalis* enzyme (Brondijk et al., 1996).

Adenylate kinase is a ubiquitous transferase found in eubacteria, archaea and in various compartments of eukaryotes: cytosol, intermembrane space and matrix of mitochondria, chloroplasts and glycosomes. Phylogenetic analysis of a hydrogenosomal *T. vaginalis* adenylate kinase revealed a grouping with mitochondrial-type adenylate kinases but an independent origin (or an unusually rapid rate of evolution) was suggested based on the long branch displayed for this enzyme (Lange et al., 1994). Thus, so far, the preliminary phylogenetic analyses of the metabolic proteins malic enzyme, ferredoxin, SCS and adenylate kinase show a possible relationship between mitochondria and the hydrogenosomes. However, the presence of typically anaerobic enzymes such as PFO and hydrogenase, which do not occur in mitochondria, challenges this idea.

## STRUCTURE OF THE TRICHOMONAD HYDROGENOSOME

### Physical structure

Hydrogenosomes of *T. vaginalis* and of *T. foetus* are bound by double membranes, like mitochondria, but do not have cristae (Honigberg, 1984). In *T. foetus*, ultrastructural studies (Benchimol et al., 1996) showed that the hydrogenosomes were concentrated near the axostyle and costae of the parasites and occupied about 4% of the total cell volume. The organelles were mostly spherical, with diameters of 0.5 to 2 μm. The two membranes, each about 6 nm thick showed some undulation, but no cristae and were closely opposed to each other, leaving a very narrow intermembrane space. Hydrogenosomes in ciliates also have double membranes, and some of them show cristae-like structures (Finlay and Fenchel, 1989). Closely opposed double membranes are also reported for *Neocallimastix* hydrogenosomes (vanderGiezen et al., 1997, Benchimol, 1997) although a previous report indicated a single membrane (Marvin-Sikkema et al., 1992), implying a microbody-like structure for these hydrogenosomes.

Some *T. foetus* hydrogenosomes showed vesicular-like structures of various forms and sizes protruding out of the organelles (Benchimol et al., 1996). These were always surrounded by concentric membranes distinct from the organelle envelope membranes, suggesting the possibility of an intra-organellar compartment. The matrix of the organelles appeared granular with electron dense spots corresponding to calcium deposits. It has been shown that trichomonad and *N. frontalis* hydrogenosomes are calcium stores, possibly in the form of calcium phosphate deposits (Biagini et al., 1997; Humphreys et al., 1994). This is similar to the role of mitochondria in storing excess cytosolic calcium as a precipitate with phosphate, forming deposits in the matrix (Carafoli, 1987).

Studies on the morphogenesis of trichomonad hydrogenosomes (Benchimol et al., 1996) revealed division by binary fission in a two-stage process involving first, growth of the hydrogenosome followed by the appearance of a central constriction, second, the development of a septum originating from the inner membrane that eventually separates the matrix into two compartments. This is very similar to division during biogenesis of mitochondria.

An electrochemical membrane potential has been detected in trichomonad hydrogenosomes (Humphreys et al., 1994) and also in *N. frontalis* hydrogenosomes (Biagini et al., 1997) where additionally a pH potential has been measured. The membrane potential could participate in active processes such as the translocation of metabolites, proteins or to drive accumulation of $Ca^{2+}$ into hydrogenosomes.

Unlike mitochondrial membranes, trichomonad hydrogenosomal membranes have been reported to contain no cardiolipin, the major component of mitochondrial inner membranes (Paltauf and Meingassner, 1982). These membranes were found to be constituted of the general membrane phospholipids phosphatidylcholine, phosphatidylethanolamine as well as small amounts of phosphatidylglycerol and N-dimethylphosphatidylethanolamine.

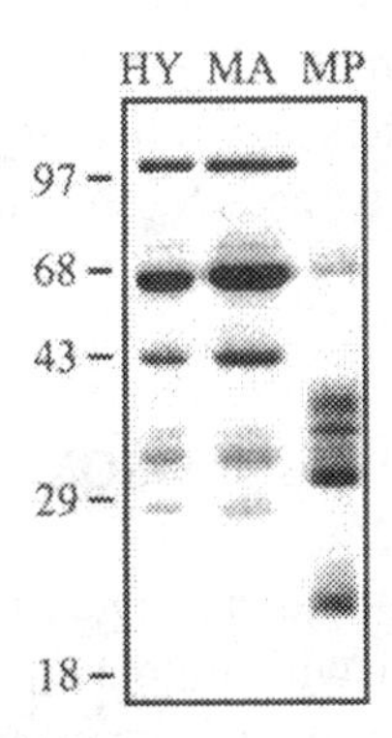


**Figure 2.** Protein profile of *Trichomonas vaginalis* hydrogenosomes. Protein markers sizes are given in kilodaltons. Isolated hydrogenosomes were subjected to sodium carbonate extraction (see detailed method in this chapter) and proteins from the two fractions were separated by SDS-PAGE, followed by Coomassie Blue

staining for visualisation. Lanes: HY=10 μg total hydrogenosomal protein, MA=10 μg soluble hydrogenosomal protein, consisting of matrix and any intermembrane space or peripheral membrane proteins, MP=10μg integral membrane proteins.

### Protein composition

The trichomonad hydrogenosome does not appear to be a complex organelle in terms of protein content (Fig. 2). The organelle consists of approximately 90% soluble proteins and 10% insoluble proteins in the form of integral membrane proteins. The major soluble polypeptides have been shown by immunolocalisation to correspond to the proteins involved in metabolism: PFO (about 120 kDa), malic enzyme and Hsp60 (about 65 kDa), β-SCS (about 43 kDa) and α-SCS (about 30 to 32 kDa). The majority of the membrane proteins range from 24 to 40 kDa with minor bands at higher molecular weights.

### Organellar DNA

No organellar-associated DNA has been found in *T. vaginalis*, *T. foetus* (Müller, 1993) or in *Neocallimastix* hydrogenosomes (vanderGiezen et al., 1997). The presence and availability of an organellar genome would have greatly facilitated phylogenetic reconstruction of the origin of these organelles, such as was the case with mitochondria (Gray et al., 1999). Recently, however, a weak labelling of hydrogenosomes with anti-DNA antiserum has been reported in *Nyctotherus ovalis*, an anaerobic heterotrichous ciliate (Akhmanova et al., 1998). Thus, ciliate hydrogenosomes (as well as those in other organisms) may ultimately be shown to have retained at least some DNA of the endosymbiont(s) that gave rise to these organelles.

## BIOGENESIS OF THE HYDROGENOSOME

Studies of biogenesis of trichomonad hydrogenosomes have revealed a number of features in common with mitochondrial biogenesis. Similarities include post-translational targeting of hydrogenosomal matrix proteins, the requirement of cytosolic proteins and similar energy and electrochemical requirements for translocation of proteins into the organelle. Furthermore, hydrogenosomal matrix proteins invariably contain N-terminal, cleavable targeting signals that greatly resemble N-terminal presequences used to target mitochondrial proteins.

## Protein synthesis in the cytosol and post-translational targeting of hydrogenosomal proteins

In *T. vaginalis*, the hydrogenosomal proteins ferredoxin and β-SCS are synthesised on free polyribosomes (Lahti and Johnson, 1991), indicating that these nuclear-encoded proteins are synthesised and released in the cytosol, then post-translationally translocated into the organelle (Fig. 3). This observation implies multiplication by binary fission as examplified by other organelles such as mitochondria, peroxisomes, glycosomes and chloroplasts (see references in Lahti and Johnson, 1991) which also acquire proteins synthesised on free polyribosomes. Proteins in other organelles such as lysosomes and vesicles that divide by budding off the rough endoplasmic reticulum, are acquired by synthesis on membrane-bound polyribosomes.

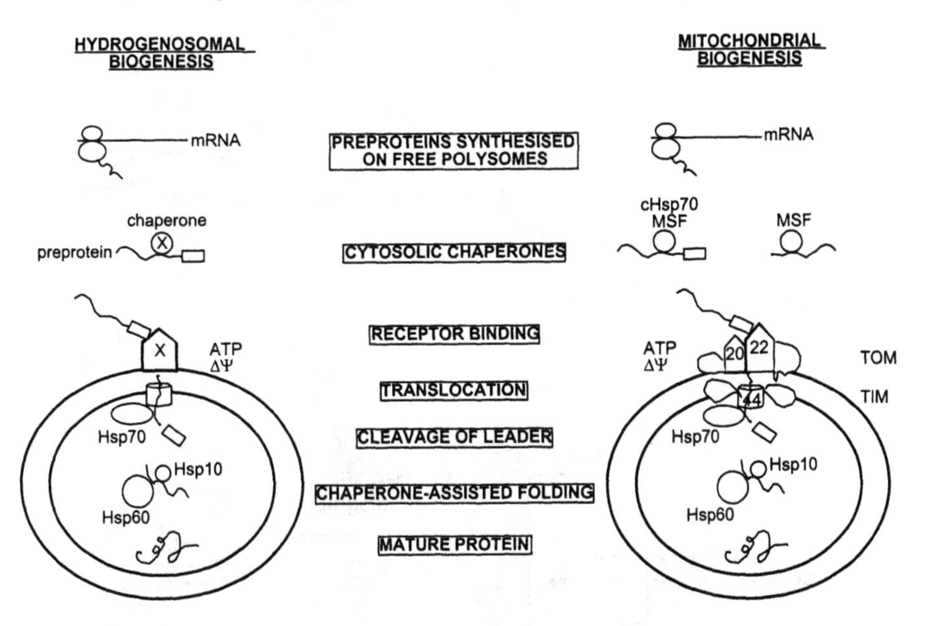


**Figure 3**. Comparison of hydrogenosomal and mitochondrial biogenesis. The N-terminal presequence of the newly synthesised protein is indicated by the rectangular box. X = unknown proteins, cHsp70 = cytosolic Hsp70, MSF = mitochondrial import stimulation factor, TOM = translocase of outer mitochondrial membrane complex, TIM = translocase of inner mitochondrial membrane complex. Tom20 , Tom22 and Tim44 are indicated by the numbers 20, 22 and 44 respectively.

**Cytosolic factor(s) required for translocation**

Recently, an in vitro import assay has been established to study the translocation of a radiolabelled precursor protein into isolated hydrogenosomes in the presence of a cytosolic extract and an ATP-regenerating system (Bradley et al., 1997). Briefly, the radiolabelled precursor is incubated with the organelles in the presence of the other components at 37°C to allow translocation. After the reaction, the organelles are protease treated to distinguish between any surface bound precursor and those that have been translocated to a protease-protected compartment. Complete translocation of the protein can be measured by the amount of the faster migrating, N-terminal-cleaved mature protein (Fig. 4).

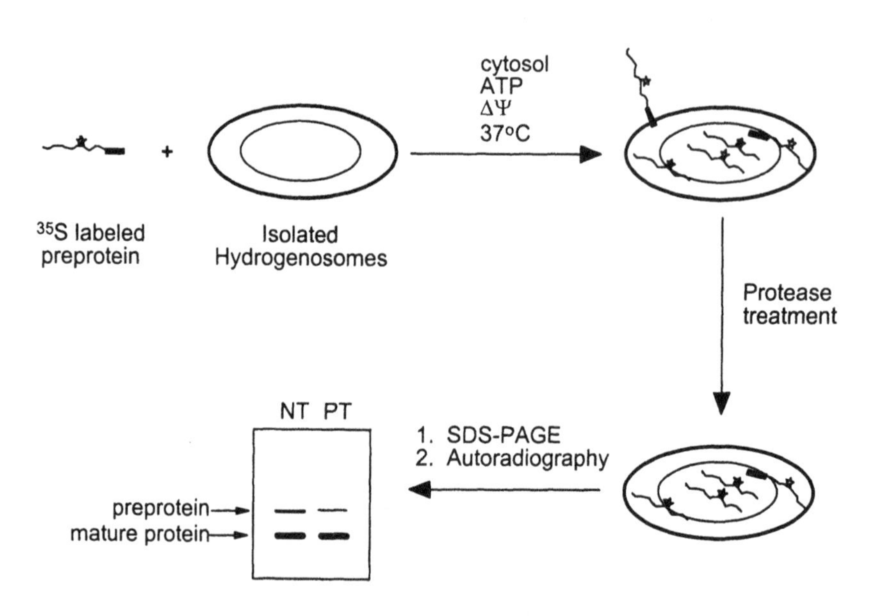


**Figure 4.** Schematic representation of in vitro protein translocation into isolated hydrogenosomes. The N-terminal presequence of the radiolabelled preprotein is indicated by the black rectangular box. NT = no protease treatment, PT = protease treatment.

The exclusion of the cytosolic extract from the in vitro system or its inactivation by an alkylating agent resulted in binding of the radiolabelled precursor but not in translocation as assayed by protease treatment. These data indicate that translocation of the precursor depends on protein(s) in the cytosol, possibly acting as chaperones (Fig. 3). Cytosolic chaperones interact with newly synthesised proteins to prevent aggregation and to keep them in a translocation-competent conformation. Two cytosolic chaperones have been implicated in translocation into mitochondria: these are the mitochondrial-import stimulating factor (MSF) which interacts with proteins destined to cross the membranes and heat shock protein 70 (Hsp70) which interacts with a variety of preproteins (Schatz and Dobberstein, 1996).

| *Trichomonas vaginalis*: Hydrogenosomal matrix proteins | | |
|---|---|---|
| Ferredoxin | MLSQVCRF↑GTI | P21149 |
| SuccinylcoA synthetase, αSCS1 | MLAGDFSRN↑LHK | P53399 |
| SuccinylcoA synthetase, αSCS2 | MLSSSFERN↑LHQ | P53400 |
| SuccinylcoA synthetase, αSCS3 | MLSSSFERN↑LHQ | P53401 |
| SuccinylcoA synthetase, βSCS | MLSSSFARN↑FNI | Q03184 |
| Adenylate kinase | MLSTLAKRF↑ASG | P49983 |
| Heat shock protein hsp60 | MSLIEAAKHFTRAF↑AKA | Q95058 |
| Malic enzyme subunit A | MLTSSVSVPVRN↑ICR | AAA92714 |
| Malic enzyme subunit B | MLTSSVNFPARE↑LSR | AAA92715 |
| Malic enzyme subunit C | MLTSVSYPVRN↑ICR | AAA92716 |
| Malic enzyme subunit D | MLTSVSLPVRN↑ICR | AAA92717 |
| Pyruvate:ferredoxin oxidoreductase A | MLRNF↑SKR | AAA85494 |
| Pyruvate:ferredoxin oxidoreductase B | MLRSF↑GKR | AAA85494 |
| Fe-hydrogenase subunit A | MLASSATAMKGFANSLRM-KDY | AAC41759 |
| Fe-hydrogenase subunit B | MLASSSRA-AANIRW-VDT | AAC41760 |
| Heat shock protein hsp70 | MLKMFNSIFARE-KNQ | AAB17251 |
| Heat shock protein hsp10 | MLATFARN-FAA | AAB17249 |
| *Trichomonas vaginalis*: Hydrogenosomal membrane protein | | |
| Hmp31 | MAQPAEQILIAT↑SPK | Unpublished |
| *Psalteriomonas lanterna*: Hydrogenosomal matrix protein | | |
| Ferredoxin | MVSGVSRN-AAR | P34806 |
| *Neocallimastix frontalis*: Hydrogenosomal matrix proteins | | |
| SuccinylcoA synthetase, βSCS | MLANVTRSTSKAAPALASIAQTAQKRF-LSV | X84222 |
| Malic enzyme | MLAPIQTIARPVSSILPATGALAAKRT↑FFA | AAC49572 |
| *Entamoeba histolytica*: Crypton protein | | |
| Heat shock protein hsp60 | MLSSSSHYNKLLSLNIDCRE↑NVL | AAC38819 |

| | | |
|---|---|---|
| *Leishmania*: Mitochondrial matrix proteins | | |
| *L. tarentolae* Heat shock protein hsp60 | MLRSAVRL↑AGK | P56281 |
| *L.major* Heat shock protein hsp70 | MFARRVCGSAAASAACLARAH↑ESQ | P12076 |
| Crown Group Eukaryotes: Mitochondrial matrix proteins | | |
| *S. cerevisiae* Heat shock protein hsp60 | MLRSSVVRSRATLRPLLRRAYSSHK↑ELK | P19882 |
| *Cucurbita sp.* Heat shock protein hsp60 | MHFRATGLASKARLARNGANQIASRSNWRRN↑AAK | X70868 |
| *H. sapiens* Heat shock protein hsp60 | MLRLPTVFRQMRPVSRVLAPHLTRAY↑AKA | P10809 |
| Plantae: Chloroplast luminal proteins | | |
| *A. thaliana* Heat shock protein hsp60 | MASTFTATSSIGSMVAPNGHKSDKKLISKLSSSSFGRRQS VCPRPRRSSSAIVCA↑AKE | P21240 |
| *P. sativum* Heat shock protein hsp70 | MASSAQIHGLGTASFSSLKKPSSISGNSKTLFFGQRLNSN HSPFTRAAFPKLSSKTFKKGFTLRVVS↑EKV | Q02028 |

**Table 1**. Comparison of N-terminal presequences of proteins from hydrogenosomes, crypton, mitochondria and chloroplasts. Known cleavage sites are indicated by '↑'and predicted cleavage sites are shown as '-'. Arginine residues found at positions –2 or –3 with respect to the cleavage site are underlined. Accession numbers for the proteins, where available, are given in the right hand column.

## Protein targeting signals

Post-translational translocation of preproteins into organelles requires the presence of targeting signals. The first putative hydrogenosomal signal sequence was characterised in the ferredoxin protein which is synthesised with an N-terminal presequence of 8 amino acids that is not present in the mature protein (Johnson et al., 1990; Lahti and Johnson, 1991). Using the in vitro translocation assay described above, Bradley *et al.* (1997) have shown that deletion of this N-terminal leader results in the disruption of binding of the leaderless ferredoxin. This implies that ferredoxin binds to the outer membrane of the hydrogenosome via a receptor or otherwise depends on the presence of the leader sequence (Fig. 3). A single amino acid mutation of leucine at position 2 of the leader (Table 1) resulted in only a small amount of protein associating with the organelle. Interestingly, the associated protein was present in a protease-protected compartment and the leader was not cleaved, as opposed to that observed using the wild type precursor. This suggests that either the mutant protein was not accessible to the peptidase, possibly because it was present in the narrow intermembrane space, or that the cleavage site was not recognised by a matrix peptidase. Similar leader sequences have been found on all *T. vaginalis* hydrogenosomal matrix proteins and recently on hydrogenosomal proteins from *N. frontalis* and *P. lanterna*. These presequences are shown in Table 1, together with N-terminal signal peptides from other systems.

The characterised leader sequences from 17 *T. vaginalis* hydrogenosomal matrix proteins are quite short, ranging from 5 to 14 residues. Similarly, hydrogenosomal leader sequences from other organisms are relatively short. Hydrogenosomal proteins from *P. lanterna* (Brul et al., 1994) and *N. frontalis* (Brondijk et al., 1996; van der Giezen et al., 1997) have leader sequences of 8 and 27 amino acids, respectively. Recently Mai et al. (1999) have reported a leader sequence of 21 amino acids on a Hsp60 of *Entamoeba histolytica*, a protein that is thought to be targeted to a newly discovered organelle, the crypton (Mai et al., 1999) or mitosome (Tovar, 1999) in this microaerophillic protist that lacks both mitochondria and hydrogenosomes. In the kinetoplast, the highly-specialized mitochondria found in Trypanosomatids, the mitochondrial N-terminal leaders are of two types: short leaders of 7 to 9 amino acids in >50% of proteins and others of 15 to 20 amino acids (Hausler et al., 1997). In the case of mitochondria from higher eukaryotes, the leader sequences are longer (Glaser et al., 1998; von Heijne et al., 1989), from 15 to 85 amino acids. The N-terminal presequences of *T. vaginalis* hydrogenosomal matrix proteins are enriched in serine (23%), leucine (14%) and arginine (12%). The mitochondrial matrix N-terminal presequences are enriched in arginine (14%), leucine (12%) serine (11%) and alanine (14%). On the other hand, chloroplast leader peptides have a different amino acid composition with serine at 19% and threonine at 9%. Both chloroplast and mitochondrial leaders lack acidic residues (von Heijne et al., 1989). However, in the known *T. vaginalis* hydrogenosomal matrix leaders, 5 out of 17 tolerate acidic residues.

Sixteen out of the 17 trichomonad hydrogenosomal leaders in Table 1 start with a Met-Leu dipeptide. The Met-Leu dipeptide at the presequence N-terminus and the Arg at –2, relative to the cleavage site, are conserved in the two *N.frontalis* hydrogenosomal proteins, the crypton hsp60 protein and in the short-type mitochondrial matrix leaders of kinetoplastids (Table I and Hausler et al., 1997). It is noteworthy that 40% of examined mitochondrial leader sequences from yeast and mammals also begin with a Met-Leu dipeptide (Hendrick et al., 1989). Moreover, as discussed above, the leucine at this position in the hydrogenosomal protein ferredoxin was shown to be necessary for translocation of the protein into *T. vaginalis* hydrogenosomes (Bradley et al., 1997), clearly demonstrating a functional role in targeting for this residue. In addition to the highly-conserved Leu at position 2 in trichomonad hydrogenosomal leaders, 12 out of 13 examined sequences on matrix proteins have either a Phe or an Asn at –1 relative to the cleavage site. Furthermore, 10 out of 13 examined proteins have an Arg at position –2 and those lacking this residue at –2 possess it at the –3 position. Interestingly, a majority of yeast mitochondrial leader sequences likewise contain Arg at –2 from the cleavage site (von Heijne et al, 1989).

Apart from a similar amino acid enrichment, a common feature of these hydrogenosomal leader peptides and of the mitchondrial matrix leaders is their ability to form amphiphilic helices (Johnson et al., 1990; Lahti et al., 1992; Pfanner et al., 1997). The amphiphilic mitochondrial N-terminal leader peptides have been shown to bind to the Tom22-Tom20 mitochondrial outer membrane receptor sub-complex (Fig. 3 and Pfanner et al., 1997) prior to insertion into the general import pore. The purified cytosolic domain of Tom22 binds presequences in an electrostatic manner whereas the interaction with Tom20 hydrophobic, suggesting that the negatively charged receptor Tom 22 recognises the positively charged face of an amphiphilic presequence and that Tom 20 binds to the hydrophobic side (Pfanner et al., 1997). Given the amphiphilicity of the *T. vaginalis* matrix leader peptide, it is possible that such a principle is utilised by the hydrogenosomal translocation apparatus.

In contrast to matrix proteins, it appears that at least some hydrogenosomal membrane proteins in *T. vaginalis* do not have a cleavable leader sequence as ascertained by N-terminal sequencing of the endogenous hydrogenosomal protein and comparison with the predicted translated gene sequence (our unpublished data). This suggests the presence of internal targeting signals that would direct them to the organelle. So far, internal membrane-targeting signals in mitochondrial proteins have not been well characterised and there does not appear to be any conserved primary sequence stretches in the membrane proteins (Pfanner et al., 1997). However, one examined hydrogenosomal membrane protein, Hmp31 (Table 1) has a 12 amino acid leader that is different from the matrix leaders in composition and in amphiphilic properties. This leader appears to have a similar composition to leaders found in plant mitochondrial membrane proteins (Glaser et al., 1998), but is shorter. Like the latter, the Hmp31 leader appears not to be necessary for translocation as assayed by in vivo transformation of *T. vaginalis* cells with a leaderless version of Hmp31, but it may participate in enhancing the efficiency of translocation (our unpublished data). The ability to generate selectable transformants in *T. vaginalis* (Delgadillo et al., 1997) allows us to overexpress epitope-tagged mutants of these hydrogenosomal proteins *in vivo* and to study their fate in the cell by immunoreaction with anti-epitope antiserum. This technique can be applied to detect whether mature proteins that cannot be isolated are cleaved upon translocation and also to study the effect of mutations on targeting.

The ability of a leaderless membrane protein to be translocated into the hydrogenosome points towards an internal membrane targeting signal. Thus, there are at least two types of translocation signals in hydrogenosomal proteins: the matrix N-terminal targeting peptides and the internal membrane targeting signal. Alternative targeting signals that have been investigated in hydrogenosomes include the C-terminal peroxisomal targeting signal (PTS-1) SKL in *N. frontalis*. It has been suggested that the hydrogenase protein has a PTS-1 signal that was detected by anti-SKL antibodies although no sequence information was available for this protein (Marvinsikkema et al., 1993). This observation together with the prior belief that the hydrogenosome in *N. frontalis* had a single membrane (Marvin-Sikkema et al., 1992) led to the association of fungal hydrogenosomal biogenesis with peroxisomal biogenesis. However, as discussed previously, it has since been reported that the *N. frontalis* hydrogenosomes do have a double membrane (vanderGiezen et al., 1997). Moreover, the two hydrogenosomal matrix proteins characterised to date from this organism have an N-terminal cleaved presequences (Brondijk et al., 1996; van der Giezen et al., 1997) that resemble mitochondrial and trichomonad hydrogenosomal targeting peptides but no PTS-1 signal. It is possible that some of the proteins in the fungal hydrogenosomes utilise peroxisomal-type signals, but no evidence has been presented to date. It has been shown, however, that trichomonad hydrogenosomes are not detected by the anti-SKL antibody (Keller et al., 1991).

**Dynamics of protein import**

The energy and electrochemical requirements for protein import have been determined by studying the translocation of proteins into hydrogenosomes using in vitro protein import assays (Bradley et al., 1997). Depletion of ATP from the in vitro system resulted in binding of the preprotein, followed by very little translocation. This implies that ATP is not required for binding to the outer membrane of the hydrogenosome (Fig. 3), but is necessary for translocation to occur. It is possible that ATP is required for cytosolic chaperone binding to keep the preprotein in a translocation-competent conformation. In addition, ATP could be required for insertion into the organelle, by interaction with a translocation/receptor component in an ATP-dependent manner. In mitochondrial translocation, ATP is required for interaction with the cytosolic factor and for translocation into the matrix across the inner membrane (Fig. 3) during interaction of the imported protein with a member of the TIM complex, Tim44 and the mitochondrial chaperone Hsp70 (Pfanner et al., 1997). Temperature also had an effect on the binding of precursor to hydrogenosomes. Binding was greatly reduced at 0°C, and the precursor was not translocated into the organelle (Bradley et al., 1997). Translocation was also demonstrated to be dependent on the membrane electrochemical potential. Dissipation of the membrane potential resulted in increased binding with about 40% of the uncleaved precursor in a protease-protected compartment (Bradley et al., 1997). This could represent an intermediate that is arrested in the intermembrane space or embedded in the membrane. This observation of dependence on membrane potential for translocation to be completed reflects the situation in mitochondria (Pfanner et al., 1997) as opposed to that of other organelles such as the chloroplast where a membrane potential is not required for translocation (Heins and Soll, 1998).

**Hydrogenosomal chaperonins**

Chaperonins or heat shock proteins (Hsp) are involved in the correct folding of mature proteins and/or translocation of newly synthesised polypeptides across membranes. They are ubiquitous proteins found in prokaryotes, archaea and various compartments of the eukaryotic cell. Phylogenetic analysis of hydrogenosomal Hsp10 (Bui et al., 1996), Hsp60 (Bui et al., 1996; Horner et al., 1996; Roger et al., 1996) and Hsp70 (Bui et al., 1996; Germot et al., 1996) with their homologues from a diverse representation of taxa indicated that these specifically grouped with mitochondrial heat shock proteins. Mitochondrial Hsp70 interacts with Tim44 to pull matrix-destined proteins through the inner membrane. It also assists in post-translocation folding in the matrix with the assistance of co-chaperones. Hsp60 and Hsp10 co-operate to correctly fold proteins to their final conformation (Pfanner et al., 1997). Both the mitochondrial-type Hsp60 and Hsp70 have been immunolocalised to the hydrogenosomes of trichomonads (Bozner, 1997; Bui et al., 1996). The N-terminal sequence of Hsp10 closely resembles the leader peptide of hydrogenosomal matrix proteins, suggesting a similar location (Fig. 3). As yet, no functional data is available on these hydrogenosomal heat shock proteins, but their co-localisation in the organelle would suggest a similar function to those of their mitochondrial homologues (Fig. 3).

**Translocation of hydrogenosomal proteins into mitochondria**
Given the similarities between the short-type leader sequences of the kinetoplastids and the hydrogenosomal leaders, Hausler et al. (1997) investigated the ability of the hydrogenosomal leader sequences from ferredoxin, β-SCS and PFO to escort a passenger peptide (mouse dihydrofolate reductase, DHFR) into *T. brucei* mitochondria in vivo. It was reported that all three leaders were capable of targeting DHFR to the organelles, albeit at low efficiency, and that some cleavage occurred in the case of the ferredoxin and β-SCS fusion proteins. This suggests that the mitochondrial processing peptidase recognises the cleavage site on these hydrogenosomal leaders. Furthermore, it has been demonstrated that the hydrogenosomal malic enzyme from *N. frontalis* can be targeted in vivo to mitochondria of the ascomycete yeast *Hansenula polymorpha*. Deletion of the 27 amino acid leader, which closely resembles yeast mitochondrial N-terminal leaders in length and in sequence (Table 1), results in no detectable translocation (vanderGiezen et al., 1998). These results show that the yeast mitochondrial translocation machinery recognises the targeting signal on the hydrogenosomal protein, providing support for similar translocation components being involved in the two organelles.

## THEORIES ON THE ORIGINS OF THE HYDROGENOSOME

The origin of the hydrogenosome involves two questions: first, what type of endosymbiont gave rise to the organelle? Second, how can we explain the occurrence of hydrogenosomes both before (trichomonads) and after (fungi and ciliates) the appearance of true mitochondria in the eukaryotic line?

Early studies on the biochemistry of the hydrogenosome that revealed the typical anaerobic enzymes PFO and hydrogenase led Müller (1980) to suggest that hydrogenosomes evolved from an endosymbiotic anaerobic bacterium related to the strict anaerobes of the *Clostridium* species. However, this theory does not explain the occurrence of anaerobic hydrogenosomal ciliates and chytrids that form sister groups with aerobic, mitochondriate relatives as assessed by small subunit RNA analyses. It seems implausible that multiple endosymbioses occurred to give rise to hydrogenosomes in these different lines. Later, as similarities in hydrogenosomal and mitochondrial biochemistry appeared, Cavalier-Smith (1987) proposed that hydrogenosomes were converted mitochondria that lost their genome and respiratory chain components. Such a theory would require the acquisition of hydrogenosome-specific genes such as hydrogenase and PFO multiple times. It would also imply that there might exist some organisms that have hydrogenosomes with mitochondrial features and yet still function as hydrogenosomes. Such organelles have been described (Finlay and Fenchel, 1989) and do provide some support for this theory. The recent isolation of a chimeric hydrogenase gene in the genome of such an organism, *N. ovalis* (Akhmanova et al., 1998) that comprises a hydrogenase-like moiety and a portion that is similar to components of the respiratory chain is consistent with a common ancestry for hydrogenosomes and mitochondria.

In the early 1990s, phylogenetic reconstructions, based on the analysis of rRNAs provided strong evidence that trichomonads diverged from the main line of eukaryotic descent prior to the appearance of true mitochondria. Taking this information into account, as well as the significant biochemical differences between mitochondria and hydrogenosomes, we proposed that these trichomonad hydrogenosomes and mitochondria evolved from a common progenitor organelle, as opposed to the notion that these hydrogenosomes were converted mitochondria (Johnson et al., 1993).

Many lines of evidence have shown that mitochondria are descendants of an α-proteobacteria that formed a stable symbiosis with some ancestral eukaryotic cell (Gray et al., 1999). As discussed above, support for a common endosymbiotic origin for hydrogenosomes and mitochondria has been provided by molecular analyses of *T. vaginalis* hydrogenosomal heat shock proteins (Bui et al., 1996; Germot et al., 1996; Horner et al., 1996; Roger et al., 1996). The results of these phylogenetic analyses obtained using three different heat shock protein sequences were exceedingly informative. A variety of inference techniques consistently put hydrogenosomal Hsp70, Hsp60 and Hsp10 on the lineage that contains mitochondria (Bui et al., 1996; Germot et al., 1996; Horner et al., 1996; Roger et al., 1996). These data provide strong evidence that hydrogenosomes and mitochondria have a common evolutionary origin. Moreover, similar phylogenetic analyses on Hsp genes isolated from three protist lineages that lack both mitochondria and hydrogenosomes, entamoebids (Clark and Roger, 1995), microsporidia (Germot et al., 1997) and diplomonads (Roger et al., 1998) also placed these gene sequences in the clade composed of mitochondrial Hsps. This indicates that these organisms once contained an endosymbiont, most likely the one that contributed to the formation of mitochondria and hydrogenosomes, which was ultimately lost from the lineages that gave rise to *Entamoeba*, *Nosema* and *Giardia*.

The phylogeny of Hsps in *Trichomonas*, *Giardia* and *Entamoeba*, as well as the observation that hydrogenosomes in free living ciliates are often found in an intimate connection with endosymbiotic methanogens, recently lead to a new hypothesis on the origin of the eukaryotic cell (Martin and Müller, 1998). This hypothesis, called the hydrogen hypothesis, takes into account the archaeal origin of the nuclear apparatus (Gupta and Golding, 1996; Lake and Rivera, 1994) versus the eubacterial origin of the core metabolism found in mitochondria, hydrogenosomes and the cytosol of eukaryotes lacking both organelles. The hydrogen hypothesis expands upon the common progenitor theory for the origin of hydrogenosomes and mitochondria (Johnson et al., 1993) and states that eukaryotes originated from the symbiotic association between an Archaea host and a hydrogen-producing proteobacterium. Proposed properties of the progenitor include both abilities to respire and to ferment (Martin and Müller, 1998). The relevant force for the symbiosis to occur was the ability of the symbiont to produce molecular hydrogen and carbon dioxide that were consumed by the autotrophic archean in an anaerobic environment. The hypothesis draws upon similar properties in some modern day methanogens that are strictly dependent on $H_2$ for their ATP production to account for origins of eukaryotic cell metabolism. It states that after establishing the symbiosis, the proteobacteria transferred its genes to the chromosomes of the host. Respiratory components of the cell then became active as levels of oxygen rose and ultimately the symbiont/progenitor organelle was converted to either mitochondria in aerobic cells or to hydrogenosomes in anaerobic cells. In cells that currently lack either organelle, such as *Giardia* and *Entamoeba*, the symbiont/progenitor organelle was secondarily lost. Consistent with the hydrogen hypothesis is the observation that some Archaea may be more closely related to eukaryotes than others based on translation elongation factor phylogeny (Embley and Hirt, 1998). Analyses of components of the machinery for replication, transcription and translation that were once thought to be eukaryote-specific and are now known in Archaea (Doolittle, 1998) are also consistent with this hypothesis for the origin of early eukaryotes.

Conceptually, the hydrogen hypothesis is an attractive theory to explain the similarities in metabolism and biogenesis of hydrogenosomes and mitochondria. However, the existence of a common progenitor does not preclude the later acquisition of specific genes by the organellar eukaryotes through unsuccesful symbioses or lateral gene transfer.

## CONCLUSION

Recent research on trichomonad hydrogenosomes has provided convincing evidence for a common origin with mitochondria. What is less well-understood is the evolutionary relationship of hydrogenosomal-like organelles in other eukaryotes and those found in trichomonads or the relationship of these organelles with mitochondria. Further studies on the biochemistry, membrane composition and components and the structure of hydrogenosome-like organelles in these organisms will be required to shed light on these relationships. It is possible that hydrogenosome-like organelles evolved independently, unlike the monophylogeny currently put forth for the evolution of mitochondria. What is clear is that studies on trichomonad hydrogenosomes have been extremely insightful, even leading to a new hypothesis for the origin of eukaryotic cells. Equally exciting evolutionary findings are likely to result from further analyses of these intriguing organelles from a variety of eukaryotic cells.

## METHOD

The following sodium extraction method derived from Fujiki et al. (1982) is used routinely in our lab to separate the hydrogenosome into two protein fractions: the soluble fraction which would consist of any matrix, intermembrane and peripheral proteins and an insoluble fraction consisting of integral membrane proteins. This method allows us first to purify membrane proteins for primary sequence analysis and second to follow in vitro or in vivo translocation experiments to determine the location of the hydrogenosomally-targeted protein under study.

Purified hydrogenosomes (Bradley et al., 1997) resuspended at 1 mg/ml in SMDI (250 mM sucrose, 10 mM MOPS-KOH, pH7.2, 5mM DTT, 50 μg/ml TLCK, 10 μg/ml leupeptin) are diluted to a final concentration of 30 μg/ml in ice-cold 0.1 M $Na_2CO_3$ and incubated on ice for 30 min. The suspension is subjected to ultracentrifugation at 200,000 g for 1 hr at 4°C to separate the disrupted, extracted membranes (pellet) from the soluble fraction (supernatant). Following neutralisation, the supernatant is precipitated with trichloroacetic acid to a final concentration of 0.5 M, centrifuged at 16,000 g and the precipitate resuspended in Laemmli buffer for SDS-PAGE analysis or for isoelectric focussing. The membrane pellet is washed with ice-cold water to remove traces of the supernatant, and resuspended in the latter gel-loading buffers. The fraction of integral membrane proteins constitutes approximately 10% of the total organelle protein content.

## ACKNOWLEDGMENTS

We thank members of our laboratory for helpful input and critical comments on this manuscript.

## REFERENCES

Akhmanova, A., Voncken, F., van Alen, T., van Hoek, A., Boxma, B., Vogels, G., Veenhuiss, M., and Hackstein, J. H. P. (1998). A hydrogenosome with a genome. Nature *396*, 527-528.

Benchimol, M., Almeida, J. C., and de Souza, W. (1996). Further studies on the organization of the hydrogenosome in *Tritrichomonas foetus*. Tissue Cell *28*, 287-99.

Benchimol, M., Johnson, P. J., and deSouza, W. (1996). Morphogenesis of the hydrogenosome: An ultrastructural study. Bio. Cell *87*, 197-205.

Biagini, G. A., Finlay, B. J., and Lloyd, D. (1997). Evolution of the hydrogenosome. Fems Microbiol. Lett. *155*, 133-140.

Biagini, G. A., vanderGiezen, M., Hill, B., Winters, C., and Lloyd, D. (1997). Ca2+ accumulation in the hydrogenosomes of *Neocallimastix frontalis* L2: A mitochondrial-like physiological role. Fems Microbiol. Lett. *149*, 227-232.

Bozner, P. (1997). Immunological detection and subcellular localization of Hsp70 and Hsp60 homologs in *Trichomonas vaginalis*. J. Parasitol. *83*, 224-9.

Bradley, P. J., Lahti, C. J., Plümper, E., and Johnson, P. J. (1997). Targeting and translocation of proteins into the hydrogenosome of the protist *Trichomonas*: similarities with mitochondrial protein import. EMBO J. *16*, 3484-93.

Brondijk, T. H., Durand, R., van der Giezen, M., Gottschal, J. C., Prins, R. A., and Fèvre, M. (1996). scsB, a cDNA encoding the hydrogenosomal beta subunit of succinyl-CoA synthetase from the anaerobic fungus *Neocallimastix frontalis*. Mol. Gen. Genet. *253*, 315-23.

Brul, S., Veltman, R. H., Lombardo, M. C., and Vogels, G. D. (1994). Molecular cloning of hydrogenosomal ferredoxin cDNA from the anaerobic amoeboflagellate *Psalteriomonas lanterna*. Biochim. Biophys. Acta *1183*, 544-6.

Bui, E. T., and Johnson, P. J. (1996). Identification and characterization of [Fe]-hydrogenases in the hydrogenosome of *Trichomonas vaginalis*. Mol. Biochem. Parasitol. *76*, 305-10.

Bui, E. T. N., Bradley, P. J., and Johnson, P. J. (1996). A common evolutionary origin for mitochondria and hydrogenosomes. Proc. Nat. Acad. Sci. USA *93*, 9651-9656.

Carafoli, E. (1987). Intracellular calcium homeostasis. Annu. Rev. Biochem. *56*, 395-433.

Cavalier-Smith, T. (1987). The simultaneous symbiotic origin of mitochondria, chloroplasts, and microbodies. Ann. N Y Acad. Sci. *503*, 55-71.

Clark, C. G., and Roger, A. J. (1995). Direct evidence for secondary loss of mitochondria in *Entamoeba histolytica*. Proc. Natl. Acad. Sci. USA *92*, 6518-21.

Delgadillo, M. G., Liston, D. R., Niazi, K., and Johnson, P. J. (1997). Transient and selectable transformation of the parasitic protist *Trichomonas vaginalis*. Proc. Natl. Acad. Sci. USA *94*, 4716-20.

Doolittle, W. F. (1998). A paradigm gets shifty [news; comment]. Nature *392*, 15-6.

Embley, T. M., and Hirt, R. P. (1998). Early branching eukaryotes? Curr. Opin. Genet. Dev. *8*, 624-9.

Finlay, B. J., and Fenchel, T. (1989). Hydrogenosomes in some anaerobic protozoa resemble mitochondria. FEMS Microbiol. Lett. *65*, 311-314.

Fujiki, Y., Hubbard, A. L., Fowler, S., and Lazarow, P. B. (1982). Isolation of intracellular membranes by means of sodium carbonate treatment: application to endoplasmic reticulum. J. Cell. Biol. *93*, 97-102.

Germot, A., Philippe, H., and Le Guyader, H. (1997). Evidence for loss of mitochondria in Microsporidia from a mitochondrial-type HSP70 in *Nosema locustae*. Mol. Biochem. Parasitol. *87*, 159-68.

Germot, A., Philippe, H., and Le Guyader, H. (1996). Presence of a mitochondrial-type 70-kDa heat shock protein in *Trichomonas vaginalis* suggests a very early mitochondrial endosymbiosis in eukaryotes. Proc. Natl. Acad. Sci. USA *93*, 14614-7.

Glaser, E., Sjöling, S., Tanudji, M., and Whelan, J. (1998). Mitochondrial protein import in plants. Signals, sorting, targeting, processing and regulation. Plant Mol. Biol. *38*, 311-38.

Gray, M. W., Burger, G., and Lang, B. F. (1999). Mitochondrial evolution. Science *283*, 1476-81.

Gupta, R. S., and Golding, G. B. (1996). The origin of the eukaryotic cell [see comments]. Trends Biochem. Sci. *21*, 166-71.

Hausler, T., Stierhof, Y. D., Blattner, J., and Clayton, C. (1997). Conservation of mitochondrial targeting sequence function in mitochondrial and hydrogenosomal proteins from the early-branching eukaryotes Crithidia, Trypanosoma and Trichomonas. Eur. J. Cell Biol. *73*, 240-251.

Heins, L., and Soll, J. (1998). Chloroplast biogenesis: mixing the prokaryotic and the eukaryotic? Curr. Biol. *8*, R215-7.

Hendrick, J. P., Hodges, P. E., and Rosenberg, L. E. (1989). Survey of amino-terminal proteolytic cleavage sites in mitochondrial precursor proteins: leader peptides cleaved by two matrix proteases share a three-amino acid motif. Proc. Natl. Acad. Sci. USA *86*, 4056-60.

Honigberg, B. M., Volkmann, D., Entzeroth, R., Scholtyseck, E. (1984). A freeze-fracture electron microscopy study of *Trichomonas vaginalis* Donne and *Tritrichomonas foetus* (Riedmuller). J. Protozool. *31*, 116-131.

Horner, D. S., Hirt, R. P., Kilvington, S., Lloyd, D., and Embley, T. M. (1996). Molecular data suggest an early acquisition of the mitochondrion endosymbiont. Proc. R. Soc. Lond. B Biol. Sci. *263*, 1053-9.

Hrdy, I., and Müller, M. (1995). Primary structure and eubacterial relationships of the pyruvate:ferredoxin oxidoreductase of the amitochondriate eukaryote *Trichomonas vaginalis*. J. Mol. Evol. *41*, 388-96.

Hrdy, I., and Muller, M. (1995). Primary structure of the hydrogenosomal malic enzyme of *Trichomonas vaginalis* and its relationship to homologous enzymes. J. Eukaryot. Microbiol. *42*, 593-603.

Humphreys, M. J., Ralphs, J., Durrant, L., and Lloyd, D. (1994). Hydrogenosomes in trichomonads are calcium stores and have a transmembrane electrochemical potential. Biochem. Soc. Trans. *22*, 324S.

Jenkins, T. M., Gorrell, T. E., Müller, M., and Weitzman, P. D. (1991). Hydrogenosomal succinate thiokinase in *Tritrichomonas foetus* and *Trichomonas vaginalis*. Biochem. Biophys. Res. Commun. *179*, 892-6.

Johnson, P. J., Bradley , P. J., and Lahti, C. J. (1995). Cell biology of trichomonads: protein targeting to the hydrogenosome. In Molecular Approaches to Parasitology, J. C. Boothroyd and R. Komuniecki, eds. (New York: Wiley-Liss, Inc.), pp. 399-411.

Johnson, P. J., d' Oliveira, C. E., Gorrell, T. E., and Müller, M. (1990). Molecular analysis of the hydrogenosomal ferredoxin of the anaerobic protist *Trichomonas vaginalis*. Proc. Natl. Acad. Sci. USA *87*, 6097-101.

Johnson, P. J., Lahti, C. J., and Bradley, P. J. (1993). Biogenesis of the hydrogenosome in the anaerobic protist *Trichomonas vaginalis*. J. Parasitol. *79*, 664-70.

Keller, G. A., Krisans, S., Gould, S. J., Sommer, J. M., Wang, C. C., Schliebs, W., Kunau, W., Brody, S., and Subramani, S. (1991). Evolutionary conservation of a microbody targeting signal that targets proteins to peroxisomes, glyoxysomes, and glycosomes. J. Cell Biol. *114*, 893-904.

Lahti, C. J., Bradley, P. J., and Johnson, P. J. (1994). Molecular characterization of the alpha-subunit of *Trichomonas vaginalis* hydrogenosomal succinyl CoA synthetase. Mol. Biochem. Parasitol. *66*, 309-18.

Lahti, C. J., d' Oliveira, C. E., and Johnson, P. J. (1992). Beta-succinyl-coenzyme-a synthetase from *Trichomonas vaginalis* is a soluble hydrogenosomal protein with an amino-terminal sequence that resembles mitochondrial presequences. J. Bacteriol. *174*, 6822-6830.

Lahti, C. J., and Johnson, P. J. (1991). *Trichomonas vaginalis* hydrogenosomal proteins are synthesized on free polyribosomes and may undergo processing upon maturation. Mol. Biochem. Parasitol. *46*, 307-10.

Lake, J. A., and Rivera, M. C. (1994). Was the nucleus the first endosymbiont? [comment]. Proc. Natl. Acad. Sci. USA *91*, 2880-1.

Lange, S., Rozario, C., and Muller, M. (1994). Primary structure of the hydrogenosomal adenylate kinase of *Trichomonas vaginalis* and its phylogenetic relationships. Mol. Biochem. Parasitol. *66*, 297-308.

Mai, Z., Ghosh, S., Frisardi, M., Rosenthal, B., Rogers, R., and Samuelson, J. (1999). Hsp60 is targeted to a cryptic mitochondrion-derived organelle ("crypton") in the microaerophilic protozoan parasite *Entamoeba histolytica*. Mol. Cell Biol. *19*, 2198-205.

Martin, W., and Müller, M. (1998). The hydrogen hypothesis for the first eukaryote [see comments]. Nature *392*, 37-41.

Marvin-Sikkema, F. D., Lahpor, G. A., Kraak, M. N., Gottschal, J. C., and Prins, R. A. (1992). Characterization of an anaerobic fungus from llama faeces. J. Gen. Microbiol. *138*, 2235-41.

Marvinsikkema, F. D., Kraak, M. N., Veenhuis, M., Gottschal, J. C., and Prins, R. A. (1993). The hydrogenosomal enzyme hydrogenase from the anaerobic fungus *Neocallimastix* sp L2 is recognized by antibodies directed against the C-terminal microbody protein targeting signal SKL. Eur. J. Cell Biol. *61*, 86-91.

Müller, M. (1980). The hydrogenosome. In The Eukaryotic Microbial Cell, G. W. Gooday, LLoyd, D. and Trinci, A.P.J., ed. (Cambridge: Cambridge University Press), pp. 127-142.

Müller, M. (1993). The hydrogenosome. J Gen Microbiol *139*, 2879-89.

Paltauf, F., and Meingassner, J. G. (1982). The absence of cardiolipin in hydrogenosomes of *Trichomonas vaginalis* and *Tritrichomonas foetus*. J. Parasitol. *68*, 949-50.

Pfanner, N., Craig, E. A., and Hönlinger, A. (1997). Mitochondrial preprotein translocase. Annu. Rev. Cell. Dev. Biol. *13*, 25-51.

Roger, A. J., Clark, C. G., and Doolittle, W. F. (1996). A possible mitochondrial gene in the early-branching amitochondriate protist *Trichomonas vaginalis*. Proc. Natl. Acad. Sci. USA *93*, 14618-22.

Roger, A. J., Svärd, S. G., Tovar, J., Clark, C. G., Smith, M. W., Gillin, F. D., and Sogin, M. L. (1998). A mitochondrial-like chaperonin 60 gene in *Giardia lamblia*: evidence that diplomonads once harbored an endosymbiont related to the progenitor of mitochondria. Proc. Natl. Acad. Sci. USA *95*, 229-34.

Schatz, G., and Dobberstein, B. (1996). Common principles of protein translocation across membranes. Science *271*, 1519-26.

Sogin, M. L. (1991). Early evolution and the origin of eukaryotes. Curr. Opin. Genet. Dev. *1*, 457-63.

Tovar, J., Fischer, A. & Clark, G. C. (1999). The mitosome, a novel organelle related to mitochondria in the amitochondrial parasite *Entamoeba histolytica*. Mol. Microbiol. *32*, 1013 - 1021.

van der Giezen, M., Rechinger, K. B., Svendsen, I., Durand, R., Hirt, R. P., Fèvre, M., Embley, T. M., and Prins, R. A. (1997). A mitochondrial-like targeting signal on the hydrogenosomal malic enzyme from the anaerobic fungus *Neocallimastix frontalis*: support for the hypothesis that hydrogenosomes are modified mitochondria. Mol. Microbiol. *23*, 11-21.

vanderGiezen, M., Kiel, J., Sjollema, K. A., and Prins, R. A. (1998). The hydrogenosomal malic enzyme from the anaerobic fungus *Neocallimastix frontalis* is targeted to mitochondria of the methylotrophic yeast *Hansenula polymorpha*. Curr. Genetics *33*, 131-135.

vanderGiezen, M., Sjollema, K. A., Artz, R. R. E., Alkema, W., and Prins, R. A. (1997). Hydrogenosomes in the anaerobic fungus *Neocallimastix frontalis* have a double membrane but lack an associated organelle genome. FEBS Lett. *408*, 147-150.

von Heijne, G., Steppuhn, J., and Herrmann, R. G. (1989). Domain structure of mitochondrial and chloroplast targeting peptides. Eur. J. Biochem. *180*, 535-45.

# 9

# LYSOSOME EXOCYTOSIS AND INVASION OF NON-PHAGOCYTIC HOST CELLS BY *TRYPANOSOMA CRUZI*

Barbara A. Burleigh
*Department of Immunology and Infectious Diseases, Harvard School of Public Health, Boston, MA 02115*

## INTRODUCTION

In recent years, significant progress has been made toward understanding the mechanisms of host cell invasion by the intracellular pathogen, *Trypanosoma cruzi*. Invasion of non-phagocytic cells by trypomastigotes, the mammalian-infective forms of *T. cruzi*, is achieved by a surprising mechanism whereby host cell lysosomes are recruited as a vehicle for parasite entry. These studies have revealed the existence of the lysosome exocytosis pathway, a previously unrecognised regulated secretion system in non-phagocytic mammalian cells. Host cell signalling pathways that regulate *T. cruzi* entry and lysosome recruitment were shown to involve the second messengers, $Ca^{2+}$ and cAMP and require secreted agonist(s) generated by the parasite. In addition, the ability of *T. cruzi* to receive signals from the host cell in a contact-dependent manner is crucial to the invasive process. In this chapter, I will outline our current knowledge of the cellular and molecular bases for host cell invasion by *T. cruzi* as well as the experimental approaches taken to study this process.

## HOST CELL INVASION: PHAGOCYTOSIS VERSUS ACTIVE PARASITE ENTRY

For over two decades, investigators have been studying the cellular basis for *T. cruzi* invasion of mammalian host cells. The early studies, which focused primarily on macrophages, generated a degree of controversy in the field. Several laboratories demonstrated that *T. cruzi* invasion of macrophages exhibited characteristics of classical phagocytosis, including pseudopodia formation and dependence on host cell actin microfilaments (Nogueira and Cohn, 1976; Meirelles et al., 1982). Other reports suggested that *T. cruzi* entered macrophages by an active parasite-driven mechanism that proceeded independently of the host cell actin cytoskeleton (Kipnis et al., 1979).

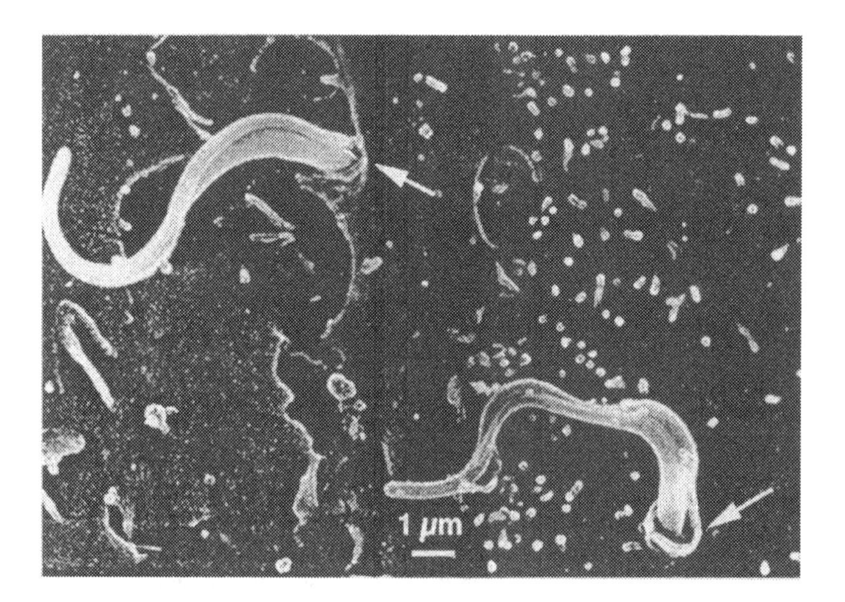


**Figure 1.** Scanning electron micrograph of *T. cruzi* trypomastigotes invading epithelial cells (E.S. Robbins and N.W. Andrews, unpublished image). The absence of pseudopodia at the point of host cell entry is indicated (arrows). Dr. Edith S. Robbins, Department of Cell Biology, NYU Medical Center, New York, New York, generously provided this micrograph.

Following invasion, *T. cruzi* trypomastigotes reside transiently within a vacuole before exiting to the cytosol where they replicate as intracellular amastigotes. The *T. cruzi*-containing (parasitophorous) vacuoles formed in both professional phagocytes (e.g. macrophages) and non-phagocytic cells (e.g. fibroblasts, epithelial cells) were shown to contain markers for host cell lysosomes (Milder and Kloetzel, 1980; Meirelles and de Souza, 1983; de Meirelles et al., 1987; Carvalho and de Souza, 1989). It was assumed that the presence of lysosomal markers in the vacuole was a consequence of phagosome maturation and suggested a common phagocytic mechanism for trypomastigote entry into mammalian cells.

The emerging picture of host cell invasion by *T. cruzi* was further complicated when a more detailed examination of trypomastigote entry of non-phagocytic cells was carried out. Scanning electron microscopy (SEM) studies of epithelial cells clearly demonstrated the absence of membrane protrusions or pseudopodia around the invading trypomastigote (Fig. 1) that would indicate a phagocytic mechanism for entry (Schenkman et al., 1988). Pre-treatment of MDCK and HeLa cells with the actin-depolymerizing agent, cytochalasin D, not only failed to inhibit *T. cruzi* entry (Schenkman et al., 1991), this treatment significantly enhanced parasite invasion (Tardieux et al., 1992). Furthermore, FITC-phalloidin staining of fibroblasts demonstrated the lack of filamentous actin associating with incoming trypomastigotes (Tardieux et al., 1992). These observations provided a strong indication that *T. cruzi* entry of fibroblasts and epithelial cells proceeded via a mechanism that is distinct from phagocytosis. If so, how is the parasitophorous vacuole formed in these cells and why does it stain positively for lysosomal markers? In the following paragraphs, I will discuss experiments that led to the elucidation of a unique pathway of host cell entry by an intracellular pathogen that involves host cell lysosome exocytosis.

## LYSOSOME RECRUITMENT AS A MECHANISM OF HOST CELL ENTRY

When the invasion of non-phagocytic mammalian cells by *T. cruzi* was examined at early time points (5 to 15 min), it became clear that association of trypomastigotes with host cell lysosomes coincided with parasite entry (Tardieux et al., 1992). Intense staining of the posterior portion of partially internalised trypomastigotes was revealed by immunofluorescence using antibodies to a major lysosomal membrane glycoprotein (lgp120) and shown in Figure 2 (Tardieux et al., 1992). The detection of horseradish peroxidase (HRP)-loaded lysosomes in clusters around invading parasites by light and electron microscopy confirmed the identity of the lgp120 staining organelles as lysosomes (Tardieux et al., 1992). The early recruitment of lysosomes by *T. cruzi* was highlighted by the striking observation that lysosomes congregated at the parasite attachment site prior to initiation of invasion (Tardieux et al., 1992). Thus, using simple cell biological methods, Tardieux et al. were the first to demonstrate that lysosomes contribute to the formation of the parasitophorous vacuole during *T. cruzi* invasion.

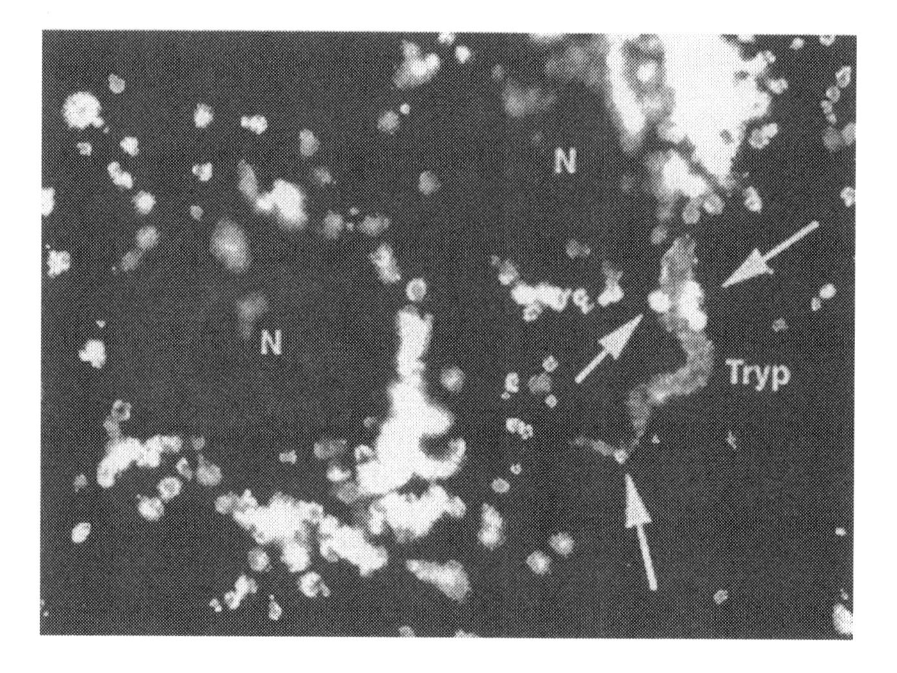


**Figure 2.** Immunofluorescence staining of NRK fibroblasts using anti-lgp120 to visualize lysosomes (E.C. Caler and N.W. Andrews, unpublished image). Co-localization of lysosomes with intracellular *T. cruzi* trypomastigotes (Tryp) is indicated by arrows; N = nucleus of NRK cell. Dr. Elisabet Caler, Yale University School of Medicine, New Haven, CT, generously provided this image.

Several treatments have been exploited to manipulate the distribution of lysosomes in mammalian cells (Heuser, 1989). These treatments were shown to have profound effects on *T. cruzi* invasion. Acidification of the cytosol promotes the redistribution of lysosomes toward the periphery of host cells and results in enhanced invasion by *T. cruzi*. Conversely, increasing the cytosolic pH causes lysosomes to aggregate in the perinuclear region of the host cell, which significantly inhibits *T. cruzi* invasion (Tardieux et al.,

1992). In addition, host cell entry by *T. cruzi* was compromised by the specific depletion of peripheral lysosomes following microinjection of NRK cells with antibodies against the cytoplasmic domain of lgp120 which promotes perinuclear aggregation of lysosomes (Rodríguez et al., 1996). In polarized MDCK cells, the predominantly basolateral distribution of lysosomes restricts trypomastigote entry to that surface (Schenkman et al. 1988). Strikingly, depolarisation of the cell monolayer following disruption of the actin microfilaments, results in a dramatic increase in lysosome-mediated invasion from the apical plasma membrane (Tardieux et al., 1992). Furthermore, functional lysosomes are required to mediate *T. cruzi* entry. Interference with the fusogenic properties of lysosomes by loading with indigestible substances (Montgomery et al., 1991) significantly impaired parasite invasion (Tardieux et al., 1992). Taken together, these experimental observations provide strong support for the early recruitment and fusion of host cell lysosomes in the mechanism of *T. cruzi* invasion.

**Lysosomes are recruited along microtubules**

The movement of lysosomes was recorded by time-lapse video microscopy in live myoblasts loaded with an endocytic tracer, bovine serum albumin-gold (Rodríguez et al., 1996). In control cells or in cells lacking attached *T. cruzi* trypomastigotes, lysosomes (and other organelles) move around in a random fashion. However, 'committed attachment' (attachment that will eventually lead to internalisation) of a trypomastigote to the cell surface results in the anterograde movement of lysosomes that are within a 12 μm radius of the parasite attachment site (Rodríguez et al., 1996). Lysosomes beyond this radius continue to move randomly. These results suggest that lysosomes recruited by trypomastigotes use the microtubule system for movement to the periphery of the cell. In support of this idea, the microtubule-depolymerizing agents, colchicine and vinblastine, had a dose-dependent inhibitory effect on *T. cruzi* invasion when used to pre-treat host cells (Rodríguez et al., 1996). Further support of a microtubule-based mechanism for lysosome recruitment comes from experiments in which microinjection of antibodies to the microtubule motor protein, kinesin, into NRK cells blocked both the acidification-induced, microtubule-dependent redistribution of lysosomes to the host cell periphery, and reduced trypomastigote entry (Rodríguez et al., 1996). The microtubule-dependence of *T. cruzi* invasion is likely to explain the observation made nearly two decades ago that mitotic cells, in which a dramatic reorganization of microtubules and organelles occurs (Moskalewski and Thyberg, 1990), are quite refractory to invasion by *T. cruzi* (Dvorak, 1981).

## *T. CRUZI*-TRIGGERED HOST CELL SIGNALLING REGULATES ENTRY

To understand how *T. cruzi* trypomastigotes communicate signals to mammalian host cells that culminate in the localized clustering and fusion of host cell lysosomes with the plasma membrane, we examined possible parasite-induced host cell signalling pathways. Since exocytosis and membrane fusion events are regulated by increases in cytosolic free calcium concentration $[Ca^{2+}]_i$ (Burgoyne and Morgan, 1993; Burgoyne and Morgan, 1998) and cAMP (Muniz et al., 1996), we investigated the role of these second messengers in host cell invasion by *T. cruzi*.

### *T. cruzi*-triggered $Ca^{2+}$-transients are required for host cell invasion

Time-lapse fluorescence microscopy was performed to examine the effect of *T. cruzi* on $[Ca^{2+}]_i$ of mammalian host cells (Tardieux et al., 1994). In these experiments, live *T. cruzi* trypomastigotes were incubated with NRK fibroblasts loaded with a $Ca^{2+}$-sensitive dye (Fluo3-AM). *T. cruzi*-host cell interactions result in the induction of repetitive $[Ca^{2+}]_i$-transients in host cells (Tardieux et al., 1994). In addition, contact-dependent $[Ca^{2+}]_i$ elevations have been observed in *T. cruzi* trypomastigotes upon interaction with mammalian host cells (Moreno et al., 1994). These observations imply that a system of reciprocal communication exists between *T. cruzi* and host cells (Burleigh and Andrews, 1998). The ability of *T. cruzi* to trigger signalling pathways in mammalian host cells is restricted to the infective trypomastigote forms and is critical for the invasion process. For example, if $Ca^{2+}$-transient formation in host cells was prevented by pre-treatment with a membrane permeant, calcium chelating agent, MAPTA-AM, or by first depleting host cell intracellular $Ca^{2+}$ stores with thapsigargin (Tardieux et al., 1994; Rodríguez et al., 1995), *T. cruzi* invasion was significantly impaired. Thus, these studies provided the first indication that *T. cruzi* trypomastigotes might produce an agonist, which promotes the mobilization of $Ca^{2+}$ from host cell intracellular stores as an essential early essential step in the invasion process.

### *T. cruzi* oligopeptidase B-dependent host cell signalling

How does *T. cruzi* trigger $Ca^{2+}$-transients in mammalian cells? We have shown that *T. cruzi* trypomastigotes, but not the non-infective epimastigotes, contains an activity in soluble extract that is able to trigger $Ca^{2+}$-transients in a wide variety of mammalian cells (Burleigh and Andrews, 1995). In addition to this soluble $Ca^{2+}$-signaling activity, a GPI-anchored surface membrane protein from metacyclic trypomastigotes (gp82) was shown to trigger $Ca^{2+}$-transients in mammalian cells (Dorta et al., 1995). These two trypomastigote $Ca^{2+}$-signaling activities are clearly distinct, since preparation of the soluble activity involves removal of membranes and concanavalin A-binding proteins (Burleigh and Andrews, 1995).

Soluble trypomastigote extracts were shown to activate a pathway in mammalian cells that leads to the mobilization of $Ca^{2+}$ from intracellular stores. This pathway is dependent upon the activation of phospholipase C and generation of $IP_3$ (Rodríguez et al., 1995). The participation of a heterotrimeric G-protein coupled plasma membrane receptor has been postulated due to the sensitivity of this pathway to pertussis toxin (Tardieux et al. 1994) and to mild trypsinization of host cells (Leite et al., 1998). In addition, heterologous expression of the mammalian $Ca^{2+}$-signaling response in *Xenopus* oocytes indicates that surface expression of the putative heterologous receptor confers sensitivity of the oocytes to the trypomastigote $Ca^{2+}$-signaling activity (Leite et al., 1998).

To our surprise, the soluble trypomastigote $Ca^{2+}$-signaling activity was found to be dependent on the activity of a *T. cruzi* peptidase, oligopeptidase B (Burleigh and Andrews, 1995; Burleigh et al., 1997). This was originally demonstrated with the use of non-specific protease inhibitors that targeted oligopeptidase B and abolished the $Ca^{2+}$-signaling activity (Burleigh and Andrews, 1995). Subsequently, a role for this peptidase was directly demonstrated using blocking antibodies produced against oligopeptidase B that also inhibited the soluble trypomastigote $Ca^{2+}$-signaling activity (Burleigh et al., 1997). Oligopeptidase B is a cytosolic enzyme expressed in all life cycle stages of *T. cruzi* (Burleigh et al., 1997). It is not secreted from trypomastigotes nor is it capable of triggering $Ca^{2+}$-transients in mammalian cells on its own. Therefore, we postulated that oligopeptidase B exerts its function in the $Ca^{2+}$-signaling pathway as a processing enzyme, which generates an active $Ca^{2+}$-agonist from a precursor molecule (Burleigh et al., 1997). While the putative $Ca^{2+}$-agonist has been elusive, we have been successful in directly demonstrating a role for oligopeptidase B in host cell signalling and invasion. *T. cruzi* oligopeptidase B knockout parasites, generated by targeted gene replacement, exhibited a dramatic decrease in infectivity of cultured mammalian cells and mice (Caler et al., 1998). Evidence to link oligopeptidase B with the $Ca^{2+}$-signaling pathway generated in mammalian cells by live trypomastigotes resulted from the observation that the residual invasion by the oligopeptidase B null mutants is no longer sensitive to thapsigargin pre-treatment of host cells. These results strongly suggest that the cytosolic enzyme oligopeptidase B produces a soluble $Ca^{2+}$-agonist, which is released by mammalian infective forms of *T. cruzi* and interacts with host cell receptors to trigger the mobilization of $Ca^{2+}$ from thapsigargin-sensitive intracellular stores (Caler et al., 1998). A summary of these signalling events is illustrated as a model in Figure 3.

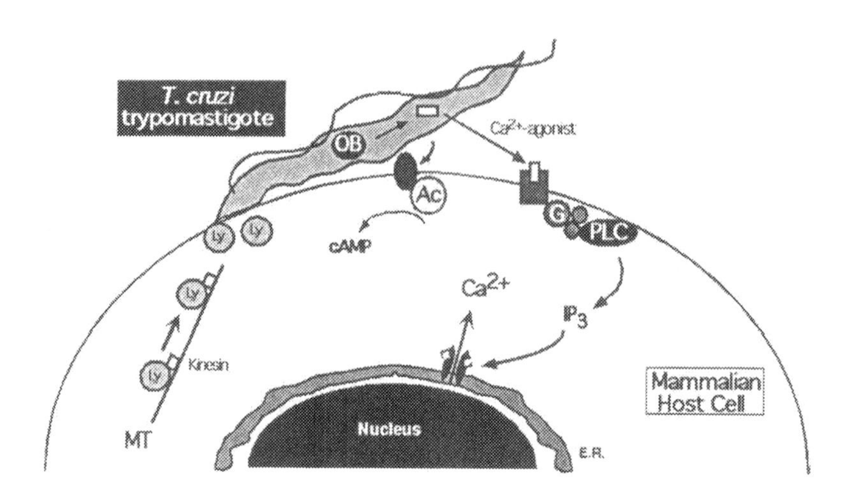


**Figure 3**. Model of host cell signalling by *T. cruzi* trypomastigotes and induction of localized lysosome exocytosis (Adapted from Caler et al., 1998). Intracellular processing of trypomastigote precursor by the cytosolic *T. cruzi* enzyme, oligopeptidase B (OB), leads to the generation of a soluble $Ca^{2+}$-agonist. Agonist binding to a heterotrimeric G-protein coupled host cell surface receptor activates phospholipase C (PLC) which leads to the generation of inositol 1,4,5-triphosphate ($IP_3$) and $Ca^{2+}$ mobilization from intracellular stores [e.g. endoplasmic reticulum (E.R.)]. Elevated levels of cyclic AMP (cAMP) follow trypomastigote-induced activation of adenylyl cyclase (Ac). Lysosome exocytosis is regulated by $Ca^{2+}$ and cAMP, where movement along microtubules (MT) toward the cell periphery is dependent on the MT motor protein, kinesin.

## LYSOSOME EXOCYTOSIS: A COMMON CELLULAR PATHWAY

In order to invade non-phagocytic mammalian cells, *T. cruzi* trypomastigotes clearly require directed fusion of host cell lysosomes with the plasma membrane at the parasite attachment site. Localized fusion of lysosomes with the plasma membrane is reminiscent of regulated exocytosis in specialized secretory cells. Our studies therefore suggest that *T. cruzi* takes advantage of an existing lysosome secretory pathway in non-specialized mammalian cells. To further investigate this putative exocytic pathway, it was desirable to set up an assay in which lysosome exocytosis could be promoted in the absence of the parasite. Rodríguez et al. demonstrated that it was possible to induce exocytosis of lysosomes in mammalian cells and monitored this event by measuring the secretion of ß-hexosaminidase (see Methods section), the lysosomally-processed form of cathepsin D, and fluid phase endocytic tracers preloaded into lysosomes (Rodríguez et al., 1997). Extracellular release of lysosomal contents coincided with the appearance of the lumenal domain of lgp120 on the plasma membrane (Rodríguez et al., 1997). The existence of the lysosome exocytosis pathway has now been demonstrated in several non-phagocytic cell types: fibroblasts, epithelial cells and myoblasts (Rodríguez et al., 1997). It is likely that this is a common pathway in mammalian cells and may provide a partial explanation for the ability of *T. cruzi* to invade a wide variety of cell types.

**$Ca^{2+}$ and cAMP regulate lysosome exocytosis and *T. cruzi* invasion**

Lysosome-plasma membrane fusion was triggered in mammalian cell lines by increasing the cytosolic free calcium concentration $[Ca^{2+}]_i$. This was achieved with the calcium ionophore, ionomycin, and using streptolysin O (SLO)-permeabilized cells with $Ca^{2+}$-containing buffers (Rodríguez et al., 1997). This process was shown to require micromolar levels of $Ca^{2+}$ and to be both temperature and ATP-dependent (Rodríguez et al., 1997). It was recently demonstrated that activation of adenylyl cyclase and elevation of cAMP in host cells potentiates the $Ca^{2+}$-dependent lysosome exocytosis pathway, however cAMP itself is insufficient to trigger the fusion of lysosomes (Rodríguez et al., 1999). Predictably, elevated levels of cAMP have been observed in mammalian cells following incubation with *T. cruzi* trypomastigotes (Rodríguez et al., 1999). The components of this *T. cruzi*-triggered cAMP-dependent pathway have not yet been defined.

**Lysosome exocytosis and the cytoskeleton**

Depolymerization of actin microfilaments with cytochalasin D or latranculin A enhanced $Ca^{2+}$-regulated lysosome exocytosis in SLO-permeabilized cells (Rodríguez et al., 1999). This result is consistent with the following observations: 1) *T. cruzi*-induced $Ca^{2+}$ transients promote the transient rearrangement of the cortical actin cytoskeleton and are required for host cell entry (Rodríguez et al., 1995); and 2) lysosome-mediated entry of *T. cruzi* is enhanced by pre-treatment of cells with actin depolymerizing drugs (Tardieux et al., 1992). Depolymerization of the cortical actin cytoskeleton is thought to provide better access of lysosomes to the plasma membrane for fusion. While host cell microtubules and the microtubule motor protein, kinesin, play important roles in lysosome-mediated *T. cruzi* invasion (Rodríguez et al., 1996), the inhibitory effect of microtubule depolymerizing agents on lysosome exocytosis in permeabilized cells is far less pronounced (Martinez, Rodríguez and Andrews, unpublished observations). This apparent discrepancy may reflect the fusion of a population of peripheral lysosomes in SLO-permeabilized cells that may not require transportation along intact microtubules.

## WHY LYSOSOMES?

Why do *T. cruzi* trypomastigotes recruit host cell lysosomes to the cell surface for invasion? Several studies suggest that transient exposure of *T. cruzi* trypomastigotes to a low pH environment in the early stages of host cell invasion may be necessary to activate parasite molecules and signalling pathways required for establishment of the intracellular cycle. Following internalisation, trypomastigotes reside within a lysosome-derived vacuole for a short period of time (20 to 60 min) before escaping to the cytosol and multiplying as amastigotes. To aide in the disruption of the parasitophorous vacuole, *T. cruzi* secretes a hemolysin into the vacuolar space, which is optimally active at pH 5.5 (Andrews and Whitlow, 1989; Andrews et al., 1990), the pH of mature lysosomes. Pre-treatment of host cells with agents that prevent acidification of the vacuole also inhibit exit of *T. cruzi* to the cytosol (Ley et al., 1990). The *T. cruzi* ecto-enzymes, trans-sialidase/neuraminidase (Schenkman et al., 1992) and protein tyrosine phosphatase (Furuya et al., 1998), are active in acidic pH ranges and both have been implicated in host cell invasion (Schenkman et al., 1991; Zhong et al., 1998). Evidence suggests that desialylation of sialic acid-rich lysosomal membrane proteins by *T. cruzi* neuraminidase renders the vacuole membrane more susceptible to lysis by secreted hemolysin and thus facilitates escape from the parasitophorous vacuole (Hall et al. 1992). Finally, an important step for continuity of the parasite life cycle is the differentiation of non-dividing trypomastigotes to the intracellular replicative form, the amastigote. Studies carried out in vitro demonstrated that trypomastigote-amastigote differentiation was accelerated by exposure to low pH (Tomlinson et al.,

1995), a process that was recently shown to involve protein serine/threonine phosphatases (Grellier et al., 1999).

While pH-dependent processes appear to be important for the establishment of the *T. cruzi* intracellular cycle, this could be achieved following phagolysosomal fusion in phagocytic cells. *T. cruzi* readily infects phagocytic cells where it is able to complete the intracellular cycle. Thus, the fact that *T. cruzi* has acquired the ability to harness the lysosome exocytosis pathway in non-phagocytic cells, a unique mechanism for pathogen entry, underscores the importance of these cells in the establishment of infections in vivo. Certainly, the persistence of *T. cruzi* in cardiac and smooth muscle of infected hosts leads to the inflammatory reactions observed in acute and chronic phases of Chagas' disease.

## CONCLUSIONS

The studies described in this chapter, while focused on the lysosome-dependent mechanism of *T. cruzi* invasion of mammalian cells, highlight an important feature of research in the area of host-pathogen interactions. By examining the mechanism of host cell entry by *T. cruzi*, we have uncovered the existence of a basic secretory pathway in mammalian cells. Prior to these studies, the capacity for lysosome exocytosis in non-specialized mammalian cells was not appreciated, although secretion of lysosomal contents by some specialized cells had been documented (LeSage et al., 1993; Silverstein and Febbraio, 1992; Hirano et al., 1991; Tapper and Sundler, 1990; Baron et al., 1988). Therefore, a careful examination of the molecular and cellular mechanisms directing entry and survival of intracellular pathogens in mammalian host cells, will not only help us to uncover critical pathways that can be exploited for disease intervention, these studies have great potential to enhance our knowledge of basic mechanisms in eukaryotic cell biology.

## METHODS

### Infection of mammalian cells with *T. cruzi* trypomastigotes

Grow mammalian cells on 12-mm cover slips. Seed cells in a 6-cm dish at a density of 7.5x$10^4$ cells/ml (4 l total), 48 hr in advance.

Wash freshly harvested trypomastigotes twice in ice-cold Ringers/BSA (10 mM HEPES pH 7.2, 155 mM NaCl, 5 mM KCl, 2 mM $NaH_2PO_4$, 2 mM $CaCl_2$, 1 mM $MgCl_2$, 10 mM glucose, 0.5 mg/ml BSA; Heuser, 1989). Resuspend parasites at a final concentration of 5x$10^6$ to 5x$10^7$/ml (this will vary depending on host cell used) in warm Ringers/BSA.

Set up 3 cover slips / 2 $cm^2$ dish for each infection condition.

If drug treatments are to be carried out, pre-treat host cells by adding appropriately diluted drug (usually in PBS) to desired final concentration. Incubate at 37°C for required time.

Remove drug from cells with two rapid washes in Ringers/BSA and replace with 2 ml diluted trypomastigotes. Incubate at 37°C for 15 min.

Stop infection by washing dishes 5 times with ice-cold $PBS^{++}$. Fix infected cells with 2% PFA/PBS at room temperature for 30 min.

### Immunofluorescence to quantitate parasite invasion*

This assay exploits the use of antibodies to *T. cruzi* surface antigens to stain extracellular parasites in non-permeabilized cells. The total number of parasites (intracellular and extracellular) and host cells are detected simultaneously by staining with 4,6-diamidino-2 phenylindole (DAPI) to visualize nuclear DNA (as well as kinetoplast DNA of *T. cruzi*). Fluorescence microscopy is used to enumerate the total number of parasites and host cells as well as the extracellular trypomastigotes. To calculate the number of intracellular parasites, subtract the number of extracellular parasites counted from the total number (DAPI-stained) per 200 to 500 host cells. Assays are carried out in triplicate.

Remove fixative and wash cover slips 3 times with PBS and quench unreacted aldehyde groups with 50 mM $NH_4Cl$/PBS for 15 min.

Block for 10 min with TBS/BSA (50 mM Tris pH 7.4, 150 mM NaCl, 1% BSA).

Dilute rabbit anti-*T. cruzi* antibody in TBS/BSA and incubate cover slips in 50 μl drops of diluted antibody on parafilm for 40 min at room temperature.

Wash cover slips in 200 μl drops of TBS/BSA: 5 washes of 5 min each.

Dilute fluorescence conjugated anti-rabbit IgG in TBS/BSA and incubate in 50 µl drops of diluted antibody for 40 min at room temperature.
Wash cover slips in 200 µl drops of TBS/BSA: 5 washes of 5 min each.
Dilute DAPI in TBS/BSA to 1 ng/ml and incubate cover slips in 50 µl drops for 2 min.
Wash 3 times for 1 min in TBS/BSA. Wash 3 times for 1 min in PBS.
Mount cover slips in anti-fade mounting medium.

**(For an alternative method to quantitate parasite invasion, see Yan and Moreno, 1998.)*

**Lysosome exocytosis assays**
Prepare the following buffers:

| **Buffer A**[a] | | **Buffer B**[b] | |
|---|---|---|---|
| Component | Concentration | Component | Concentration |
| HEPES | 20 mM | HEPES | 20 mM |
| $Na_2HPO_4$ | 0.9 mM | $K^+$-glutamate | 100 mM |
| NaCl | 110 mM | KCl | 40 mM |
| KCl | 5.4 mM | EGTA | 5 mM |
| $MgCl_2$ | 10 mM | | |
| $CaCl_2$ | 2 mM | | |
| Glucose | 11 mM | | |

[a]pH to 7.2 with 1 M NaOH [b]pH to 7.2 with 1 M KOH

ß-hexosaminidase substrate buffer

88 mM $Na_2HPO_4$
60 mM citric acid

prepare a 2.3 mg/ml solution of 4-methyl-umbelliferyl-N-acetyl-ß-D-glucosaminide in Na citrate-$PO_4$ buffer (pH 4.5) containing 0.1% TX-100.

Permeabilization buffer

1. Prepare stock of Streptolysin O (4U/ml): dilute bottle of 40 units SLO with 10 ml of Buffer A. To dissolve, warm briefly to 37°C. Quickly cool on ice and freeze in 750 µl aliquots.
2. Dilute SLO stock solution to a final concentration of 0.5 U/ml Buffer A (e.g. 750 µl of SLO in 6 ml Buffer A for 12 dishes). ALWAYS KEEP ON ICE.

$Ca^{2+}$-loading buffer

1. B-ATP: dissolve 60 mg of ATP in 50 ml Buffer B.
2. B-ATP-Ca: (Solution X): prepare 100 mM $MgCl_2$ in B-ATP.
   (Solution Y): prepare 100 mM $CaCl_2$ in B-ATP.
   For a solution containing 1 μM $Ca^{2+}$, mix 750 μl of solution X with 405 μl of solution Y and add to 8.8 ml of B-ATP.
3. B-ATP-Mg: dilute 960 μl of solution X in 10 ml Buffer B-ATP.

Procedure:

Plate mammalian cells 24 hr prior to experiment in 60-mm dishes at appropriate density to obtain confluent monolayers (approximately $8.8x10^5$ cells/60-mm dish).

Place plates on an ice platform. Wash twice with ice-cold Buffer A. Add 600 μl SLO diluted in Buffer A and incubate on ice for 6 min.

Wash once (gently) with 4 ml cold Buffer B. Add 600 μl pre-warmed Buffer B-ATP-Mg (for background level of exocytosis) or Buffer B-ATP-Ca (to stimulate exocytosis) and incubate at 37°C for 10 min (or time points up to 10 min). Set up triplicates for each condition.

Collect supernatant and lyse cells in 600 μl Buffer B/1% NP-40 for 10 min. Centrifuge all samples for 5 min at top speed in microfuge. Store on ice.

**ß-Hexosaminidase assay**

Place 350 μl of supernatant fraction (sup) in fresh microfuge tube. For the detergent soluble extracts, add 35 μl to 315 μl Buffer B-ATP.

Add 50 μl of ß-hexosaminidase substrate to supernatant or detergent soluble fraction. Incubate at 37°C for 15 min. Stop reaction by adding 100 μl stop buffer (200 mM $Na_2CO_3$, 110 mM glycine).

Measure fluorescence at Ex. = 365nm; Em. =450nm.

Calculate % ß-hexosaminidase released [sup/(sup + detergent extract x 10)].

## ACKNOWLEDGEMENTS

The contribution of unpublished micrographs by Drs. Edith S. Robbins and Elisabet V. Caler is greatly appreciated. I would also like to thank Dr. Norma W. Andrews for critical reading of the chapter.

## REFERENCES

Andrews, N.W., Abrams, C.K., Slatin, S.L. and Griffiths, G. (1990). A *T. cruzi*-secreted protein immunologically related to the complement component C9: evidence for membrane pore-forming activity at low pH. Cell *61,* 1277-87.

Andrews, N.W. and Whitlow, M.B. (1989). Secretion by *Trypanosoma cruzi* of a hemolysin active at low pH. Mol. Biochem. Parasitol. *33,* 249-256.

Baron, R., Neffe, L., Brow, W., Courtoy, P.J., Louvard, D. and Farquhar, M.G. (1988). Polarized secretion of lysosomal enzymes: co-distribution of cation-independent mannose-6-phosphate receptors and lysosomal enzymes along the osteoclast exocytic pathway. J. Cell Biol. *106,* 1863-1872.

Burgoyne, R.D. and Morgan, A. (1993). Regulated exocytosis. Biochem. J. *293,* 305-316.

Burgoyne, R.D. and Morgan, A. (1998). Calcium sensors in regulated exocytosis. Cell Calcium *24,* 367-376.

Burleigh, B. and Andrews, N.W. (1995). A 120 kDa alkaline peptidase from *Trypanosoma cruzi* is involved in the generation of a novel $Ca^{2+}$ signaling factor for mammalian cells. J. Biol. Chem. *270,* 5172-5180.

Burleigh, B.A. and Andrews, N.W. (1998). Signaling and host cell invasion by *Trypanosoma cruzi.* Curr. Opin. Microbiol. *1,* 461-465.

Burleigh, B.A., Caler, E.V., Webster, P. and Andrews, N.W. (1997). A cytosolic serine endopeptidase from *Trypanosoma cruzi* is required for the generation of $Ca^{2+}$ signaling in mammalian cells. J. Cell Biol. *136,* 609-620.

Caler, E.V., Vaena, S., Haynes, P.A., Andrews, N.W. and Burleigh, B.A. (1998). Oligopeptidase B-dependent signaling mediates host cell invasion by *Trypanosoma cruzi.* EMBO J. *17,* 4975-4986.

Carvalho, T.M.V. and de Souza, W. (1989). Early events related with the behavior of *Trypanosoma cruzi* within an endocytic vacuole in mouse peritoneal macrophages. Cell Struct. Funct. *14,* 383-392.

de Meirelles, M.N., de Araujo Jorge, T.C., de Souza, W., Moreira, A.L. and Barbosa, H.S. (1987). *Trypanosoma cruzi:* phagolysosomal fusion after invasion into non-professional phagocytic cells. Cell Struct. Funct. *12,* 387-393.

Dorta, M.L., Ferreira, A.T., Oshiro, M.E.M. and Yoshida, N. (1995). $Ca^{2+}$ signal induced by *Trypanosoma cruzi* metacyclic trypomastigote surface molecules implicated in mammalian cell invasion. Mol. Biochem. Parasitol. *73,* 285-289.

Dvorak, J.A. and Crane, M.S. (1981). Vertebrate cell cycle modulates infection by protozoan parasites. Science *214,* 1034-1036.

Furuya, T., Zhong, L., Meyer-Fernandes, J.R., Lu, H.G., Moreno, S.N. and Docampo, R. (1998). Ecto-protein tyrosine phosphatase activity in *Trypanosoma cruzi* infective stages. Mol. Biochem. Parasitol. *92,* 339-348.

Grellier, P., Blum, J., Santana, J., Bylen, E., Mouray, E., Sinou, V., Teixeira, A.R. and Schrevel, J. (1999). Involvement of calyculin A-sensitive phosphatase(s) in the differentiation of *Trypanosoma cruzi* trypomastigotes to amastigotes. Mol. Biochem. Parasitol. *98,* 239-252.

Hall, B.F., Webster, P., Ma, A.K., Joiner, K.A. and Andrews, N.W. (1992). Desialylation of lysosomal membrane glycoproteins by *Trypanosoma cruzi*: a role for the surface neuraminidase in facilitating parasite entry into the host cell cytoplasm. J. Exp. Med. *176,* 313-325.

Heuser, J.E. (1989). Changes in lysosome shape and distribution correlated with changes in cytoplasmic pH. J. Cell Biol. *108,* 855-864.

Hirano, T., Saluja, A., Ramarao, P., Lerch, M.M., Saluja, M. and Steer, M.L. (1991). Apical secretion of lysosomal enzymes in rabbit pancreas occurs via a secretagogue regulated pathway and is increased after pancreatic duct obstruction. J. Clin. Invest. *87,* 865-869.

Kipnis, T.L., Calich, V.L.G. and Dias da Silva, W. (1979). Active entry of bloodstream forms of Trypanosoma cruzi into macrophages. Parasitol. *78,* 89-99.

Leite, M.F., Moyer, M.S., Nathanson, M.H. and Andrews, N.W. (1998). Expression of the mammalian calcium signaling response to *Trypanosoma cruzi* in *Xenopus laevis* oocytes. Mol. Biochem. Parasitol. *92,* 1-13.

LeSage, G.D., Robertson, W.E. and Baumgart, M.A. (1993). Bile acid-dependent vesicular transport of lysosomal enzymes into bile in the rat. Gastroenterology *105,* 889-900.

Ley, V., Robbins, E.S., Nessenzweig, V. and Andrews, N.W. (1990). The exit of *Trypanosoma cruzi* from the phagosome is inhibited by raising the pH of acidic compartments. J. Exp. Med. *171,* 401-413.

Meirelles, M.N.L., Araujo Jorge, T.C. and de Souza, W. (1982). Interaction of *Trypanosoma cruzi* with macrophages in vitro: dissociation of the attachment and internalization phases by low temperature and cytochalasin B. Z. Prasitenkd *68,* 7-14.

Meirelles, M.N.L. and de Souza, W. (1983). Interaction of lysosomes with endocytic vacuoles in macrophages simultaneously infected with *Trypanosoma cruzi* and *Toxoplasma gondii*. J. Submicrosc. Cytol. Pathol. *17,* 327-334.

Milder, R. and Kloetzel, J. (1980). The development of *Trypanosoma cruzi* in macrophages *in vitro*. Interaction with host cell lysosomes and host cell fate. Parasitology *80,* 139-145.

Montgomery, R., Webster, P. and Mellman, I. (1991). Accumulation of undigestible substances reduces fusion competence of macrophage lysosomes. J. Immunol. *147,* 3087-3093.

Moreno, S.N.J., Silva, J., Vercesi, A.E. and Docampo, R. (1994). Cytosolic free calcium elevation in *Trypanosoma cruzi* is required for cell invasion. J. Exp. Med. *180,* 1535-1540.

Moskalewski, S. and Thyberg, J. (1990). Disorganization and reorganization of the Golgi complex and the lysosomal system in association with mitosis. J. Submicrosc. Cytol. Pathol. *22,* 159-171.

Muniz, M., Alonso, M., Hidalgo, J. and Velasco, A. (1996). A regulatory role for cAMP-dependent protein kinase in protein traffic along the exocytic route. J. Biol. Chem. *271,* 30935-30941.

Nogueira, N. and Cohn, Z. (1976). *Trypanosoma cruzi*: mechanism of entry and intracellular fate in mammalian cells. J. Exp. Med. *143,* 1402-1420.

Rodríguez, A., Martinez, I., Chung, A., Berlot, C.H. and Andrews, N.W. (1999). cAMP regulates $Ca^{2+}$-dependent exocytosis of lysosomes and lysosome-mediated cell invasion by trypanosomes. J. Biol. Chem. *274,* 16754-16759.

Rodríguez, A., Rioult, M.G., Ora, A. and Andrews, N.W. (1995). A trypanosome-soluble factor induces IP3 formation, intracellular $Ca^{2+}$ mobilization and microfilament rearrangement in host cells. J. Cell Biol. *129,* 1263-1273.

Rodríguez, A., Samoff, E., Rioult, M.G., Chung, A. and Andrews, N.W. (1996). Host cell invasion by trypanosomes requires lysosomes and microtubule/kinesin-mediated transport. J. Cell Biol. *134,* 349-362.

Rodríguez, A., Webster, P., Ortego, J. and Andrews, N.W. (1997). Lysosomes behave as $Ca^{2+}$-regulated exocytic vesicles in fibroblasts and epithelial cells. J. Cell Biol. *in press.*

Ruiz, R.C., Favoreto, S., Dorta, M.L., Oshiro, M.E.M., Ferreira, A.T. and Manque, P.M. (1998). Infectivity of *Trypanosoma cruzi* strains is associated with differential expression of surface glycoproteins with differential $Ca^{2+}$ signaling activity. Biochem. J. *330,* 505-511.

Schenkman, S., Andrews, N.W., Nussenzweig, V. and Robbins, E.S. (1988). *Trypanosoma cruzi* invade a mammmalian epithelial cell in a polarized manner. Cell *55,* 157-165.

Schenkman, S., Jiang, M.S., Hart, G.W. and Nussenzweig, V. (1991). A novel cell surface trans-sialidase of *Trypanosoma cruzi* generates a stage-specific epitope required for invasion of mammalian cells. Cell *65,* 1117-1125.

Schenkman, S., Pontes, de, Carvalho, L and Nussenzweig, V. (1992). *Trypanosoma cruzi* trans-sialidase and neuraminidase activities can be mediated by the same enzymes. J. Exp. Med. *175,* 567-75.

Schenkman, S., Robbins, E.S. and Nussenzweig, V. (1991). Attachment of *Trypanosoma cruzi* to mammalian cells requires parasite energy, and invasion can be independent of the target cell cytoskeleton. Infect. Immun. *59,* 645-654.

Silverstein, R.L. and Febbraio, M. (1992). Identification of lysosome-associated membrane protein-2 as an activation-dependent platelet surface glycoprotein. Blood *80,* 1470-1475.

Tapper, H. and Sundler, R. (1990). Role of lysosomal and cytosolic pH in the regulation of macrophage lysossomal enzyme secretion. Biochem. J. *272,* 407-414.

Tardieux, I., Nathanson, M.H. and Andrews, N.W. (1994). Role in host cell invasion of *Trypanosoma cruzi*-induced cytosolic free $Ca^{2+}$ transients. J. Exp. Med. *179,* 1017-1022.

Tardieux, I., Webster, P., Ravesloot, J., Boron, W., Lunn, J.A., Henser, J.E. and Andrews, N.W. (1992). Lysosome recruitment and fusion are early events required for trypanosome invasion of mammalian cells. Cell *71,* 1117-1130.

Tomlinson, S., Vandekerckhove, F., Frevert, U. and Nussenzweig, V. (1995). The induction of *Trypanosoma cruzi* trypomastigote to amastigote transformation by low pH. Parasitology *110,* 547-554.

Yan, W. and Moreno, S.N. (1998). A method to assess invasion and intracellular replication of *Trypanosoma cruzi* based on differential uracil incorporation. J. Immunol. Meth. *220,* 123-128.

Zhong, L., Lu, H.G., Moreno, S.N. and Docampo, R. (1998). Tyrosine phosphate hydrolysis of host proteins by *Trypanosoma cruzi* is linked to cell invasion. FEMS Microbiol. Lett. *161,* 15-20.

# 10

# CELL BIOLOGICAL APPROACHES TO THE STUDY OF INTRACELLULAR PATHOGENS: MOTILITY, INVASION, SECRETION AND VESICULAR TRAFFICKING

David G. Russell, Dana G. Mordue, Wandy Beatty, Olivia K. Giddings, Jennie L. Lovett, Andreas Lingnau, Maren Lingnau, Jaime Dant and L. David Sibley

*Department of Molecular Microbiology, Washington University School of Medicine, St. Louis, MO 63110.*

## OVERVIEW AND BACKGROUND INFORMATION

Gaining an appreciation of the interplay between intracellular pathogens and their host cells requires the exploitation of an ever-broadening range of techniques. These techniques are bundled under the blanket term "Cell Biology", but are in reality a bastardized mixture of biochemistry, molecular biology and physiology. This chapter describes general protocols and specific experiments that have been developed and applied to several different intracellular pathogens.

The pathogens that we use in the course include the apicomplexan *Toxoplasma gondii*, the trypanosomatid *Leishmania mexicana*, and the bacterium *Mycobacterium avium*. The reason for this selection (and the inclusion of a prokaryote) is that they represent the broad range on differing styles of intracellular infection, see (Russell, 2000). *Toxoplasma* is an active invader of mammalian cells that builds its own unique vacuole sequestered outwith the normal trafficking machinery of the host cell. In contrast, both *Leishmania* and *Mycobacterium* are phagocytosed by their host macrophage and remain within the endosomal/lysosomal continuum of the cell. *Leishmania* survives and replicates in an acidic, hydrolytically-competent lysosome-like compartment, while *Mycobacterium* arrests the normal progression of their phagosome and is retained within the early, or recycling endosomal system of the cell.

The techniques discussed in this chapter cover maintenance of parasites and host cells, the use and analysis of fluorescent tracers and antibodies against intracellular markers, motility, secretion and invasion assays for *Toxoplasma*, and different cell fractionation protocols for the isolation and analysis of intracellular vesicles and phagosomes. The techniques are valid for a range of probes and markers but only a limited range of these applications are explored in depth in this chapter.

**Isolation and maintenance of host cells**

**Bone marrow-derived macrophages**

Reagents:

L-cell-conditioned medium: Contains the growth factors monocyte colony stimulating factor (M-CSF) and granulocyte/monocyte colony stimulating factor (GM-CSF): L929 fibroblasts are grown in DMEM + 10% fetal calf serum in T160 flasks until the monolayer starts to break down. The medium is harvested, pooled and filtered through a 0.2 μm pore media sterilization filter. It is then aliquoted into 50 ml sterile screw cap centrifuge tubes (45 mls in each) and frozen at -20$^{0}$C.

Bone marrow-derived macrophage (BMMO) medium: is prepared from DMEM + 10% fetal calf serum, 5% horse serum, and 10 to 20% L-cell-conditioned medium.

Procedure:

1. Following euthanization, trim the fur from the two hind legs of the mouse. Remove the foot from one leg and peel the muscle off the tibia. Cut the tibia just below the "knee" and place in medium on ice. Repeat procedure with the femur, cutting just below the "hip" joint. Do the other leg, and additional mice as needed.
2. Place the bones in a Petri dish. Ensure that virtually all tissue has been removed from the bones, then cut off the unbroken end of the bone.

Using a 5 ml syringe with a 23 gauge needle, flush the bone with BMMO medium to wash out the marrow plug.

3. Once all the bones have been flushed through take a 10 or 20 ml syringe with an 18 gauge needle and gently aspirate the cell suspension until all clumps have been disrupted.
4. Count the cells and dilute to 4 x $10^5$ per ml in BMMO medium and place in 10 ml aliquots in 10 cm diameter sterile microbiological plasticware Petri dishes. We usually use Ultradish Petri dishes from Valmark.
5. Macrophages are left to differentiate and fed on either day 4 or 5 by addition of a further 10 ml of BMMO medium.
6. Once the Petri dishes are confluent (approximately 6 to 7 days) the macrophages are harvested by removal of the medium and addition of 10 mls ice cold sterile PBS. The dish is placed at $4^0$C for 10 to 20 min, and the cells removed by gently scraping (single passes across the dish) with a Sarstedt cell scraper. The cells are counted and plated in tissue culture flasks at the required density.

**Fibroblasts culture and reagents**

Reagents:

Human foreskin fibroblasts cells (HFF): are grown in DMEM containing 10% fetal bovine serum (FBS), 10 mM HEPES, 2 mM L-glutamine, and 10 µg/ml gentamicin (referred to as D10). When confluent, they are ready to use for *Toxoplasma* infection. The main advantage of HFF cells is that they are contact inhibited allowing for synchronous infection and efficient lysis of the monolayer which facilitates parasite harvest. HFF cells generally work well only for the first 25 passages. A variety of readily available mammalian cell lines also work well for this purpose including mouse 3T3 and MRC5 cells (obtained from ATCC, Rockville, MD).

Procedure:

1. Expand HFF cultures by trypsinization, seeding the appropriate size culture flasks by dilution of 1:10 to 1:20 of the original stock. Keep track of the passage number each time you split the cells.
2. Wash cells twice with 10 ml warm calcium-magnesium free (CMF) PBS to remove serum and remaining media. Dilute trypsin 1:1 with CMF-PBS. Add 2.5 ml diluted trypsin to each T75 flask. Mix gently for a few seconds to cover entire cell surface with the solution.
3. Incubate at 37°C for 2 to 10 min (time depends on confluency of cells, passage number, etc.) until cells release from the flask surface. Tapping on the bottom or side of the flask will help release any "stubborn" cells.

4. Add desired volume of warm D10 to flask and use pipette to run cells up and down flask side to break up any cell clumps. In general, a 1:10 split will reach confluency in 3 to 4 days. Resuspend the contents of 1 T75 in 10 ml, use 1 ml cells per new flask plus 4 mls of D10 medium. Incubate at 37°C in $CO_2$ incubator. Once confluent, they are good for two weeks; however, the media must be changed after one week.

**Mycoplasma testing and treatment**

Mycoplasma are common colonizers of the human upper respiratory tract where they are often commensal. Contamination with human mycoplasma is a common problem in cell culture. Such contamination can lead to significant problems in interfering with biochemical assays designed to monitor enzymatic pathways, induction of immune responses and in the generation of gene probes or DNA libraries. Fortunately contamination is easily detected and usually easily treated. Experience has taught that regular testing combined with strict quarantine of infected or newly obtained cultures is imperative to control the spread of mycoplasma between cell cultures. Along with these precautions, it is essential to use only the highest quality cell culture reagents that have been screened and certified as negative for mycoplasma contamination.

Reagents:

Mycoplasma testing kit: Is based on oligonucleotide hybridization to a conserved segment of the SSU RNA is available from GenProbe (Cat 1591, GenProbe). An effective and non-toxic treatment is available using mycoplasma removal agent (MRA) that contains 4-oxo-quinoline-3-carboxylic acid (Cat. 33-500-44, ICN Biomedical Inc.).

Procedures

**Testing**

1. Remove all antibiotics for 5 to 6 days prior to testing. HFF cells must be removed from antibiotics 2 days prior to being used for *Toxoplasma* growth. Replace the media again when you inoculate the culture with parasites.
2. The final passage should be a 3-day spent culture, reduce the inoculum accordingly (i.e. 2 drops of RH per T25). Allow the flask to fully lyse out, harvest the supernatant for testing.
3. Test for mycoplasma using the GenProbe kit.

**Treatment**

1. Pass *Toxoplasma* in T25s containing confluent monolayers of HFF cells to achieve lysis in 2 days (i.e. dilute 1:20 into new flask/plate containing drug).
2. Treat with MRA (100x stock: 50 μg/ml, protect from light!) for 6 to 8 days. This normally requires 2 to 3 successive passages where the standard dose of *Toxoplasma* is used to inoculate a new flask of HFF cells in D10 supplemented with drug.
3. After treatment, remove all antibiotics for 1 week and retest for presence of mycoplasma.

**Parasite cell cultures**

**Leishmania promastigotes and amatigote-like forms**

Reagents:

Promastigote Medium: We grow all our species and strains of *Leishmania* in a medium developed for the procyclic stages of *Trypanosoma brucei*, Semi-Defined medium '79 (SDM79) (Brun and Schonberger, 1979) at 25°C. The media is prepared from powder premixed by JRH Biosciences (Cat. 57453-10L), dissolved in double distilled water, filter sterilized and supplemented with 10% fetal calf serum, and hemin to 7.5 mg/l (from a 7.5 mg/ml stock solution in 0.05 M NaOH).

Amastigote Medium: Amastigote-like forms are cultured in modified UM54 containing 1 Packet ME199 (Hank's salts), 2.5 g glucose, 5 g Trypticase (Cat. 11921, BBL Microbiology Systems), 7.5 mg hemin (from stock detailed above), 25 mM Hepes, 0.75 g glutamine, 100 ml fetal calf serum (heat inactivated at 56°C for 4 hr), and 100 ml Serum plus. The medium is pH'ed 5.5 and filter-sterilized.

Procedure:

For transformation of promastigotes into tissue-culture amastigotes take 5 mls from a late log phase culture of *L. mexicana* promastigotes and add to 25 mls of UM54 medium. Place the culture at 32 to 33°C and leave for at least 48 hr. It should be ready for sub culture after 4 to 5 days and at that time the transformation will be complete in the entire population of parasites. (The majority of cells will have differentiated after 36 hr).

***Mycobacterium avium* bacilli**

Reagent:

*Mycobacterium* Medium: We grow all our *Mycobacterium avium* cultures in Middlebrook 7H9 medium (Difco, cat # 0713-01-7) or on Middlebrook nutrient agar plates (Cat. 0627-01-2, Difco), both supplemented with OADC (Cat. 0722-64-0, Difco).

Procedure:

*M. avium* attenuates in culture and its loss of virulence correlates with an alteration in colony formation on agar plates. To overcome this loss of virulence we isolated and plated *M. avium* from the spleen of an infected mouse, then expanded and froze aliquots of the bacteria. Fresh aliquots are thawed every two weeks. These aliquots have a >99% smooth transparent (the virulent phenotype) colony morphology.

***Toxoplasma* growth and harvest**

Parasites are maintained by serial passage in confluent monolayers of HFF cells. Use a high multiplicity of infection (MOI) to achieve complete infection of the monolayer and subsequent synchronous lysis. Experience as shown that a two day passage schedule is the most efficient at achieving high yields and high viability. This balance is important as once the parasites lyse the monolayer they begin to die with the $T_{1/2}$ of 6 hr. Lower MOIs can be used to maintain a parasite line by weekly passage; however, these cultures do not yield parasites of sufficient viability for experimentation. Both the host cell lines and the parasite-infected cell lines can be cryopreserved by routine procedures, allowing for the long-term storage of specific strains.

Reagents:

Harvesting Medium: Parasites are harvested in Hank's Balanced Salt Solution (minus $Mg^{+2}/Ca^{+2}$), 0.1 mM EGTA, and 10 mM HEPES (referred to as HHE). Parasite cells are separated from host cell debris by centrifugation and filtration through a 3 micron pore membrane (Nucleopore). Parasites are resuspended in HHE or in invasion medium (referred to as IM) consisting of MEM (low bicarbonate), 10 mM HEPES and 3% FBS.

Procedure:

1. With a sterile transfer pipette, resuspend the contents of a totally lysed flask of HFF cells that was infected two days previously.

2. Add the desired volume of inoculum from a lysed out culture to new T-25 flasks containing a confluent monolayer of HFF cells and label accordingly. For a 2 day passage (complete lysis in 36 to 48 hr), set up separate flasks with 6 drops, 8 drops, and 10 drops. To scale up, use 1.5 to 2.0 ml of lysed out parasites to inoculate fresh HFF cells grown in D150 dishes. Recap and place in 37°C $CO_2$ incubator.
3. After two days (36 to 48 hr), the monolayer should have lysed completely. Resuspend the material in the flask by pipetting and transfer to a 50 ml polystyrene tube.
4. Syringe passage using a 20 gauge needle attached to a 10 or 20 ml syringe, drawing the parasites in and forcing out three times. Be careful not to cause foaming. For difficult to lyse cells, it may help to passage the lysate sequentially through 20, 23, and 25 gauge needles.
5. Assemble the filter apparatus consisting of a 47 mm diameter Nucleopore SwinLock filter (Cat. 420400) which contains a 3 micron pore polycarbonate membrane (Cat. 111112). Place the filter over the top of a 15 ml polystyrene tube. Using the syringe barrel, inject the lysate into the top of the assembled filter and flush with 15 ml of HHE. Typically, the harvest from one D-150 can be filter-purified using a single filter.
6. Centrifuge at 1,800 rpm for 10 min and resuspend the cell pellet in HHE. Parasites may be counted using a standard hemacytometer chamber. Typical yields after purification for the RH strain are $5 \times 10^7$ parasites from a lysed out T25 and $5 \times 10^8$ from a lysed out D150. Slower growing strains, such as PLK, CEP, typically yield less and require higher inoculum to achieve confluent lysis.

## EXPERIMENTAL SECTION

### Motility and invasion by *Toxoplasma*

Like other Apicomplexa, *Toxoplasma* is actively motile and invades its host cell by direct penetration (Dobrowolski and Sibley, 1996). *Toxoplasma* gliding motility can be easily monitored by examining the trails left by parasites migrating on serum-coated slides (Håkansson et al., 1999). Previous observations have shown that these trails contain all the major surface proteins (SAGs) as well as membrane lipids (Håkansson et al., 1999). Invasion is an active process wherein the parasite attaches to the host cell and penetrates into a vacuole that forms by invagination of the host cell plasma membrane (Suss-Toby et al., 1996). The mechanism of cell penetration is similar to that used in gliding motility; however, adhesion to host cell receptors is also required for entry.

The following protocols can be used to evaluate potential inhibitors that affect *Toxoplasma* gliding motility, cell attachment and invasion. Freshly isolated parasites will be allowed to glide on bovine serum albumin (BSA)-coated glass slides and trails will be visualized by immunofluoresence (IF) staining of the major surface protein SAG1. Host cell binding will be monitored by detecting the presence of the heterologous reporter enzyme β-Galactosidase (β-Gal) expressed by transgenic parasites (Seeber and Boothroyd, 1996). Invasion will be monitored by two-color immunofluorescence staining (IFA) to distinguish intracellular from extracellular parasites.

**Gliding of Toxoplasma**

Reagents:

Antibodies: Gliding trails will be stained using the mAb DG52 to the major tachyzoite surface antigen SAG1 (Burg et al., 1988) followed by fluorescently conjugated secondary antibodies (obtained from Jackson ImmunoResearch Laboratories, Inc., West Grove, PA). Glass multi-chamber LabTek slides are obtained from Nunc.

Procedure:

1. Coat two empty glass LabTek chamber slides with 50 μg/ml BSA by adding 0.25 ml and incubating at 37°C for at least 30 min. Rinse with PBS. Add 0.25 ml of parasite suspension, containing inhibitor dilutions, to each well of the slides. Place at 37°C in the $CO_2$ incubator for 10 min. Remove the suspension from the slide by dumping the contents into a beaker.
2. Fix the slide by adding 0.5 ml 2.5% formalin in PBS for 10 min at room temperature. Rinse gently in PBS by changing 3 times 2 min.
3. Block with 5% normal goat serum (NGS), 5% FBS, 0.05% saponin for 10 min. Rinse 3 times 2 min with PBS containing 1% NGS, 0.01% saponin.
4. Add 0.25 ml of mAb anti-SAG1 (diluted 1:1000) and incubate for 10 min., rinse 3 times, followed by 3 times 5 min changes of PBS, 1% NGS, 0.01% saponin.
5. Add 0.25 ml of goat anti-IgG-FITC (diluted 1:500) and incubate for 45 min. Rinse 3 times 5 min changes of PBS, 1% NGS, 0.1% saponin.

6. Remove the wells and mount using a No. 1 glass coverslip in Prolong antifade (No. P-7481, Molecular Probes).

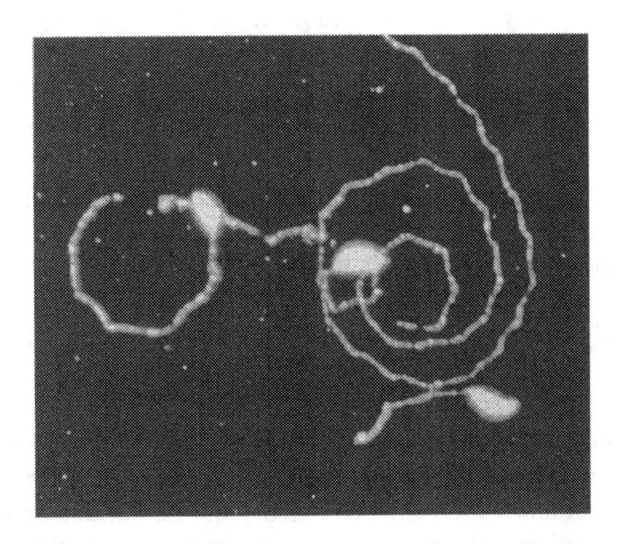

**Figure 1**. Immunofluorescence staining of trails left by gliding *Toxoplasma*. Trails were stained with mAb DG52 to the surface protein SAG1 followed by goat anti-mouse IgG conjugated to FITC.

**Analysis**

Examine by epifluorescence and record the average trail length in control and treated samples. Inhibition of gliding will be apparent as reduced or absent trails. An example of the trails typically formed during this assay is shown in Figure 1.

**Determination of extracellular versus intracellular parasites**

Parasite attachment versus invasion is most easily quantified by two-color IF staining to distinguish parasites that are still extracellular from those which are fully within the host cell. The principle of this assay is that once a parasite has fully entered the host cell, it is inaccessible to antibodies unless the cell is first permeabilized with detergent. Only extracellular parasites will be detected by the first antibody to SAG1 that is added after fixation but before permeabilization. Subsequently, the monolayer will be fully permeabilized and a second antibody added to detect all parasites. The use of two different fluorochromes allows ready identification of parasites that are inside versus outside.

Reagents:

Substrates and Antibodies: Prepare two, 4-chamber permanox LabTek chamber slides that have sub-confluent monolayers of HFF cells. These work best when plated the day before. Purify mAb to SAG1 (i.e. DG52) using protein G chromatography (Harlow and Lane, 1988). Directly conjugate the fluorochromes Oregon Green (OG) or Texas Red to the purified mAb using FluoReporter protein labeling kits for Texas Red (Cat. F-6162) and Oregon Green (Cat. F-6153) (Molecular Probes). Oregon Green is a photostable fluorochrome with a similar excitation and emission spectra as FITC. Once conjugated, the fluorescently labeled antibodies can be aliquoted and stored at -70°C.

Procedure:

1. Remove medium from wells and immediately add 0.25 ml of parasite suspension containing inhibitors or control to each well of the chamber slides. Do not allow the monolayer to dry out during this process. For efficient infection, it is important to rapidly warm the slide chamber. Submerge the chamber slide in 37°C water bath during the pulse-invasion. Incubate for 5 min or 15 min at 37°C in the $CO_2$ incubator. Remove the parasite suspension by inverting the slide and dumping the contents into the beaker in the hood. Rapidly wick away the excess liquid by touching the slide wells to a piece of bench-coat contained within a Petri plate (don't allow the slide to dry out!). Rinse 3 times with warm IM. During rinsing, add the medium gently to the side of the well.
2. Fix for 15 min in 2.5% formalin-PBS. Wash monolayers with PBS/1% NGS. Block for 15 min in 5% NGS/5% FBS in PBS. Wash with PBS/1% NGS.
3. Incubate with Texas-Red-conjugated DG52 in PBS/1% NGS for 45 min RT. Rinse 3 times 5 min with PBS/1% NGS. Permeabilize cells with PBS/1% NGS and 0.05% saponin, 10 min at room temperature. Rinse in PBS/1% NGS and 0.01% saponin. Block for 15 min in PBS/1% NGS & 0.01% saponin. Wash with PBS/1% NGS & 0.01% saponin.
4. Incubate in OG-conjugated DG52 in PBS/1% NGS & 0.1% saponin for 45 min at room temperature. Wash w/ PBS/1% NGS & 0.01% saponin. Rinse 3 times 5 min in PBS and 0.01% saponin. Mount using a No. 1 coverslip in 10% glycerol/PBS.

**Analysis**
Examine by epifluorescence using FITC and Texas Red filters in the microscope. Green staining cells are intracellular, red (or orange-yellow depending on the ratio of antibodies used) staining parasites are extracellular. Quantify results by counting the number of parasites (inside or out) per 100 host cells or per ten 40x fields (assuming a reasonable attachment rate). The principle advantage of this assay is that it allows discrimination of agents that disrupt invasion but still allows attachment of the parasite to the monolayer, for example cytochalasin D. Drugs which block both attachment and entry will be revealed by a decrease in cell-associated parasites (total parasites per 100 host cells).

**Quantitation of cell-associated parasites by β-galactosidase activity**
A convenient and sensitive method for monitoring parasite cell numbers is the heterologous reporter provided by the *E. coli lacZ* gene that confers expression of β-Gal in *Toxoplasma*. Expression of the heterologous reporter β-Gal was initially described for *Toxoplasma* using a genetic construct harboring the *lacZ* gene driven by *SAG1*, a stage specific tachyzoite gene (Seeber and Boothroyd, 1996). Independently, we have constructed a *GRA1* promoter driven construct (Howe and Sibley, 1997) and generated stable lines of RH that contain this plasmid integrated into the genome (Dobrowolski and Sibley, 1996). Monitoring β-Gal expression allows for rapid quantification of parasite numbers that can be used for attachment assays, growth in vitro and dissemination in vivo. β-Gal can be monitored enzymatically using a variety of substrate and by Western blot using the mAb 40a-1 (obtained from Joshua Sanes, Washington University).

Reagents:

The β-galactosidase expressing 2F line of *Toxoplasma* is propagated in HFF cells the same as the normal RH strain. Harvest parasites and resuspend in HHE as described above.

Lysis buffer: 100 mM HEPES (pH 8.0), 1 mM $MgSO_4$, 1% Triton X-100, 1 mM DTT. Prepare and store at $4^0$C, add DTT fresh.

Assay buffer: 100 mM Phosphate buffer (pH 7.3), 102 mM β-mercaptoethanol, 9 mM $MgCl_2$. Prepare and store at $4^0$C, add β-mercaptoethanol fresh (stock is 14.4 M).

CPRG: The substrate is freshly made as a 6.25 mM stock in phosphate buffer (chlorophenolred-β-D-galactopyranoside) (Cat. 4521220, Boehringer Mannheim).

Procedure:

**Standard curve**

To relate enzyme activity to parasite cell numbers, it is necessary to generate a new standard curve for each experiment.

1. Add 50 μl of stock cells ($5 \times 10^6$ cells/ml) to 200 μl of 1x lysis buffer to generate a $10^6$/ml stock lysate.
2. Make serial 10-fold dilution of the $10^6$ stock to obtain $10^5$, $10^4$, $10^3$ cells/ml (i.e. 100 μl lysate plus 900 μl of lysis buffer).
3. Place 50 μl of cell suspensions in an empty 96 well plate for the standards $10^6$/ml ($5 \times 10^4$ cells) through $10^3$/ml ($5 \times 10^1$ cells). Run each standard in quadruplicate.

**Binding/invasion assay based on β-Gal activity**

1. Transfer 100 μl of parasites at $5 \times 10^6$ cells/ml in IM to each well of a 96 well plate containing confluent HFF cells (include four wells per sample group). Inhibitors can be incubated with the parasites and or host cells prior to and/or during incubation, to assess their effects on attachment. Incubate at 37°C for 30 min in $CO_2$ incubator.
2. Include wells that have no HFF cells as a control for washing and HFF cells which are not challenged with *Toxoplasma* as a control for endogenous activity.
3. Wash by adding 200 μl PBS containing 1 mM $Ca^{+2}$ (37°C) per well. Repeat washing six times by gently flicking the contents from the plate into a tray containing disinfectant. Check the plate using the inverted microscope to be sure you have removed the nonadherent parasites without dislodging the monolayer.
4. Add 50 μl of 1x lysis buffer per well. Remove 5 μl for total protein assay, (used to normalize for effects on cell monolayer density).

**β-Gal Assay**

1. Incubate 96 well plate at 50°C for 10 min to destroy endogenous activity. Add 160 μl of assay buffer per well. Incubate plate at 37°C for 5 to 10 min.
2. Add 40 μl of 6.25 mM CPRG per well (final concentration = 1 mM). Incubate at 37°C. Read plate at 570 nm at 30 min. (Incubation time may need to be adjusted from between 30 min and a few hours in order to observe the optimal range.)
3. Calculate the mean for the quadruplicate wells for each treatment vs. controls. The wells which had no HFF cells serve as a control for

washing. HFF cells with no parasites serve as a control for endogenous activity in the host cells.

4. Compare the number of attached/invaded parasites in control (no treatment) versus treated groups (use diluted standards to determine cell numbers).
5. Analyze the total protein content per well using the Pierce Protein Assay (Cat. 23235, Pierce). This monitors the intactness of the cell monolayer (parasites make a negligible contribution) and can be used to normalize for loss of cell numbers.

**Analysis**

This assay allows evaluation of agents that decrease (or increase) cell-associated parasites. Because it measures both attached and intracellular parasites, agents that only effect internalization may not show a decrease in this assay. Such effects are best determined using the 2-color immunofluorescence assay described above. The advantage of the β-Gal assay is that it is rapid and easily quantified. As described here, the assay is designed to monitor cell-associated parasites. The assay can also be prolonged by incubating the plate at 37°C in complete culture medium for one to several days prior to assaying for β-Gal activity. Modified in this way, the assay can be used to monitor the effects of drugs on growth, here the compounds are typically added after infection and allowed to remain during the duration of the assay.

**Transient transfection of *Toxoplasma* using heterologous reporters**

Molecular genetic tools have recently been developed for *Toxoplasma* allowing both transient and stable transfections that have been useful for a variety of genetic manipulations (Roos et al., 1994). The ability to analyze protein expression by transient transfection allows for rapid assessment of protein trafficking and function in *Toxoplasma.* Transient transfection of heterologous genes merely requires a convenient means of detecting the product, for example using a specific antibody. In contrast, transfection of genes that encode proteins which are naturally expressed in *Toxoplasma* requires the ability to distinguish the introduced version of the protein by epitope tagging. A variety of epitope tags are available for this purpose, the one illustrated here is a 9 amino acid sequence defined in the influenza hemagglutinin called HA9 or flu tag (Wilson et al., 1984). This has been used previously to study the membrane partition of the *Toxoplasma* secretory proteins GRA2 (Mercier et al., 1998) and GRA5 (Lecordier et al., 1999). It is important to emphasize, that whenever an epitope tag is utilized, it is imperative that the investigators show that its addition does not change the trafficking or function of the protein in question. This also applies to over-expression, which may be a consequence of the transient expression construct that is chosen. More natural levels of expression can generally be achieved using the homologous promoter sequences. The following protocol illustrates the utility of heterologous expression using β-Gal or HA9-tagged proteins in *Toxoplasma*, both of which can be evaluated by enzyme assay, Western blotting or immunofluorescence staining

Reagents:

β-Gal expression: Heterologous expression requires the construction of a plasmid vector that utilizes flanking sequences from a *Toxoplasma* gene, used to drive the expression of a foreign gene of interest. Based on the original strategy used for transient expression in *Toxoplasma* (Soldati and Boothroyd, 1993), we have engineered a series of constructs that utilize the NsiI site to generate an ATG and PacI site to generate a stop codon. The open reading from a gene of interest is cloned into these sites by PCR to generate an in-frame expression cassette. We typically rely on the GRA1 promoter (Mercier et al., 1996) for high-level, constitutive expression and the SAG1 promoter (Soldati and Boothroyd, 1995) for stage specific expression. Using the *E. coli lacZ* gene, we generated in-frame expression constructs called pSAG1/lacZ and pGRA1/lacZ. The expression of β-Gal is detected by immunostaining (either IFA or western blot) using the mAb 40a-1 (provided by Joshua Sanes, Washington University) or enzymatically as described above.

HA epitope tag: The HA9 tag consists of the linear amino acid sequence of YPYDVPDYA (Wilson et al., 1984). It is recognized by the mAb 12CA5 or by rabbit polyclonal sera called HA11 (Berkeley Antibody Co.). It can be conveniently placed at the C terminus or internally, although the epitope is somewhat conformationaly dependent and detection may vary. The epitope is generally added to the protein of interest by PCR-based cloning which offers the advantage of tailor-made junctions to splice the epitope in frame in the protein sequence.

Cytomix buffer: 10 mM $KPO_4$, 120 mM KCl, 0.15 mM $CaCl_2$, 5 mM $MgCl_2$, 25 mM HEPES, 2 mM EDTA, pH 7.6. Sterile filter and store at $4^0$C.

Procedure:

1. Harvest freshly lysed cultures of *Toxoplasma* tachyzoites and resuspend at $10^7$ to $10^8$/ml in Cytomix buffer supplemented with 5 mM ATP and 5

mM glutathione. Solutions and procedure should be performed at room temperature.

2. Add 100 to 200 μg of plasmid DNA prepared by Quiagen maxiprep and desalting.
3. Add the mixture to a 4 mm gap disposable cuvette and place in the BTX Cell Micromanipulator 600 electroporater. The settings should be 2.5 kV, 24 ohms (range of 13 to 48 ohm), with a charging voltage of 2.0 kV.
4. Keep the parasites in the cuvette for 5 min. Centrifuge at 1,800 rpm for 10 min and resuspend in 5 ml of D10. Transfer to an empty 6 well plate or a T25 containing a monolayer of confluent HFF cells.
5 At intervals following infection, harvest cells and prepare for analysis. Extracellular parasites can be harvested at any time, they begin showing expression at 6 to 8 hr which generally peaks from 20 to 30 hr before falling. Intracellular parasites are most conveniently harvested when they naturally lyse the monolayer which takes from 20 to 36 hr.

**Analysis**

Following harvest, parasites can be analyzed by air-drying onto glass slides followed by a standard IFA fixation and staining protocol. Transient expression generally leads to easily detected signals in 5 to 10% of tachyzoites. Alternatively, expression can be detected by Western blot. In order to maximize the signal, it is generally necessary to load $1/10^{th}$ of the total transfected population per lane when detecting by ECL. Finally, in the case of β-Gal, the activity can be monitored by enzymatic activity using dilutions of lysate from the transfected parasites. Constructs bearing epitope tags or polyhistidine tags also offer the potential to purify proteins that are expressed in *Toxoplasma*. Such a situation may be important to maintain proper folding and or post-translational modifications.

**Protein secretion by *Toxoplasma***

Apicomplexan parasites are highly developed for regulated secretion and contain three different classes of apically-located secretory organelles: micronemes, dense granules and rhoptries (Carruthers and Sibley, 1997). Microneme proteins are released on contact with the host cell and participate in cell attachment (Carruthers et al., 1999). The triggers that cause microneme protein release are unknown, but may involve the initial interaction with the cell which involves both lateral association and apical reorientation. Microneme secretion occurs spontaneously once parasites are warmed to 37°C in the presence of serum, thus timing is quite critical to observing a decrease or stimulation of secretion. We will use ethanol to stimulate secretion, this is a potent trigger that acts by raising intracellular calcium (Carruthers et al., 1999). The protocol is designed to pretreat *Toxoplasma* with inhibitors, then stimulate them with ethanol and check for secretion of proteins into the supernatant.

Reagents:

Antibodies: Micronemes will be tracked using antibodies to the protein MIC2 (mAb 6D10, recognizes a 116 kDa form in the cell pellet and a 90 kDa form in the supernatant). Cell lysis will be evaluated using the cytoplasmic marker β-Gal (mAb 40a-1, recognizes a 120 kDa protein). Antibodies to GRA1 (mAb Tg17-43, 26 kDa protein, obtained from Marie-France Cesbron, Lille, France) and ROP1 (mAb TG49, 65 kDa protein (obtained from Joe Schwartzman, Dartmouth, US) will be used to check for release from dense granules and rhoptries, respectively.

SDS PAGE gels are run using standard buffers and conditions (Laemmli, 1970) and Western blots are developed using enzyme chemiluminescence (ECL).

Blocking solution: 5% skim milk powder, 5% normal goat serum (NGS), 0.05% Tween-20, 0.05% Triton X114 in PBS.

Wash solution: 0.1% skim milk powder, 0.1% NGS, 0.05% Tween-20, PBS.

Secondary antibodies: Goat anti-mouse IgG or goat anti-rabbit IgG conjugated to horseradish peroxidase (Jackson). Dilute 1:10,000 to 1:20,000 in washing solution.

Procedure:

Parasites are kept at 18°C prior to initiating the secretion assay. The tubes must be warmed quickly to 37°C, and the timing of additions is critical.

1. Prepare a set of 100x inhibitor stocks in DMSO or $H_20$. Prepare two sets of parallel tubes for plus (+) and minus (-) ethanol. Add 1.0 µl of each 100x inhibitor stock to the appropriate tubes. Add an equal volume of DMSO to the control tube.
2. Dispense cells (1 x $10^8$ cell/ml) at 100 µl per tube. Incubate for 20 min at 18°C.
3. Add 1 µl of absolute ethanol (final 1%) to each of the + tubes. Transfer immediately to 37°C bath. Incubate at 37°C for exactly 2 min. Chill by immersion in wet ice bath.

4. Spin at 2,000 rpm (800g) at 4ºC and remove 80 μl of supernatant to a new tube. Spin the supernatant at 2,500 rpm (1,000g) at 4ºC, and remove 60 μl of supernatant to a new tube.
5. Add 15 μl of 5x SDS PAGE sample buffer with protease inhibitor cocktail and 2% β-mercaptoethanol (total volume 75 μl). Boil 5 min, spin briefly. Resolve by SDS PAGE using a 10% acrylamide gel.
6. Load 10 μl of the control followed by the various concentrations of inhibitors. Load one lane of MW standards per gel. Each gel should include standards for the $10^4$, $10^5$, and $10^6$ cell equivalents (10 μl per lane). Run the gels at 160 volts for 1 to 1.5 hr (or 15 to 20 mA per gel). Remove the gel, transfer to nitrocellulose filters and stain the filter with Ponceau S to check the transfer (0.2% acetic acid, 0.2% ponceau S (Cat. P3504, Sigma).
7. Develop Western blots by ECL: incubate blots in 25 ml blocking solution for 30 min at room temperature. Drain off blocking solution and replace with 10 ml of diluted primary antibodies. Incubate 45 min at room temperature. Wash 4 times 5 min then add 10 ml secondary antibody diluted 1:10,000. Incubate 45 min at room temperature. Wash 4 times 5 min then add 5 ml ECL substrate solution (Cat. 0034080, Pierce): 1:1 mix of peroxide and luminol solutions). Incubate 2 to 3 min at room temperature. Briefly, blot dry between sheets of 3 MM filter paper then tape into sheet protector. Perform exposures on Kodak XAR sheet film for 5 sec to 30 min depending on signal strength.

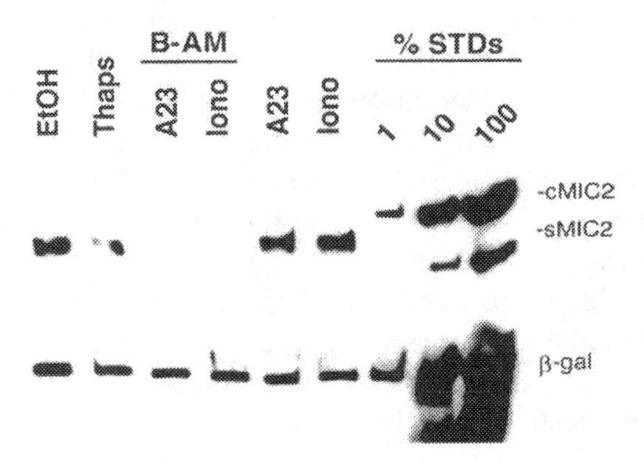


**Figure 2.** Effect of agents which modulate intracellular calcium on *T. gondii* MIC2 secretion. Purified tachyzoites were pretreated with the calcium ionophore BAPTA-

AM (B-AM, 100 μM), then stimulated with ethanol (EtOH, 1%), thapsigargin (Thaps, 1 μM), A23187 (A23, 400 nM), or ionomycin (Iono, 400 nM) for 2 min at 37 C. Serial loadings of parasite lysates indicate the degree of MIC2 secretion or parasite lysis (β-gal). Ethanol, thapsagorgin and the calcium ionophores (ionomycin and A23187) raise intracellular calcium thus stimulating discharge. Chelation of intracellular calcium with BAPTA-AM results in inhibition of secretion.

**Analysis**

Express secretion levels as a percentage of total content for total cells used in the assay after subtraction for lysis (derived from the β-Gal samples). The gels should be loaded so that the 1% level of cell standards represents secretion of 1% of the micronemes from the number of cells that were used in the assay. Likewise, 10% and 100% standards correspond to those levels of secretion. Gels can be scanned by Phosphorimager or films by densitometry for quantitative analyses. An example of the results obtained by this protocol is shown in Figure 2.

## INTRACELLULAR TRAFFICKING IN INFECTED CELLS

All cells actively internalize material by fluid-phase and receptor-mediated mechanisms. The intracellular trafficking of this cargo is indicative of the use to which the cell puts this internalized material. The majority of macromolecules are trafficked to the cell's lysosomes, an acidic, hydrolytic environment in which molecules are digested and used as nutrients for the cell. This pathway is easily delineated through analysis of the trafficking of fluorescent markers such as FITC dextran or Texas red dextran (Russell et al., 1992; see Fig. 3). Other molecules, notably the iron-carrying serum protein transferrin, traffic in and out of the cell via recycling pathways (Sturgill-Koszycki et al., 1996). Transferrin is bound by its receptor at the cell surface, internalized and delivered to the sorting endosome. Here, at pH 6.2, the iron comes off transferrin and is transported into the cell. Transferrin, still complexed with its receptor, is selected out of the sorting endosome and into the recycling endosomal pathway for return to the cell surface. At the surface, at neutral pH, the iron-free transferrin disengages from its receptor and is replaced by fresh, iron-loaded transferrin.

The differential trafficking of these molecules in cells infected by pathogens that remain within membrane-bound compartments has been invaluable in determining the position of the vacuole within the endosomal/lysosomal continuum, and therefore the physical characteristics of that compartment (Clemens and Horwitz, 1996; Russell, 1994; Russell et al., 1996; Russell et al., 1992; Schaible et al., 1999; Schaible et al., 1998; Sturgill-Koszycki et al., 1996; Sturgill-Koszycki et al., 1994; Via et al., 1997; Via et al., 1998; Xu et al., 1994). This approach, in conjunction with immunofluorescent studies exploiting antibodies against markers of known intracellular distribution has formed the basis of the majority of studies on intracellular pathogens and will be explored in *Mycobacterium avium* and *Leishmania mexicana* infected macrophages. In contrast to the above pathogens, *Toxoplasma* actively invades the host cell to form a compartment that excludes host cell proteins and resists fusion with endocytic marker (Joiner et al., 1990; Mordue et al., 1999; Mordue and Sibley, 1997). The nature of this compartment and its dependency on mode of entry is explored in the section following this one.

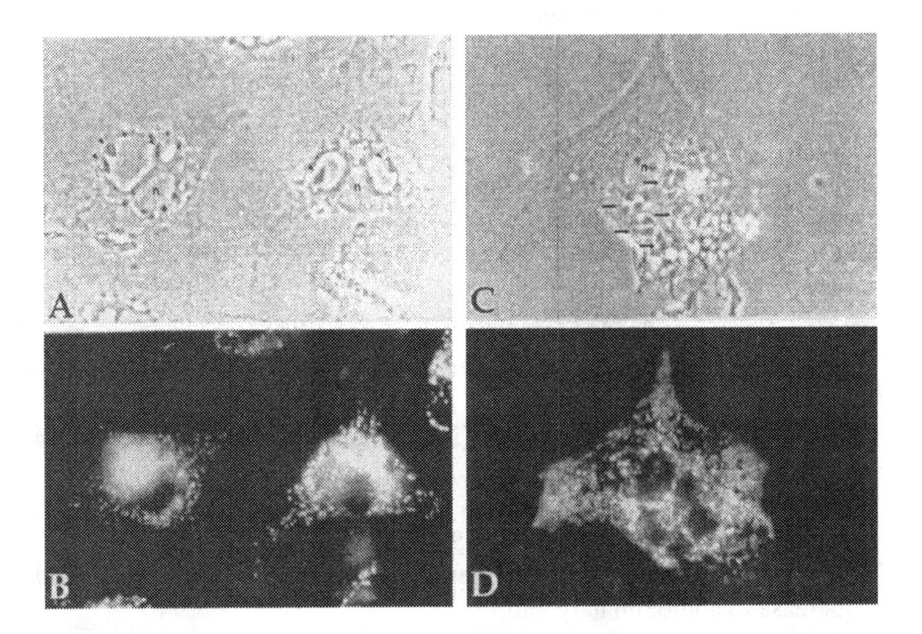


**Figure 3**. Macrophages infected with *Leishmania mexicana* (A and B) or *Toxoplasma gondii* (C and D) incubated in Texas red dextran (10 kDa) at 200 µg/ml for a 45 min pulse and 45 min chase. The fluorescence views (B and D) show that while the tracer has free access to the *Leishmania* amastigote-containing vacuoles it does not gain entry to the vacuoles containing *Toxoplasma*.

***Mycobacterium-*, *Toxoplasma-* and *Leishmania*-infected macrophages**
In the trafficking experiments detailed below macrophage monolayers were established in either Labtek multi-chamber slides (Nunc) for fluorescent studies or T25 flasks for electron microscopy and then infected with either *Leishmania mexicana*, *Toxoplasma gondii* or *Mycobacterium avium* the day before the experiment. The cells were washed and placed in medium with Texas red dextran (10 kDa; cat. D-1863, Molecular Probes) at 2 mg /ml and fluorescein transferrin at 20 ug/ml or biotin dextran (10 kDa; cat D-1956, Molecular Probes) and digoxigenin-transferrin at comparable concentrations. The duration of the pulse and chase periods were varied and the samples were fixed and processed for immunofluorescence or immunoelectron microscopy.

**Preparation of "tagged" transferrin**
Reagents:
Transferrin + Label: Transferrin can be exploited as a marker for trafficking when tagged with fluorescent markers for fluorescence microscopy or tagged with biotin or digoxigenin for immunoelectron microscopy. For these experiments we use human transferrin (Cat. T4778, Sigma) tagged with either of the following reagents; n-hydroxysuccinimide (NHS)-carboxyfluoresein (Cat. 1 386 093, Boehringer Mannheim), NHS-sulfo-SS-biotin (Cat. 21331, Pierce) or NHS-aminodigoxigenin (Cat. A-2952, Molecular Probes).

Procedure:
1. Dissolve 5 mg iron-loaded transferrin in 500 μl PBS, pH adjusted to 7.8. This pH is suboptimal for NHS coupling but it doesn't denature or release the iron from the transferrin.
2. Add a 20-fold molar excess of label (NHS-digoxigenin) dissolved in 50 μl DMSO. Cover the tube in aluminum foil and place on ice for 30 min.
3. Apply to a gel filtration column (Kwiksep 40 to 100 μm, Cat. 20449, Pierce) already equilibrated in PBS pH 7.2.
4. Collect the 1st colored peak to come off the column. The volume should increase to approximately 1ml.
5. Sterile filter, run a protein concentration assay and store at 4$^0$C.
6. The trafficking of transferrin through the recycling pathway is easily disturbed by either addition of too much transferrin, or over-derivatization of the protein with reporter (i.e. FITC). Transferrin trafficking experiments are routinely performed with 20 μg/ml labelled transferrin. Aberrant trafficking is controlled for by adding a 5 times

excess of unlabelled transferrin and ensuring that it blocks uptake of the labelled transferrin.

**Immunoelectron microscopy**

Reagents:

Pipes/Mg buffer: 200 mM Pipes, 0.5 mM $MgCl_2$, pH 7.0.

Fixative: 4% paraformaldehyde dissolved in Pipes/Mg buffer, boiled on a heated stir plate until dissolved, cooled on ice and filtered through a 0.4 μm pore filter-sterilizing unit. We store the fixative at $4^0$C and use it for up to 5 days.

10% gelatin: gelatin (225 bloom from calf skin, Cat. 27,162-4, Aldrich) is dissolved in hot Pipes/Mg buffer. The solution is dispensed into 500 μl aliquots and stored at -20°C.

1.86 M sucrose and 20% polyvinyl pyrrolidone in Pipes/Mg buffer: 63.67g of sucrose is added to 20 ml 1M Pipes, 2.5 mM MgC12, pH 7.0. The volume is made up to 80 ml and the solution heated and mixed until the sucrose is dissolved. 20 g of polyvinyl pyrrolidone is placed in a 150 ml glass beaker, approximately 5 ml of the sucrose solution is added and mixed into a paste with a glass rod. The sucrose solution is added gradually until a smooth mixture is achieved. The volume is increased to 100 ml and the mixture is covered with Parafilm and placed on a stir plate overnight. This facilitates removal of all the air bubbles. The clear, yellow PVP/sucrose solution is dispensed into 1 ml aliquots and stored at -20°C.

Block Buffer: is prepared from Pipes/Mg buffer plus 5% fetal calf serum and 5% goat serum.

Procedure:

1. Infected cultures of BMMO in T25 tissue culture flasks are processed as follows: Fix cells in 5ml of cold 4% paraformaldehyde Pipes/Mg buffer in the flask. Leave on a rocking table for 30 min (this is better kept cold if possible), pour off the medium, add 1 ml of fresh fixative, then GENTLY scrape the cells free from the flask and transfer to a 1.5 ml microfuge tube. Leave for a further 30 min on the rocking table.
2. Pellet the cells by centrifugation (gently, for 5 to 10 sec). Wash the cells in the Pipes/Mg buffer, pellet again and remove all the buffer. Keep the cells at room temperature. Then resuspend the cell pellet in 50 μl of warm 10% gelatin in Pipes/Mg buffer, quickly centrifuge the cells to pellet them before the gelatin sets. Place the gelatin pellets on ice for 10 min to set the gelatin.

3. Cut the tip off the tube with a razor blade, and place directly in 1 ml of 4% paraformaldehyde Pipes/Mg buffer for 15 min at room temperature. Then remove the gelatin-containing tip and place in 1 ml of 10% polyvinylpyrrolidone in 2.3 M sucrose in Pipes saline. Leave the tube on a rocking table (preferably a Nutator) overnight at 4°C or for 60 to 120 min at room temperature.
4. Gently remove the gelatin tip from the tube, trim the block with a razor blade, and place on the roughened tip of an aluminum stud. Freeze by plunging directly into liquid nitrogen. The block can then be cut or stored in liquid nitrogen until needed.
5. Following sectioning in an RMC MT7 cryoultramicrotome operating at -95°C, the sections are separated using an eyelash, lifted off the knife with a 1 mm diameter wire loop holding a drop of 2.3 M sucrose in 200 mM Pipes, 0.5 mM $MgCl_2$ pH 7.0, and gently thawed to allow stretching of the sections prior to touching onto freshly glow discharged, formvar-carbon coated grids. The grids are floated off the sucrose droplet by placing on the surface of Pipes/Mg buffer.
6. The frozen sections are blocked for 20 min on Pipes/Mg buffer with 5% FCS/ 5% goat serum (block buffer) in a 24 well plate, filtered prior to use.
7. The grids and sections are placed face down on 1st antibody diluted in filtered block buffer in 15 μl in the wells of a Terasaki plate for 30 to 60 min at room temperature or overnight at 4°C. Double labelling is carried out by mixing different species of primary and relevant secondary antibody, see next step. The grids are washed on 2 ml of block buffer in a 24 well plate on a rocking table for 10 min at room temperature.
8. The grids are then placed on second antibody (Jackson Immunoresearch labs gold conjugated antibodies). The grids are incubated in second antibodies as above for the primary antibody.
9. Grids are washed in the wells of a 24 well plate in 2 ml of block buffer for 10 min, then Pipes/Mg buffer for 10 min, followed by 2 times 5 min changes in $dH_2O$.
10. Grids are then stained/embedded in 2% PVA (or 2% methyl cellulose) with 0.3% uranyl acetate by placing the grids face down on a drop of stain on Parafilm on ice for 10 min. The grid is then lifted off with a wire loop, and the excess stain drained from the loop.
11. The grid is then air-dried and examined by electron microscopy.

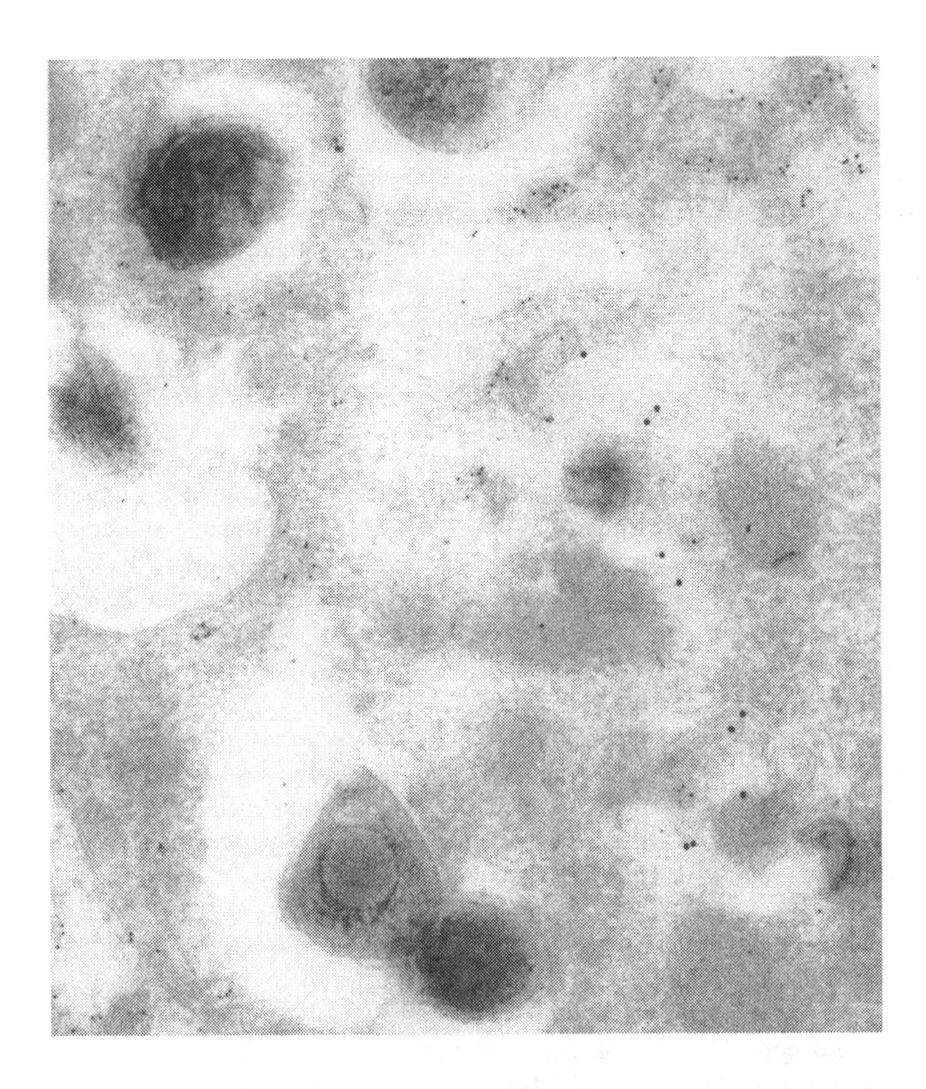

**Figure 4**. An immunoelectron micrograph of a *Mycobacterium avium*-infected murine macrophage following incubation with biotinylated Cholera toxin B subunit (detected by streptavidin, rabbit anti-streptavidin antibody, and 6 nm gold-conjugated goat anti-rabbit IgG) and digoxigenin dextran (10 kDa) (detected with mouse anti-digoxigenin antibody and 18 nm gold-conjugated goat anti-mouse IgG). The overlapping distribution of label is clearly visible. At later time points the two markers segregate and the Cholera toxin accesses the mycobacterial vacuoles while the dextran accummulates in lysosomes (Russell et al., 1996).

**Analysis**

Analysis of the distribution of the fluorescent tracers, dextran and transferrin, should demonstrate the gross cellular localization of the endocytic network. The transferrin should be confined to the early and recycling endosomal pathway, while dextran will percolate through the system and accumulate in lysosomes. The pathogen-containing vacuoles should, in the case of *Leishmania* and *Toxoplasma*, be clearly visible. If the experiment is successful the leishmanial vacuoles should be positive for dextran and negative for transferrin while the vacuoles containing *Toxoplasma* should be negative for both tracers. The distribution of dextran in infected macrophages is illustrated in Figure 3.

Two obvious drawbacks for examining markers by this method are first, the evaluation is highly subjective and two, the resolution is limited, particularly in the case of *Mycobacterium*. Nonetheless it is important to have a "feel" for the overall cellular distribution of the tracers so although immunoEM will improve resolution and allow some quantitation it is limited because the sampled regions are more restricted.

When conducting and analyzing immunoelectron microscopy images it is important to include both a negative control, irrelevant first antibody of the correct species, and to use co-labelling with another, well-characterized antibody of known distribution as an internal control. This point is illustrated in the immunoelectron micrograph shown in Figure 4.

**Intracellular fate of *Toxoplasma***

**Opsonization of *Toxoplasma***

This experiment will examine the formation and fate of the parasitophorous vacuole occupied by *Toxoplasma* versus phagosomes that are formed by Fc-receptor mediated phagocytosis. *Toxoplasma* actively invades the host cell to form a compartment that excludes host cell proteins and resists fusion with endocytic marker (Joiner et al., 1990; Mordue et al., 1999; Mordue and Sibley, 1997). However, when opsonized with specific antibodies to the cell surface, the parasite is engulfed by phagocytosis. The fate of the vacuole is evident from the acquisition of host markers through fusion with endosomes versus the active secretion of parasite proteins within the compartment. Finally, the association of host cell mitochondria with the parasitophorous vacuole versus phagosomes containing opsonized parasites will be examined.

Parasites are opsonized (coated for phagocytic uptake by Fc receptors) by incubation with specific antibodies to the cell surface protein SAG1. When engulfed by phagocytosis, the parasites are rapidly targeted for endocytic fusion and acidification (Mordue and Sibley, 1997). The entry and fate of opsonized parasites provides a convenient control for examining the nonfusigenic vacuole created by active parasite invasion.

Reagents:

Antibodies: A variety of mAbs are available to SAG1 including DG52 (Burg et al., 1988) and rabbit polyclonal antibodies to SAG1 have been described previously (Mineo et al., 1993).

Procedure:

1. Harvest *Toxoplasma* parasites from two-day passage as described above. Resuspend parasites in HHE to an approximate concentration of 5 x $10^7$/ml.
2. Add anti-SAG1 antibody to a final dilution of 1:1000. Use an opsonizing antibody to a surface protein that will not be recognized by secondary antibodies in subsequent immunofluoresence detection protocols. Incubate at room temperature for 30 min.
3. Wash opsonized parasites twice with HHE (centrifuge at 1,800 rpm). Resuspend opsonized parasites in room temperature invasion media (IM). Pass the opsonized parasites through a 25 gauge syringe twice being careful not to introduce air bubbles which will make the media alkaline.
4. For infection of macrophages, use opsonized parasites at a 5- to 10-fold lower concentration than non-opsonized parasites.

**Endosome/lysosome fusion**

The fate of parasitophorous vacuoles containing live parasites will be compared with phagosomes containing antibody-opsonized parasites. BMO cultured on LabTek chamber slides will be used to examine the intracellular fate of intracellular parasites by immunofluorescence. The extent of fusion with host endosomes/lysosomes will be evaluated by co-staining for parasites and the host marker LAMP1. The active secretion of rhoptries that occurs during formation of the parasitophorous vacuole will be detected by staining for ROP2.

Reagents:

| Antibodies: | Protein | Ab | Location |
|---|---|---|---|
| | LAMP1 | 1D4B (rat) | endosomal/lysosomal |
| | ROP2 | rabbit | Rhoptries/PVs |
| | SAG1 | rabbit | Parasite Surface |

*(ROP2 antibodies were obtained from Con Beckers, UAB, US; SAG1 antibodies were obtained form Lloyd Kasper, Dartmouth, US)*

Prepare two 4-chamber LabTek chamber slides containing BMO that are sub-confluent. Isolate *Toxoplasma* cells from freshly lysed cultures and prepare one set that is opsonized and one set that is not.

Procedure:

1. Take the two LabTeks containing BMO from the $CO_2$ incubator. Remove the medium and add 0.25 ml of parasite suspension to each well. Alternate the parasites so that you have two sets of nonopsonized followed by opsonized.
2. Warm rapidly to 37°C by submersion in the $H_2O$ tray in the $CO_2$ incubator. Allow invasion to proceed for 5 min. Rinse 3 times in PBS.
3. Fix the first slide in 2.5% formalin, 0.01% glutaraldehyde-PBS for 15 min at 4°C. Add fresh medium and return the other slide to the $CO_2$ incubator for 25 min
4. Remove the second slide and fix as above. Rinse 3 times 1 min in PBS. Incubate both slides for 10 min in PBS-5% NGS/5% FBS and 0.02% saponin. Rinse 3 times 1 min in PBS-1% NGS and 0.02% sap.
5. Add 0.25 ml of combined rat anti-LAMP mAb 1D4B and rabbit anti-ROP2 primary antibodies to one well of opsonized and one well of nonopsonized parasites on each slide. Add 0.25 ml of combined rat anti-LAMP1 (1:2 dilution of culture supernatant) (1:500 dilution) and rabbit anti-SAG1 antibodies to the remaining wells infected with opsonized or nonopsonized parasites on each slide. Incubate for 45 min at room temperature. Rinse 3 times 2 min PBS-1% NGS and 0.02% saponin.
6. Add 0.25 ml of combined secondary antisera dilutions to each well. (Goat anti-rat FITC plus goat anti-rabbit-Texas Red diluted 1:500 each). Incubate for 45 min at room temperature. Rinse 3 times 5 min PBS-1% NGS and 0.02% saponin.

**Analysis**

Co-staining with ROP2 and LAMP1 is designed to illustrate the mutually exclusive nature of parasite secretion versus lysosome fusion as described previously (Mordue and Sibley, 1997). When parasites enter the cell by active penetration, they secrete the contents of the rhoptries in the process of generating a parasitophorous vacuole which resists fusion with endosome and lysosomes. In contrast, when opsonized and taken up by phagocytosis, the parasite does not discharge rhoptries and the vacuole undergoes rapid fusion with endosomes and lysosomes. Consequently, vacuoles should be ROP2 positive or LAMP1 positive but not both. Costaining with SAG1 allows easy detection of the parasite in order to quantify the extent of fusion that occurs in cells pulsed with non-opsonized (typically less than 10%) versus opsonized (typically more than 50%) parasites.

**Protein secretion and organelle recruitment**

The parasitophorous vacuole formed by active invasion not only avoids endosome fusion, it recruits host organelles including mitochondria and endoplasmic reticulum (Jones et al., 1972; Sinai et al., 1997). The association of the vacuole with host cell mitochondria will be examined by co-staining for parasites and organelles using MitoTracker Red.

Reagents:

Hoechst 33342 (Cat. H-1398, Molecular Probes): Make stock solution at 1 mg/ml in $H_20$, dilute to 1 ug/ml in PBS for use. Hoechst is a DNA binding fluorochrome that fluoresces bright blue with UV excitation. The excitation/emission maximum are about 350/460 nm.

MitoTracker Red CMXRos (Cat. M-7512, Molecular Probes): Make fresh stock at 1 mM in DMSO. MitoTracker Red is a cell permeant, mitochondrion selective dye with excitation/emission maxima at 578/599 nm. Cells are labeled by passive diffusion, the dye flows across the cell and mitochondrion membranes and accumulates in the mitochondria. The dyes are fixable, allowing them to be used in combination with indirect fluorescence antibody localization studies.

Prepare two 4-chamber LabTek slides with subconfluent monolayers of BMO.

Procedure:

1. Take the two LabTeks containing BMO from the $CO_2$ incubator. Add 0.5 ml of diluted Mitotracker Red (100 to 250 nM) to each well. Incubate slides at 37°C for 30 min. Rinse 3 times 1 min in D10 at 37C. Add fresh

D10 and incubate at 37°C for 30 min. Rinse 3 times 1 min in D10 at 37°C

2. Add 0.25 ml of parasite suspension to each well. Alternate the parasites so that you have two sets of nonopsonized followed by opsonized. Warm rapidly to 37°C by submersion in the $H_2O$ tray in the $CO_2$ incubator. Allow invasion to proceed for 5 min. Rinse 3 times in PBS.
3. Fix the first slide in 2.5% formalin, 0.01% glutaraldehyde in PBS for 15 min at 4°C. Add fresh medium and return the other slide to the $CO_2$ incubator for 25 min. Remove the second slide and fix as above. Rinse 3 times 1 min in PBS. Incubate both slides for 10 min in PBS/5% NGS/5% FBS & 0.02% saponin. Rinse 3 times in PBS/1% NGS and 0.02% sap.
4. Add 0.25 ml of diluted anti-LAMP (mAb 1D4B, 1:2 culture supernatant) to one well of opsonized and one well of nonopsonized parasites on each slide. Add 0.25 ml of rabbit anti-ROP2 (1:500) to the other wells of each slide. Incubate 45 min at room temperature. Rinse 3 times 2 min in PBS/1% NGS and 0.02% sap.
5. Add 0.25 ml of diluted goat anti-rat-FITC (1:500) to the wells which received anti-LAMP1 on each slide. Add 0.25 ml of goat anti-rabbit-FITC (1:500) to wells that received anti-ROP2 on each slide. Incubate for 45 min at room temperature. Rinse 3 times 5 min PBS/1% NGS and 0.02% sap.
6. Add 0.25 ml of diluted Hoechst dye to each well. Incubate at room temperature for 5 min. Rinse 3 times 1 min in PBS. Remove the wells from the slide, mount in Antifade.

**Analysis**

You should observe that mitochondria are associated with the parasitophorous vacuoles but not phagosomes. That is, a given parasite containing vacuole should stain either for LAMP1, indicating lysosome fusion, or show surrounding mitochondria. The extent of mitochondrial association various with the maturity of the vacuole as shown previously (Sinai et al., 1997). LAMP1 positive vacuoles and mitochondria provide a rigorous negative control for colocalization studies as they are near by in the cytoplasm but should not colocalize.

## SUBCELLULAR FRACTIONATION OF PARASITES AND HOST CELLS: ISOLATION OF PHAGOCYTIC COMPARTMENTS FROM MACROPHAGES

Both *Leishmania* parasites and *Mycobacterium* bacilli are internalized by phagocytosis by macrophages, nonetheless the vacuoles in which they reside behave very differently inside the cell. Our appreciation of the differences between these compartments has been aided greatly by our ability to isolate and purify the phagosomes from the macrophages. In this section we detail methods for isolating phagosomes containing amastigote-like forms of *Leishmania mexicana* and IgG-coated polystyrene beads. The IgG-beads, which bind to the $Fc_{\gamma}RII$ receptors, are used as a control preparation to show how the phagosomes formed during uptake through defined receptor-mediated pathway differentiate following internalization.

### IgG-bead phagosomes

Reagents:

Homogenization buffer: 250 mM (8.6%) sucrose, 0.5 mM EGTA, 0.5 mM EDTA, 0.05% gelatin (Cat. G7765, Sigma) 20 mM Hepes, pH 6.8. This is prepared as a 5 times working strength stock solution to facilitate easier preparation of Ficoll and sucrose solutions.

Sucrose solutions: 50% sucrose, and 12% sucrose are prepared by dissolving 50 g or 12 g of sucrose in 20 ml 5 times homogenization buffer made up to 100 ml with $H_2O$.

Protease inhibitor cocktail: Each of the following are stored as 100 times stock solutions at -20°C and added to the homogenization buffer prior to use. The final concentrations are TLCK 100 µg/ml, pepstatin A 50 µg/ml, leupeptin 50 µg/ml and E64 50 µg/ml.

Ficoll solution; 15% high molecular weight Ficoll (400,000 kDa) in 5% sucrose, 0.5 mM EGTA, 0.5 mM EDTA, 0.05% gelatin, 20 mM Hepes, pH 6.8.

Plasticware: all polyproplylene from Sarstedt; 50 ml conical screw cap centrifuge tubes, 15 ml conical screw cap centriguge tubes, and 1.5 ml screw cap microfuge tubes. All plastic ware is siliconized by rinsing with Sigmacote (Sigma) and drying. It may appear trivial but we have found that different sources of plasticware have very different surface properties and some are "sticky" to membranes.

IgG-coated iron-loaded beads: Dynabeads M280, tosylactivated polystyrene beads (Cat. 142.03, Dynal) are covalently coupled to human IgG as follows. We allow 20 to 50 µg of protein per mg of bead. The beads are washed by centrifugation with a microfuge (use Sarstedt screw cap microfuge tubes) and resuspended in 100 mM phosphate buffer (20 mM $NaH_2PO_4$, 80 mM $Na_2HPO_4$, pH 8.5). An equal volume of buffer containing IgG at 500 μg/ml is added and the tube is placed on a mixer overnight at room temperature. The bead suspension is then washed 4 times and resuspended in 0.1% albumin. We use transfusion grade human serum albumin because bovine serum albumin is frequently contaminated with endotoxin. The bead suspension is stored at 4°C following addition of sodium azide to 0.02%.

47 mm 5μm pore Nucleopore filter (Millipore).

Procedure:

1. Add 75 μl (20 mg/ml beads) of beads per T75 flask of macrophages and incubate at 37°C for the time period required. A more synchronous population of vacuoles may be generated by incubating the IgG-beads with the cells at 4°C, gently washing the monolayer with cold medium, and then adding 10 ml of warm (37°C) medium to each flask for the required time window.
2. After phagocytosis, the flask of macrophages rinsed with cold PBS, scraped into 5 ml homogenization buffer with protease inhibitors. The cells are centrifuged 10 min at 300 g (1,200 rpm in Beckman tabletop) and the pellet resuspended in 1 ml of homogenization buffer.
3. Passage of the cells through a tuberculin syringe with a 23 gauge needle about 8 to 10 times results in complete lysis. Monitor carefully by microscope until >95% of the cells are lysed.
4. The lysate is layered onto a 2 ml 15% Ficoll (400 kDa) cushion and centrifuged at 800 g (2,000 rpm) for 10 min, followed by resuspension of the pellet in 250 ml homogenization buffer. This step removes unlysed cells.
5. The crude phagosome prep is layered onto 100 μl of a 30% sucrose cushion in a microcentrifuge tube. The tube is placed on end in the Magnetic Particle Collector so that all the beads concentrated at the bottom, this takes around 30 min the first time. The concentrated beads are washed in fresh homogenization buffer and the sucrose cushion step repeated for a total of 4 washes. These subsequent washes take around 10 min each time. Crossover contamination analysis (see below) of phagosomes isolated this way indicates that contamination varies between 3 to 5 %.

***Leishmania* amastigote-like form containing phagosomes**

Reagents:

Homogenization buffer: 20 mM Hepes, pH 6.8, 0.5 mM EGTA, 40 mM NaCl, 5% sucrose, 0.05% gelatin, and protease inhibitors as detailed above.

Gradient buffer: 20 mM Hepes, pH 6.8, 0.5 mM EGTA, 100 mM NaCl, 5% sucrose, 0.05% gelatin and protease inhibitors.

Sucrose solutions: 60% sucrose, 40% sucrose and 20% sucrose all made up in gradient buffer.

Procedure:

1. 2x T75 tissue culture flask of confluent bone marrow derived MO used per isolation. Infected with ALFs (10:1) for 2 hr at 37°C. MO washed twice in infection medium (DMEM + 10% FCS + Pen/Strep) and incubated for a further 30 min to ensure complete internalisation of the ALFS.
2. The cells were then washed twice in warm PBS and scraped off the flasks into 2 ml of chilled homogenization buffer.
3. The homogenization buffer is then removed from the flasks and transferred to a 1 ml tuberculin syringe on ice. Cells are disrupted by passage through a syringe with 23 gauge needle as above. This usually takes 20 to 30 passages. (alternatively a tight fitting stainless steel Dounce homogenizer works well.
4. Following homogenization the suspension is made up to 10 ml with chilled gradient buffer (20 mM Hepes, pH 6.8, 0.5 mM EGTA, 100 mM NaCl, 5% sucrose, 0.05% gelatin, TLCK and leupeptin as above) and centrifuged at 1000 rpm for 10 mins to pellet intact cells and nuclei.
5. The supernatant from the low speed spin is loaded onto two discontinuous sucrose gradients poured in 15 ml Sarstedt tubes (siliconized).

   3 ml 60% sucrose in gradient buffer
   3 ml 40% sucrose in gradient buffer
   3 ml 20% sucrose in gradient buffer
   5 ml of the supernatant from step 4.

6. The gradient is centrifuged at 1,700 rpm at 4°C for 25 min (No brake on the centrifuge). The vacuoles are harvested from the 40 to 60% interface using a pasteur pipette.
7. Hepes saline (20 mM Hepes, 100 mM NaCl, pH 6.8) is added to the vacuole solution to dilute out the remaining sucrose and the vacuoles

can be pelleted by centrifuging at 15,000 rpm for 25 min at 4°C in the Beckman J2-21 centrifuge. These phagosomes are not so clean, contamination runs at 10 to 12%.

**Contamination: How does one evaluate it?**

In our early experiments we used electron microscopy to examine the cleanliness of our phagosome preparations. This was inadequate, because it was rather subjective, and did not allow quantitation or identification of contaminants. We have developed a more rigorous approach to assaying contamination based on a cross-over contamination assay conducted as described below. This assay allows one to assess contamination in radiolabelled preparations, or in preparations where infected cells have been fed endocytic tracers.

To evaluate the purity of isolated vacuoles 4 T75 flasks of macrophages are prepared, 2 of which are metabolically labelled with $^{35}S$ methionine overnight. Particles are added to one labelled and one unlabelled flask. The cells from all 4 flasks are scraped and combined for processing as follows. The labelled cells with particles are combined with unlabelled cells with no particles, and the unlabelled cells with particles are combined with the labelled cells with no particles. After the isolation procedures are complete an equivalent aliquot from the two samples is measured for radioactivity. The % contamination was calculated as c.p.m. from particles in unlabelled cells / c.p.m. from particles in labelled cell + c.p.m. from particles in unlabelled cells x 100. In the protocols given the levels of contamination varied between 3 to 8%. 2D-PAGE analysis of labelled proteins copurified with phagosomes from unlabelled cells indicated that there was no selection for specific contaminating proteins (Sturgill-Koszycki et al., 1997). In trafficking experiments involving transferrin or cholera toxin trafficking (Russell et al., 1996; Sturgill-Koszycki et al., 1996), the tracer is used in place of the radioactivity.

**Analysis**

Analysis of isolated cellular fractions is conducted by 2D-SDS PAGE, which allows one to take a "global" view of the range of polypeptides that are present in the isolated preparation (Sturgill-Koszycki et al., 1997). However, this approach does not allow one to identify the proteins directly, although the field of proteomics will change this quite rapidly. An alternative that we use is immunoblot and probing with antibodies against either known macrophage proteins, or against tracers added to the cell during the course of the experiment. This lends itself more readily to direct experimentation on the biology of the intracellular compartment and the factors important to the maintenance of infection (Russell et al., 1996; Schaible et al., 1998; Sturgill-Koszycki et al., 1996; Sturgill-Koszycki et al., 1994).

## SUBCELLULAR FRACTIONATION OF PARASITES AND HOST CELLS: FRACTIONATION OF INTRACELLULAR COMPARTMENTS BY DENSITY GRADIENT ELECTROPHORESIS

Isolation of intracellular compartments can be achieved through different criteria. The procedures described in the preceding section rely on the unique bouyant density of the ALF-phagosomes, or the use of an iron-loaded particle. Obviously, these methods have limited general application. Our interest in exploring alternative strategies has come from our wish to study the trafficking of pathogen-derived molecules within infected cells. Obviously, such molecules have considerable inherent interest due to their potential effects on host cell function or their possible properties as antigens presented by infected cells. In our studies (Beatty et al., 2000) we used density gradient electrophoresis to separate lysosomes from early and late endosomes to delineate the progression of mycobacterial lipids through the host cell. This method was developed by Pieters and colleagues to examine antigen processing compartments inside cells (Engering et al., 1997; Tulp et al., 1997; Tulp et al., 1994; Tulp et al., 1993).

Reagents:

Density gradient electrophoresis (DGE) buffer: 250 mM sucrose, 1 mM EDTA, 0.5 mM EGTA, 10 mM triethanolamine, pH to 7.4 with acetic acid.

Ficoll solutions: Use 70,000 Da Ficoll (Cat. F2878, Sigma) and make up 12% (10 ml per gradient), 10% (200 μl per gradient) and 8% (6 ml per gradient) Ficoll solutions in DGE buffer.

Enzyme assays for verification of cellular fractions:

β-hexaminidase (lysosomal marker) is detected using 4-methylumbelliferyl-2-acetamido-2-deoxy-β-D-glucopyranoside as a substrate (4 mM in $dH_2O$) (Cat. M6634, Sigma). The reaction mixture is prepared by combining 2.5 ml substrate, 2.5 ml sodium acetate (0.4 M, pH 4.4), 125 μl 10% Triton X-100, and 4.6 ml $dH_20$. The solution is warmed to 37°C. 50 μl of each DGE fraction and 150 μl of reaction mixture are added to wells of a 96 well plate and incubated at 37°C for 1 hr. The assay is stopped with 50 μl of stop buffer (1 M glycine, 1 M $Na_2CO_3$). The reaction is read spectrophotometrically at excitation 364/emission 448.

α-mannosidase II (Golgi marker) is detected using 4-methylumbelliferyl-α-D-mannopyranoside as a substrate (4 mM in $dH_2O$) (Cat. M0905, Sigma). The reaction mixture consists of 5 ml substrate, 5 ml PBS, 250 μl 10% Triton X-100, and 9.75 ml $dH_2O$. The reaction is carried out as described above.

Alkaline phosphodiesterase (plasma membrane marker) is detected using Na-thymidine 5'-monophosphate, p-nitrophenylester as a substrate (10 mM in $dH_2O$) (Cat. T4510, Sigma). The reaction mixture is prepared by combining 2 ml substrate, 2 ml 0.1 M Tris-HCl (pH 9), and 4 ml dH2O. 50 μl of each DGE fraction and 150 μl of reaction mixture are added to wells of a 96 well plate and incubated at 37°C for 2 hr or until yellow color is evident in positive samples. The reaction is stopped as described above and absorbance is read at 410 nm.

Procedures:

**Preparation of cell lysate**

1. Prepare macrophage cultures in T160 flasks. (1 confluent T160 flask/gradient). Wash 1 time with cold PBS and place cell monolayer on ice. Rinse cell monolayer with cold homogenization buffer
2. Add 6 ml of homogenization buffer to each flask and scrape monolayers with cell scraper to remove cells. Collect cells in a 15 ml centrifuge tube and wash flask with additional 4 ml buffer.
3. Centrifuge cell suspension (10 ml total) at 300 g for 10 min at 4°C and discard the supernatant. Resuspend the pellet in 1 ml homogenization buffer and lyse cells as described above.
4. Centrifuge the lysate at 100 g for 6 min to remove nuclei. Retain the supernatant on ice, subject the pellet to step 3 for an additional 5 passes through the syringe or floating ballbearing homogenizer and repeat the post nuclear spin. Collect and combine supernatants.
5. The supernatants are layered onto a step gradient of 2 ml of 12% sucrose in homogenization buffer over 2 ml of 30% sucrose in homogenization buffer and centrifuged for 800 g for 60 min at 4°C. The bacteria and large cell debris go to the lowest phase of the gradient.
6. The supernatant from the upper region of the gradient which contains the intramacrophage vesicles were collected and subjected to

ultracentrifugation in a Beckman tabletop ultracentrifuge at 55,000 rpm (100,000 g) for 45 min at 4°C.

7. For improved separation, treat with trypsin prior to centrifuging. Add 25 µl/mg protein for 5 min at 37°C. Place on ice immediately and add 100 µg/mg protein soybean inhibitor. Repeat the above centrifugation step. The resulting pellet is VERY GENTLY resuspended in 200 µl 10% Ficoll in DGE buffer using a tuberculin syringe.

**Density gradient electrophoresis**

1. The density gradients are set up and run in the cold room. DGE columns are set up in upper buffer chamber with prewet dialysis membrane (6 to 8,000 MW) in place at bottom of columns and secured with rubber O-ring.
2. The lower chamber is filled with DGE buffer and the upper chamber containing columns is placed onto lower chamber.
3. 10 ml of 12% Ficoll is put in the bottom of each column and the sample (in 200 µl of 10% Ficoll in homogenization buffer) is carefully layered onto the 12% Ficoll cushion using tuberculin syringe.
4. With a two chamber gradient maker, a continuous gradient of 8% to 0% Ficoll is overlaid on top of the sample (the chambers contain 6 ml DGE buffer and 6 ml 8% Ficoll to make 12 ml gradient). This is carefully run down the side of the DGE column to avoid disrupting the sample or gradient itself.
5. Create a meniscus with DGE buffer at top of column and place a piece of Nitex cloth on top of column. This reduces disruption of gradient when filling upper buffer chamber. (Nitex will float and is removed after upper buffer chamber is filled.)
6. The reservoir chamber is then filled and connected to the upper chamber which fills back into the lower chamber and returns to the reservoir. The buffer circulation is mediated by a peristaltic pump.
7. The anode (+ve) terminal of the power pack is connected to the upper electrode of the DGE apparatus which is run at 15 mAmp constant current (starts at approximately 100 V) for 4 gradients.
8. After approximately 2 hr, 3 distinct bands should be evident. The upper band is comprised of late endosomes and lysosomes, the middle contains intermediate endocytic compartments, Golgi, and ER, and the lower band contains early endosomes and plasma membrane.
9. Fractions from the entire gradient are harvested from the top of the column by siphon. The vesicular material in the fractions are concentrated by ultramicrofugation at 55,000 rpm (100,000 g) for 45 min at 4°C.

10. The subcellular fractions are characterized by analysis of enzymatic activity (described above), analysis of known biochemical markers by Western blot, or immunoelectron microscopy.

**Analysis**

This technique is extremely suitable for examination of molecules released inside the host cell, and in common with the isolated phagosomes, these can be studied by 2D-SDS PAGE, or by immunoblot. In addition, tracers such as labeled transferrin, or horseradish peroxidase can be added to the cells for various times prior to processing to facilitate further the characterization of the vesicular fractions.

## SUBCELLULAR FRACTIONATION OF PARASITES AND HOST CELLS: CELL FRACTIONATION OF *TOXOPLASMA* AND PURIFICATION OF MICRONEMES

*Toxoplasma* is specially equipped for protein secretion and harbors three distinct exocytic organelles, the micronemes, rhoptries and dense granules (Carruthers and Sibley, 1997). During invasion, the contents of these organelles are discharged sequentially beginning with micronemes which mediate cell attachment, followed by rhoptries which are involved in vacuole biogenesis, and finally dense granules that modify the intracellular vacuole. Purification of these compartments would greatly aid in studying their composition, mechanisms of release, and function of their contents. The following protocol provides a partial purification of micronemes from *Toxoplasma*.

Reagents:

Start with extracellular tachyzoites that are purified after growth in fibroblast monolayers. Parasite cells will be disrupted by sonication using conditions which have proven effective at breaking the cell membrane but not in disrupting the micronemes. Micronemes will then be semi-purified using differential centrifugation. Finally, micronemes will be broken using harsh sonication and centrifuged to separate a membrane pellet and supernatant. The fractions will be tested for purity by Western blot using markers for the cell surface (SAG1), the micronemes (MIC2), rhoptries (ROP1) and dense granules (GRA1).

Protease inhibitors (PI): Prepare a 1000x cocktail containing the following. 1 mg/ml APMSF in DMSO, 1 mg/ml E64 in DMSO, 1 mg/ml TLCK in DMSO, 1 mM pepstatin in DMSO. Store in aliquots at -20°C.

Markers:

| Protein | mAb | Location | Size |
|---|---|---|---|
| MIC2 | 6D10 | micronemes | 110/90 kDa |
| SAG1 | DG52 | surface | 30 kDa |
| ROP1 | Tg49 | rhoptry | 60 kDa |
| GRA1 | Tg17-43 | dense granule | 26 kDa |

Procedure:

1. Harvest freshly lysed tachyzoites of *Toxoplasma* in HHE as described above. Resuspend at $10^9$/ml in HHE. Keep the cells cold and on ice the entire time! Sonicate cells on medium setting using a macroprobe for 3 times 15 sec bursts. Remove 50 µl of the supernatant as fraction 1.
2. Spin the sonicate at 2,000 g to pellet broken cells, nuclei etc. The supernatant should be cloudy, if not, resuspend the pellet and repeat the sonication at higher intensity. Transfer to another tube and spin at 2,000 g.
3. Save the sup to a new tube and spin at 8,000 g x 20 min at 4°C. Remove the 8K sup to a new tube. Resuspend the 8 K pellet, which contains the dense granule and rhoptry organelles, in 1.0 ml of PBS pH 6.0 + PI (1000x cocktail). Remove 50 µl for fraction 2.
4. Spin the 8 K sup at 30,000 g for 30 min and save 50 µl of the sup for fraction 3. Save the 30 K pellet which contains micronemes.
5. Resuspend the 30 K pellet in 1.0 ml of PBS, pH 6.0 + PI and subject to 3 rapid cycles of freeze/thaw (alternate between an ethanol dry ice bath and a 37°C water bath). Sonicate on high intensity using a microprobe for 3 times 15 sec bursts. Spin at 100,000 g for 1 hr at 4°C.
6. Save the 100K sup which contains soluble microneme proteins. Remove 50 µl of sup for fraction 4.
7. Resuspend the 100K pellet which contains membrane proteins and unbroken micronemes in 1.0 ml of PBS, pH 6.0 + PI. Remove 50 µl for fraction 5.

**Prepare the fractions for SDS PAGE**

*Run two gels, one with reduction, the other without. The antibodies to SAG1 (DG52) and GRA1 (Tg17-43) should be used to blot the non-reduced gel.*

1. Divide each fraction in half, separate to two tubes. Add 50 μl of cold acetone to each of the fractions in a microfuge tube. Put on wet ice or at –20°C for 30 min to overnight.
2. Spin in the microfuge on high speed (12,000 g) for 20 min (keep cold!). Decant the acetone, without disturbing the pellet and wash the pellet with 70% ethanol, microfuge 5 sec on high. Decant the ethanol, air dry.
3. Resuspend one pellet in 22 μl 1x SDS PAGE sample buffer containing PI. Resuspend the other pellet in 22 μl of sample buffer containing 2% β-mercaptoethanol and PI. Boil samples 2 min and load 10 μl per lane on the gel. Load one gel with the reduced samples, one gel with the non-reduced samples. Include two sets of loadings per gel with molecular weight (MW) markers (10 μl) in the center of each gel.
4. Transfer to NCP, cut the membrane in half so that you have four sets of samples. Blot each half of the membrane with one of the antibodies.

**Analysis**

You should observe that MIC2, GRA1 and ROP1 are in all the fractions, but that there is a general enrichment of the MIC2 protein and a depletion of GRA1 versus the starting material.

## ACKNOWLEDGEMENTS

The authors are grateful to the generations of students (most notably Classes of '97, '98 and '99) at the Woods Hole Biology of Parasitism Course, who suffered at our hands in the fine-tuning of these protocols! The laboratories of DGR and LDS are supported by funds from Burroughs Wellcome and the National Institutes of Health.

## REFERENCES

Beatty, W. B., Rhoades, E. R., Ullrich, H. J., Chatterjee, D., and Russell, D. G. (2000). Trafficking of mycobacterial lipids in infected macrophages: From the phagosome to the exosome. Traffic, *in press*.

Brun, R., and Schonberger, M. (1979). Cultivation and in vitro cloning of *Trypanosoma brucei* in semi defined medium. Acta Trop. *36*, 289-291.

Burg, J. L., Perlman, D., Kasper, L. H., Ware, P. L., and Boothroyd, J. C. (1988). Molecular analysis of the gene encoding the major surface antigen of *Toxoplasma gondii*. J. Immunol. *141*, 3584-3591.

Carruthers, V. B., Giddings, O. K., and Sibley, L. D. (1999). Secretion of micronemal proteins is associated with *Toxoplasma* invasion of host cells. Cell. Micro. *in press*.

Carruthers, V. B., Moreno, S. N. J., and Sibley, L. D. (1999). Ethanol and acetaldehyde elevate intracellular calcium and stimulate microneme discharge in *Toxoplasma gondii*. Biochem. J. *342*, 379-386.

Carruthers, V. B., and Sibley, L. D. (1997). Sequential protein secretion from three distinct organelles of *Toxoplasma gondii* accompanies invasion of human fibroblasts. Eur. J. Cell Biol. *73*, 114-123.

Clemens, D. L., and Horwitz, M. A. (1996). The *Mycobacterium tuberculosis* phagosome interacts with early endosomes and is accessible to exogenously administered transferrin. J Exp Med *184*, 1349-55.

Dobrowolski, J., and Sibley, L. D. (1996). The role of the cytoskeleton in host cell invasion by *Toxoplasma gondii*. Boeh. Inst. Mitt. *99*, 90-96.

Engering, A., Lefkovits, I., and Pieters, J. (1997). Analysis of subcellular organelles involved in major histocompatibility complex (MHC) class II-restricted antigen presentation by electrophoresis. Electrophor. *18*, 2523-30.

Håkansson, S., Morisaki, H., Heuser, J. E., and Sibley, L. D. (1999). Time-lapse video microscopy of gliding motility in *Toxoplasma gondii* reveals a novel, biphasic mechanism of cell locomotion. Molec. Biol. Cell *in press*.

Harlow, E., and Lane, D. (1988). Antibodies: A laboratory manual (Cold Spring Harbor, NY: Cold Spring Harbor Laboratory).

Howe, D. K., and Sibley, L. D. (1997). Development of Molecular Genetics for *Neospora caninum* A complementary system to *Toxoplasma gondii*. Meth. Comp. Meth. Enzym. *13*, 123-133.

Joiner, K. A., Furhman, S. A., Miettinen, H. M., Kasper, L. H., and Mellman, I. (1990). *Toxoplasma gondii* : Fusion competence of parasitophorous vacuoles in Fc-receptor transfected fibroblasts. Science *249*, 641-646.

Jones, T. C., Yeh, S., and Hirsch, J. G. (1972). The interaction between *Toxoplasma gondii* and mammalian cells. I Mechanism of entry and intracellular fate of the parasite. J. Exp. Med. *136*, 1157-1172.

Laemmli, U. K. (1970). Cleavage of structural proteins during the assembly of the head of bacteriophage T4. Nature (Lond.) *227*, 680-685.

Lecordier, L., Mercier, C., Sibley, L. D., and Cesbron-Delauw, M. F. (1999). Transmembrane insertion of the *Toxoplasma gondii* GRA5 protein occurs following soluble secretion into the host cell. Mol. Biol. Cell *in press*.

Mercier, C., Lefebvre-Van Hende, S., Garber, G. E., Lecordier, L., Capron, A., and Cesbron-Delauw, M. F. (1996). Common cis-acting elements critical for the expression of several genes of *Toxoplasma gondii*. Molecular Microbiology *21*, 421-428.

Mercier, C. M., Cesbron-Delauw, M. F., and Sibley, L. D. (1998). The amphipathic alpha-helices of the *Toxoplasma* protein GRA2 mediate post-secretory membrane association. J. Cell Science *111*, 2171-2180.

Mineo, J. R., McLeod, R., Mack, D., Smith, J., Khan, I. A., Ely, K. H., and Kasper, L. H. (1993). Antibodies to *Toxoplasma gondii* major surface protein (SAG-1, P30) inhibit infection of host cells and are produced in murine intestine after peroral infection. J. Immunol. *150*, 3951-3964.

Mordue, D., Håkansson, S., Niesman, I., and Sibley, L. D. (1999). *Toxoplasma gondii* resides in a vacuole that avoids fusion with host cell endocytic and exocytic vesicular trafficking pathways. Exp. Parasitol. *92*.

Mordue, D. G., and Sibley, L. D. (1997). Intracellular fate of vacuoles containing *Toxoplasma gondii* is determined at the time of formation and depends on the mechanism of entry. J. Immunol. *159*, 4452-4459.

Roos, D. S., Donald, R. G. K., Morrissette, N. S., and Moulton, A. L. (1994). Molecular tools for genetic dissection of the protozoan parasite *Toxoplasma gondii*. Meth. Cell Biol. *45*, 28-61.

Russell, D. G. (1994). Immunoelectron microscopy of endosomal trafficking in macrophages infected with microbial pathogens. Meth. Cell Biol *45*, 277-88.

Russell, D. G. (2000). Where to stay inside the cell: A homesteader's guide to intracellular parasitism. Cellular Microbiology *(ed. Cossart, Boquet, Normark, Rappouli)* ASM Press.

Russell, D. G., Dant, J., and Sturgill-Koszycki, S. (1996). *Mycobacterium avium*- and *Mycobacterium tuberculosis*-containing vacuoles are dynamic, fusion-competent vesicles that are accessible to glycosphingolipids from the host cell plasmalemma. J Immunol *156*, 4764-73.

Russell, D. G., Xu, S., and Chakraborty, P. (1992). Intracellular trafficking and the parasitophorous vacuole of *Leishmania mexicana*-infected macrophages. J Cell Sci *103*, 1193-210.

Schaible, U. E., Schlesinger, P. H., Steinberg, T. H., Mangel, W. F., Kobayashi, T., and Russell, D. G. (1999). Parasitophorous vacuoles of *Leishmania mexicana* acquire macromolecules from the host cell cytosol via two independent routes. J Cell Sci *112*, 681-93.

Schaible, U. E., Sturgill-Koszycki, S., Schlesinger, P. H., and Russell, D. G. (1998). Cytokine activation leads to acidification and increases maturation of *Mycobacterium avium*-containing phagosomes in murine macrophages. J Immunol *160*, 1290-6.

Seeber, F., and Boothroyd, J. C. (1996). *Escherichia coli* β-galactosidase as an in vitro and in vivo reporter enzyme and stable transfection marker in the intracellular protozoan parasite *Toxoplasma gondii*. Gene *169*, 39-45.

Sinai, A. P., Webster, P., and Joiner, K. A. (1997). Association of host cell endoplasmic reticulum and mitochondria with the *Toxoplasma gondii* parasitophorous vacuole membrane: a high affinity interaction. J. Cell Sci. *110*, 2117-2128.

Soldati, D., and Boothroyd, J. C. (1995). A selector of transcription initiation in the protozoan parasite *Toxoplasma gondii*. Mol. Cell. Biol. *15*, 87-93.

Soldati, D., and Boothroyd, J. C. (1993). Transient transfection and expression in the obligate intracellular parasite *Toxoplasma gondii*. Science *260*, 349-352.

Sturgill-Koszycki, S., Haddix, P. L., and Russell, D. G. (1997). The interaction between *Mycobacterium* and the macrophage analyzed by two-dimensional polyacrylamide gel electrophoresis. Electrophor. *18*, 2558-65.

Sturgill-Koszycki, S., Schaible, U. E., and Russell, D. G. (1996). *Mycobacterium*-containing phagosomes are accessible to early endosomes and reflect a transitional state in normal phagosome biogenesis. EMBO J *15*, 6960-8.

Sturgill-Koszycki, S., Schlesinger, P. H., Chakraborty, P., Haddix, P. L., Collins, H. L., Fok, A. K., Allen, R. D., Gluck, S. L., Heuser, J., and Russell, D. G. (1994). Lack of acidification in *Mycobacterium* phagosomes produced by exclusion of the vesicular proton-ATPase Science *263*, 678-81.

Suss-Toby, E., Zimmerberg, J., and Ward, G. E. (1996). *Toxoplasma* invasion: The parasitophorous vacuole is formed from host cell plasma membrane and pinches off via a fusion pore. Proc. Natl. Acad. Sci. USA *93*, 8413-8418.

Tulp, A., Verwoerd, D., Benham, A., and Neefjes, J. (1997). High-resolution density gradient electrophoresis of proteins and subcellular organelles. Electrophor. *18*, 2509-15.

Tulp, A., Verwoerd, D., Dobberstein, B., Ploegh, H. L., and Pieters, J. (1994). Isolation and characterization of the intracellular MHC class II compartment. Nature *369*, 120-6.

Tulp, A., Verwoerd, D., and Pieters, J. (1993). Application of an improved density gradient electrophoresis apparatus to the separation of proteins, cells and subcellular organelles. Electrophor. *14*, 1295-301.

Via, L. E., Deretic, D., Ulmer, R. J., Hibler, N. S., Huber, L. A., and Deretic, V. (1997). Arrest of mycobacterial phagosome maturation is caused by a block in vesicle fusion between stages controlled by rab5 and rab7. J Biol Chem *272*, 13326-31.

Via, L. E., Fratti, R. A., McFalone, M., Pagan-Ramos, E., Deretic, D., and Deretic, V. (1998). Effects of cytokines on mycobacterial phagosome maturation. J Cell Sci *111*, 897-905.

Wilson, I., Niman, H. L., Houghten, R. A., Cherenson, A. R., Connolly, M. L., and Lerner, R. A. (1984). The structure of an antigenic determinant in a protein. Cell *37*, 767-778.

Xu, S., Cooper, A., Sturgill-Koszycki, S., van Heyningen, T., Chatterjee, D., Orme, I., Allen, P., and Russell, D. G. (1994). Intracellular trafficking in *Mycobacterium tuberculosis* and *Mycobacterium avium*-infected macrophages. J Immunol *153*, 2568-78.

# 11

# CYTOKINE REGULATION OF FILARIA: INDUCED AIRWAY AND CORNEAL DISEASE

Eric Pearlman, Rajeev K. Mehlotra, Musa A. Haxhiu and Laurie R. Hall
*Case Western Reserve University, Cleveland, OH 44106*

## INTRODUCTION

Approximately 200 million individuals are infected with filarial helminths, the parasitic helminths that cause lymphatic filariasis and onchocerciasis (river blindness). Filariae are thread-like nematodes that are transmitted by the bite of blood-sucking arthropods (mosquitoes transmit lymphatic filariae and black flies transmit onchocerciasis). Adult males and females are viviparous, and the early first stage larvae, termed microfiariae, are present in either the blood (lymphatic filariae) or the skin (onchocerciasis).

This review will focus on two animal models of disease caused by microfiariae in which eosinophils are prominent, and will examine the differential outcomes of IL-5 deficiency and recombinant IL-12 treatment. The murine models are of Tropical Pulmonary Eosinophilia (TPE), caused by *Brugia malayi* and *Wuchereria bancrofti* (reviewed in Ottesen and Nutman, 1992), and *Onchocerca volvulus*: mediated corneal disease, which is a major cause of river blindness (reviewed in Ottesen, 1995; Hall and Pearlman, 1999).

### Tropical Pulmonary Eosinophilia

Much of the pathology associated with the lymphatic filariasis is attributed to the presence of adult worms causing blockage of the lymphatics. First stage larvae (microfilariae), which are present in the blood, generally do not induce pathological sequelae. However, in many individuals, the presence of microfilariae in the lungs is associated with a severe asthmatic response. This reaction, termed Tropical Pulmonary Eosinophilia (TPE), can be distinguished from allergic asthma by the effectiveness of anthelminthics in relieving clinical symptoms (Ottesen and Nutman, 1992). Patients with acute TPE have a pronounced eosinophilic alveolitis and elevated blood eosinophilia (>3,000/μl) and serum IgE (>2,000 ng/ml) (Ottesen and Nutman, 1992; Pinkston et al., 1987). Ultrastructural analysis of eosinophils from these patients show that these cells are activated, based on their highly vacuolated appearance and loss of granule content (Pinkston et al., 1987). In that study, the number of eosinophils recovered from the lungs of patients with acute TPE was 20-fold greater than in a group of asthmatics (Pinkston et al., 1987). TPE patients also demonstrate increased airway hyper responsiveness to inhalation of cholinergic agonists (Chhabra and Gaur, 1988; Ottesen and Nutman, 1992).

***Onchocerca volvulus*- mediated corneal disease (river blindness)**

In contrast to lymphatic filariasis, adult male and female *O. volvulus* are found in subcutaneous host-derived nodules. The adults cause no pathology, although blood eosinophilia and serum IgE are elevated (Ottesen, 1995). Microfilariae are present throughout the dermis, and migrate through the conjunctiva into the cornea. So long as the parasites remain alive, they elicit little or no inflammatory response, and motile worms can be detected in the cornea by slit lamp examination (Abiose, 1998; Hall and Pearlman, 1999). However, when the microfilariae die, parasite antigens are released into the microenvironment of the corneal stroma and trigger a local inflammatory response. In individuals who have been sensitised by chronic exposure to the parasites, the immediate inflammatory response is characterized by discrete areas of corneal inflammation (seen as opaque areas in the otherwise transparent cornea) termed punctate keratitis (Abiose, 1998; Hall and Pearlman, 1999). Histological examination of punctate keratitis shows local edema with infiltrating lymphocytes and eosinophils (WHO 1987; Abiose, 1998). These lesions resolve spontaneously with minimal visual impairment. In contrast, in heavily infected individuals where there is prolonged invasion of the cornea, the inflammatory response progresses through a stage of bilateral corneal opacification to stromal keratitis, where much of the cornea is opaque and vascularized. These individuals develop severe visual impairment and eventually become completely blind (WHO 1987; Abiose, 1998).

**Filarial antigens induce a pronounced Th2 response**

As IL-4 is essential for B cell switching to IgE production, and IL-5 is required for eosinophil development and maturation, elevated production of both of these cytokines would be anticipated in patients with TPE or onchocerciasis. Nutman and colleagues compared responses of TPE patients with filaria-infected individuals with no clinical disease. He found that the frequency of IL-4-producing cells in peripheral blood was elevated about 20-fold in TPE patients, and the frequency of IL-5-producing cells was elevated about 10-fold. (T.B. Nutman, personal communication). Similarly, Freedman and co-workers compared the cytokine responses in the peripheral blood of *O. volvulus*-infected individuals and found that patients with ocular disease had higher IL-4, IL-5 and IL-10 production than infected individuals with no apparent disease (Plier et al., 1996).

Experimental animals have been used to characterize immune responses associated with filarial disease. For example, repeated immunization of immunocompetent mice with *B. malayi* microfilariae or with a soluble extract of *B. malayi* or *O. volvulus* antigens results in production of CD4+ cell dependent IL-4 and IL-5, both systemically and at sites of inflammation (Lawrence et al., 1994; Mehlotra et al., 1998; Pearlman et al., 1993; Pearlman et al., 1995). Although IFN-γ is also produced, these animals have elevated IgE and eosinophilia, indicating that IL-4 and IL-5 have biological activity in these models.

**IL-5 and pulmonary eosinophilia are essential for development of B. malayi - induced airway disease**

In a murine model for TPE, mice are given weekly subcutaneous immunizations with 100,000 killed (frozen) microfilariae, followed 10 days later by tail vein injection of 200,000 live microfilariae. Parasites can then be detected in lung capillaries (Fig. 1), which is the most likely site for initiating the inflammatory response.

An earlier study demonstrated that these mice develop clinical symptoms similar to TPE, including elevated serum IgE, blood eosinophilia and the presence of eosinophils in bronchoalveolar lavage fluid (Egwang and Kazura, 1990). In more recent studies in our laboratory, we have further defined the characteristics of airway disease, including: 1) development of a Th2 response, especially IL-5; 2) recruitment of eosinophils to the lung parenchyma and the airways; 3) deposition of eosinophil major basic protein in the airways; and 4) hyperresponsiveness of tracheal smooth muscle to cholinergic agonists. Immunocompetent mice that are immunized subcutaneously with microfilarial antigens and then injected intravenously with live microfilariae develop all of these characteristics.

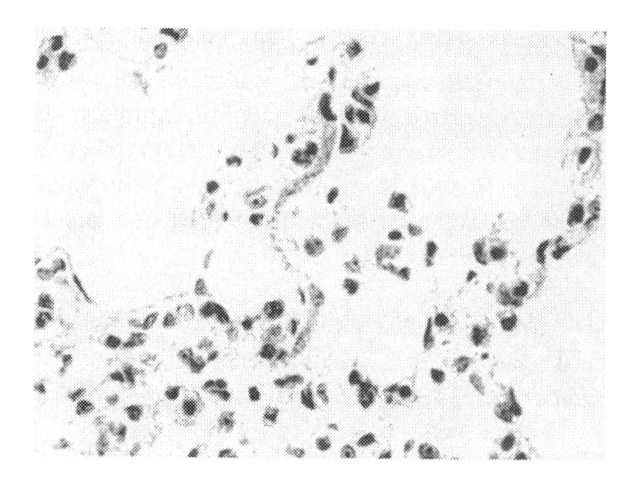

**Figure 1**. *B. malayi* first stage larva (microfilaria) in a lung capillary of a C57Bl/6 mouse 1 day after intravenous injection. Photograph by Alan Higgins.

To determine if IL-5 and eosinophils are essential for development of airway disease, we utilized mice in which the IL-5 gene was disrupted (IL-$5^{-/-}$ mice) and which do not produce eosinophils (Kopf et al., 1996). C57Bl/6 and IL-$5^{-/-}$ mice (on a C57Bl/6 background) were immunized and injected as described above. One week after intravenous injection of *B. malayi* microfilariae, lungs were lavaged, and inflammatory cells in the bronchoalveolar lavage (BAL) were counted. The number of eosinophils in the BAL fluid increased to 5.3 x $10^5$/ml (84%) total cells by day 10 (Fig. 2; Hall et al., 1998). In contrast, no eosinophils were recovered from BAL of IL-$5^{-/-}$ mice at any time, and the cellular infiltrate was primarily composed of lymphocytes (58.2%) and macrophages (39%) (Fig. 2; Hall et al., 1998). Consistent with these observations, eosinophil major basic protein was detected on the surface of bronchial epithelial cells of C57Bl/6 mice by immunohistochemical analysis (see Methods), but not in the lungs of IL-$5^{-/-}$ mice (Hall et al., 1998).

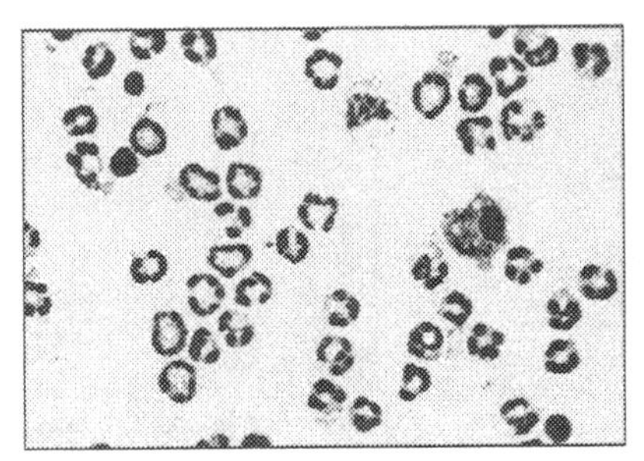
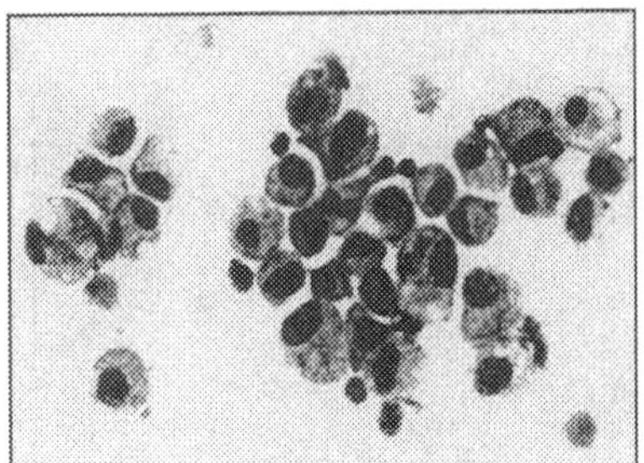

**Figure 2.** BAL from C57Bl/6 (left) and IL-5$^{-/-}$ mice (right) 10 days after intravenous injection of microfilariae. Note the predominant eosinophil infiltrate in C57Bl/6 mice compared with the mononuclear cell infiltrate in IL-5$^{-/-}$ mice. Reprinted with permission from Hall et al., (1998).

To determine if the eosinophil infiltrate to the lung and deposition of major basic protein was associated with physiologic changes in lung function, airway hyperresponsiveness was examined by addition of cholinergic agonists (carbachol and acetylcholine) to tracheal smooth muscle. This is a standard method to examine airway dysfunction. Tracheas isolated from naive C57Bl/6 mice responded to carbachol in a dose-dependent manner (Fig. 3A). However, the contractile force generated by tracheas from sensitised-challenged mice was greater at all doses of carbachol tested (Fig. 3A), and the maximal contractile force of tracheas isolated day 8 post-challenge was significantly higher than naive mice ($p=0.002$) (Hall et al., 1998). These data demonstrate that intravenous injection of microfilariae in sensitised mice induces airway hyperresponsiveness in these animals.

In contrast to immunocompetent mice, tracheas from sensitised IL-5$^{-/-}$ mice did not exhibit hyperresponsiveness to carbachol (Fig. 3B), and the maximal contractile force was not significantly different from naive animals. Therefore, there is an absolute requirement for IL-5 and eosinophils in filaria-induced airway hyperresponsiveness. These observations are in agreement with models of allergic asthma in which mice are immunized and exposed to an aerosol challenge of ovalbumin (Foster et al., 1996; Hogan et al., 1997).

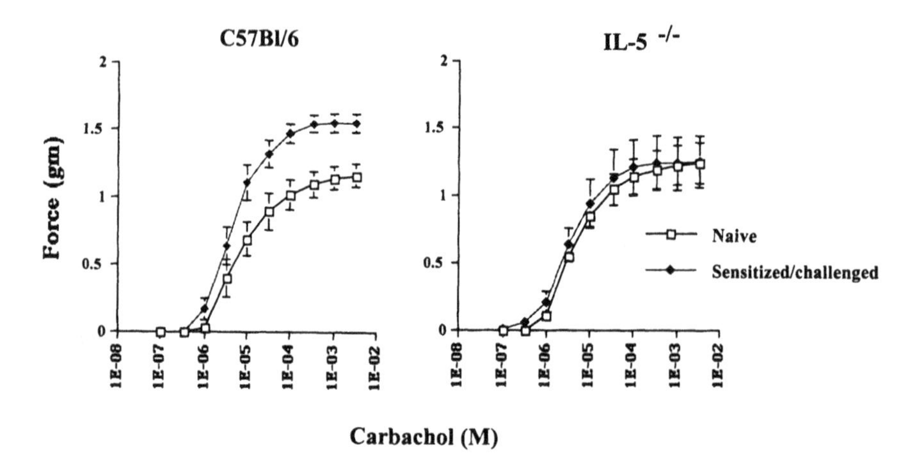


**Figure 3.** Contractile responses of tracheal smooth muscle in naïve and challenged/microfilaria-challenged C57Bl/6 and IL-5$^{-/-}$ mice. C57Bl/6 and IL-5 $^{-/-}$ mice were immunized subcutaneously and injected intravenously with *B. malayi* microfilariae. On day 8 post-challenge, tracheas were dissected and tracheal cylinders were submerged in an organ bath and exposed to cumulative doses of carbachol. Naïve mice were used as controls. The data points represent force generated in grams (mean ± SE of 8 animals per group). Data were analysed by nonlinear regression using PRISM. Statistical significance was determined using unpaired t-test. Maximal contractile force in tracheas from sensitised C57Bl/6 mice increased significantly compared to naive controls (1.6 ± 0.07 g versus 1.2±0.08 g, p=0.002). In contrast, tracheal smooth muscle responses of sensitised/challenged IL-5 $^{-/-}$ mice were similar to naive IL5 $^{-/-}$. Reprinted with permission from Hall et al., (1998).

## IL-12 abrogates filaria-induced airway hyperresponsiveness

Given our observations on the role of IL-5 and eosinophils in airway hyperresponsiveness, and our earlier studies that administration of recombinant IL-12 modulates the Th response to *B. malayi* antigens from Th2 to a predominant Th1 response (Pearlman et al., 1995), we predicted that rIL-12 would diminish airway disease in the murine model for TPE. To test this hypothesis, animals were injected intraperitoneally with 1.25 μg rIL-12 during the first week of sensitization. Injection of recombinant murine IL-12 modulated the predominant Th2- response in the lungs and spleen to a Th1 response, with elevated IFN-γ and IgG2a, and decreased antigen-specific IL-4 and IgG1, and total serum IgE (Mehlotra et al., 1998). In addition, IL-5 production was decreased and BAL eosinophilia was significantly reduced.

Consequently, there was no detectable MBP on respiratory epithelial cells (Mehlotra et al., 1998). In addition, there was no difference in tracheal reactivity to cholinergic agonists between IL-12 treated animals and naïve mice (Fig. 4), indicating that IL-12 also suppresses airway hyperresponsiveness. IL-12 therefore suppresses filaria-induced airway disease by skewing the Th response, and down-regulating IL-5 and eosinophil production. As with IL-5 deficiency, the effect of rIL-12 on murine TPE is consistent with reports of aeroallergen–induced airway hyperresponsiveness in which rIL-12 down-modulates Th2 responses eosinophil migration to the lungs and airway hyperresponsiveness (Gavett et al., 1995; Sur et al., 1996). These observations are also consistent with pulmonary granuloma formation by eggs of *Schistosoma mansoni*. In that model, rIL-12 down modulates the Th2 response and results in significantly smaller granulomas (Wynn et al., 1994; Wynn et al., 1995).

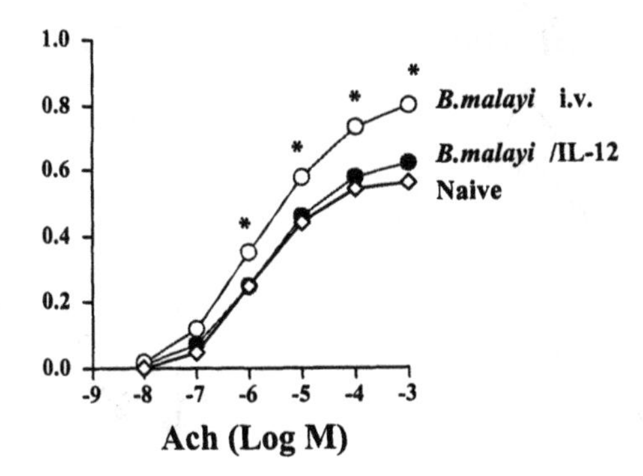


**Figure 4**. Inhibition of AHR in IL-12-treated animals. Tracheal smooth muscle response to cholinergic agonists was measured in immunized, challenged mice injected i.p. with either saline (control) or IL-12. Each data point represents mean value of contraction force generated by isolated tracheas from 8 mice in each group. Hyperresponsiveness to acetylcholine (Ach) was detected in immunized, challenged mice given saline rather than IL-12, whereas mice injected with IL-12 had no greater response than naive mice. Asterisks represent statistical significance between control and IL-12 treated animals ($p < 0.05$). Reprinted with permission from Mehlotra et al., (1998).

## *ONCHOCERCA VOLVULUS*-INDUCED KERATITIS (RIVER BLINDNESS)

Several studies have utilized animal models to examine the immune response associated with development of corneal disease (reviewed in Pearlman, 1996). Among these, Gallin and co-workers showed that clinical manifestations similar to human disease (corneal opacification and neovascularization) could be induced in sensitised, but not naïve guinea pigs by direct injection of soluble *O. volvulus* antigens into the corneal stroma (Gallin et al., 1988). Similarly, Chakravarti and co-workers found that prior immunization with *O. volvulus* antigens is essential for maximal severity of disease in A/J mice (Chakravarti et al., 1993). These workers also demonstrated by immunohistochemical analysis that CD4+, but not CD8+ cells, were in the corneal stroma of these animals (Chakravarti et al., 1994).

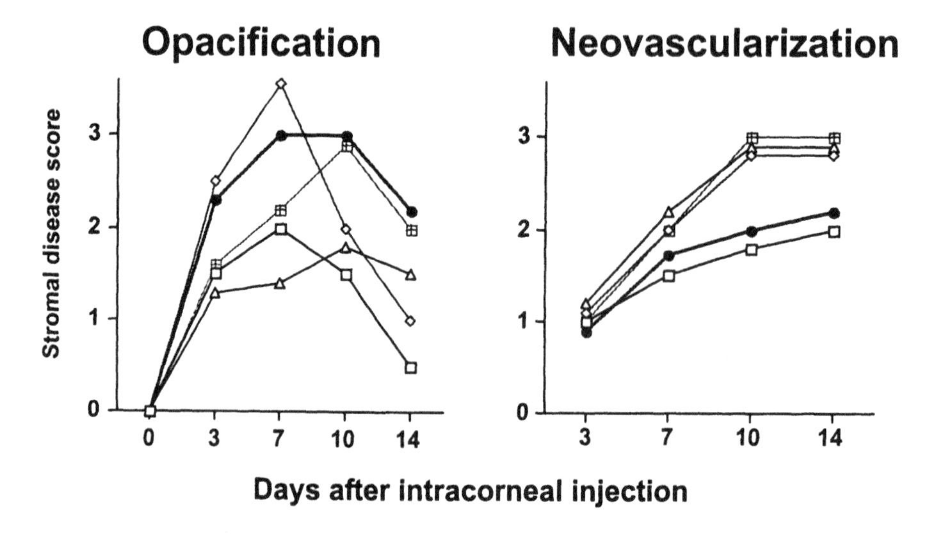


**Figure 5.** Temporal profile of OvAg-mediated stromal keratitis in immunized BALB/c mice. Mice were immunized subcutaneously and injected intrastromally with *O. volvulus* antigens. Eyes were examined by slit lamp microscopy, and corneal opacification was scored in increments of 0.5 as follows: 0: no pathology; 1.0: slight opacity; 2.0: moderate opacity; 3.0: severe opacity; 4.0: total opacity. Neovascularization scores were graded 0 to 3.0 based on the number of vessels and the distance from the limbus. Results presented are of 5 individual animals from a single representative experiment. Similar results were obtained in 8 additional experiments. Reprinted with permission from Pearlman et al., (1995).

Data from our laboratory also showed that mice immunized subcutaneously and then injected intrastromally with soluble *O. volvulus* antigens develop pronounced corneal opacification and neovascularization (Fig. 5). We demonstrated that prior sensitisation is essential for the development of pathology, and that T cells are required for the development of keratitis, as athymic mice that were immunized and challenged intrastromally with *O. volvulus* antigen do not develop keratitis (Pearlman et al., 1995). Furthermore, adoptive transfer of spleen cells from immunized, but not naive mice reconstituted the development of keratitis in these animals after intrastromal challenge (Pearlman et al., 1995). Although these studies did not identify specific sub-populations of cells, CD4 and CD8 mRNA were detected in recipient corneas. Mice develop a predominant Th2 response to repeated immunization with *O. volvulus* antigen, with *O. volvulus* stimulated lymph node and spleen cells from immunized mice producing IL-4 and IL-5, and little IFN-γ (Pearlman et al., 1995). This response is also manifested locally, as mRNA for IL-4 and IL-5 is up-regulated in the corneas of immunized mice upon intrastromal challenge (Chakravarti et al., 1996; Pearlman et al., 1995). Although T cells were detected, they represent a minority of the total inflammatory cells in the cornea (Pearlman, unpublished observations). At 12 hr and 24 hr after injection, neutrophils were the predominant cell type in the corneas. However, after this time eosinophils were recruited and by 7 days were the predominant cell type (Fig. 6B; Pearlman et al., 1998).

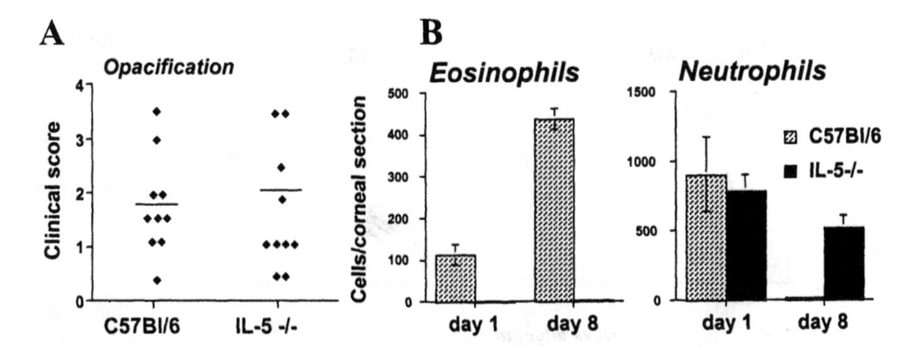


**Figure 6.** *O. volvulus* -mediated corneal disease in IL-$5^{-/-}$ mice is due to recruitment of neutrophils rather than eosinophils to the cornea. A. Corneal opacification in C57Bl/6 and IL-$5^{-/-}$ mice 8 days after intrastromal injection. Data points represent individual animals. B. Eosinophils and neutrophils were detected in the cornea by immunostaining with antibody to eosinophil major basic protein or anti-neutrophil

7/4 (see appendix). Note the presence of neutrophils rather than eosinophils on day 8 in IL-5$^{-/-}$ mice (adapted with permission from Pearlman et al., 1998).

**IL-5 and eosinophilia are not essential for development of keratitis**

To examine the role of eosinophils and neutrophils in onchocercal keratitis, C57Bl/6 and IL-5$^{-/-}$ mice were immunized subcutaneously and injected into the corneal stroma with soluble *O. volvulus* antigens. In contrast to observations with the murine model for TPE in which IL-5$^{-/-}$ mice did not develop disease, IL-5$^{-/-}$ developed severe keratitis after intrastromal injection (Fig. 6A). Except for the absence of IL-5, there was no difference in cytokine production by splenocytes, or cytokine expression in the corneas of IL-5$^{-/-}$ mice, indicating that IL-5 is not essential for production of other Th-associated cytokines (Pearlman et al., 1998). However, histological examination showed that eosinophils were absent in these animals and neutrophils were able to cause inflammation in the absence of eosinophils (Fig. 6B). Furthermore, the distribution of neutrophils was different from that of eosinophils, with neutrophils accumulating at the epithelial cell layer. In many cases, this led to loss of the epithelium and formation of corneal ulcers in these animals (Pearlman et al., 1998). Depletion of neutrophils using anti-GR1 Ab RB6 8C5 abrogated this effect (LR Hall, unpublished observations), indicating that neutrophils mediate disease in IL-5$^{-/-}$ mice.

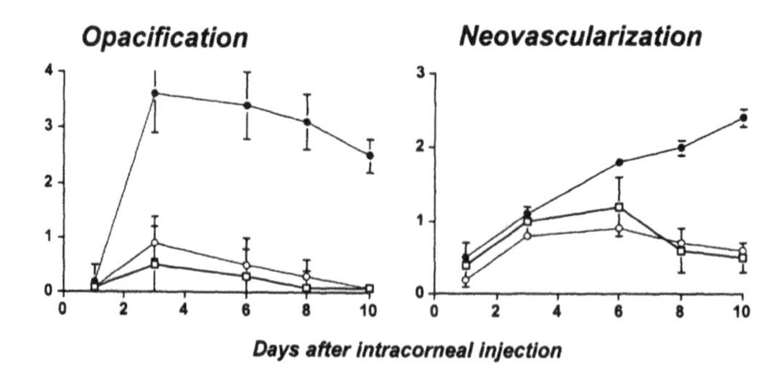


**Figure 7.** Temporal analysis of development of keratitis after IL-12 administration BALB/c mice received 2 subcutaneous immunizations and were injected intrastromally with *O. volvulus* antigens. Animals were injected intraperitoneally with either saline (control, open squares), or IL-12 at the time of initial sensitization

(closed circles) or at the time of intrastromal injection (open circles). 0.5 μg IL-12 was given on day 0, and 0.25 μg IL-12 on days 1, 3, 5 and 7. Corneal opacification and neovascularization scores were determined by slit lamp microscopy. Note that the mild response in control animals is due to receiving two rather than 3 s.c. immunizations. Similar results were obtained in 4 repeat experiments with varying levels of pre-sensitization (from Pearlman et al., 1997, with permission).

**IL-12 exacerbates *O. volvulus* : mediated keratitis**

Since recombinant IL-12 modulates the cytokine response to filarial antigens and down-regulates filaria-induced airway hyperresponsiveness, we anticipated that administration of IL-12 would also abrogate onchocercal keratitis. Corneas of IL-12 treated animals had elevated expression of IFN-γ and diminished expression of IL-4, IL-5, IL-10 and IL-13 (Pearlman et al., 1997). However, in contrast to expected results, IL-12 caused significant exacerbation of corneal pathology (Fig. 7). Histological examination of the corneas revealed increased eosinophil infiltration into the corneal stroma, and RT-PCR analysis showed elevated expression of chemokines with known activity for these inflammatory cells, including MIP1-α, RANTES and eotaxin (Pearlman et al., 1997). Unlike the regulatory mechanism of *Brugia* - induced AHR, in which IL-12–mediated down-modulation of IL-5 is sufficient to inhibit lung pathology, rIL-12, although it reduced IL-5 expression, did not diminish *O. volvulus*-induced corneal pathology. Instead, there was still sufficient IL-5 present to stimulate eosinophil production, and exacerbated keratitis appeared to be associated with elevated expression of chemotactic cytokines (chemokines) in the corneas of IL-12 treated mice (Pearlman et al., 1997). Consistent with this hypothesis, eosinophil recruitment to the cornea was found to be significantly inhibited in mice in which the gene for the eosinophil chemokine eotaxin was deleted (Rothenberg et al., 1997). Although neutrophils are also present in the cornea in *O. volvulus* – mediated keratitis, we have yet to determine if IL-12 also stimulates neutrophil migration to this site.

## SUMMARY AND CONCLUSIONS

Results of all of these studies are summarized in Figure 8. In *Brugia*-induced airway disease, AHR is dependent on the presence of IL-5 and eosinophils. IL-12 treatment or IL-5 deficiency reduces eosinophil numbers to a level where there are insufficient cells to induce disease. AHR can therefore be regulated at the level of induction of the Th2 response. In contrast, in *Onchocerca*-induced corneal pathology, modulation of IL-5 by recombinant IL-12 or by the use of IL-5 deficient mice does not have the predicted effect. Firstly, because IL-12 also regulates chemokine expression in the cornea, leading to increased inflammatory cell migration to that site. Secondly, in the absence of IL-5 and eosinophils, neutrophils are able to migrate and can mediate corneal disease. Regulation of corneal pathology appears to be at the level of cell recruitment, as cytokine regulation that affects the ability of inflammatory cells to migrate into the cornea also regulates the level of disease. In support of this concept, IL-4 gene knockout mice do not develop disease compared with control animals (Pearlman et al., 1995). As IL-4$^{-/-}$ mice produce IL-5 and other Th2 associated cytokines except IL-4, but have fewer inflammatory cells in the cornea (Pearlman et al., 1996), it appears that IL-4 functions to regulate cell recruitment to this site.

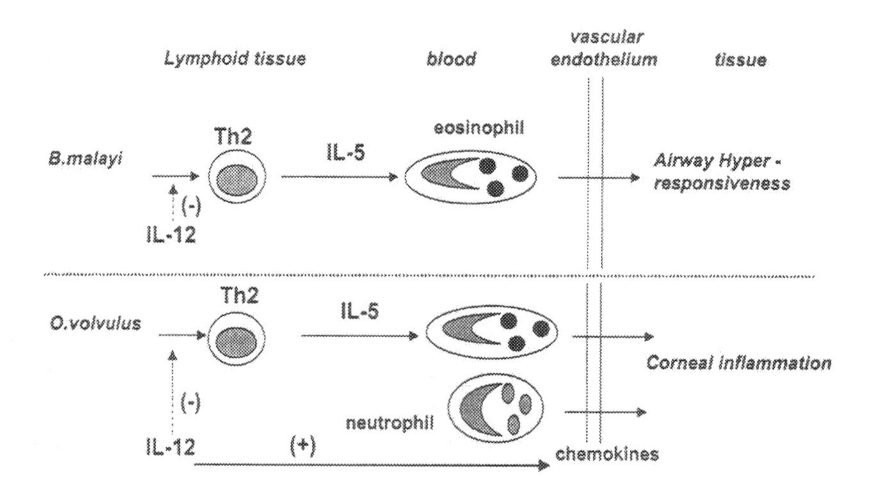


**Figure 8.** Proposed sequence of events in Filaria–induced airway and corneal disease. *B. malayi* and *O. volvulus* antigens induce a Th2 response, notably production of IL-5 and eosinophils. Intravenous injection of *B. malayi* microfilariae results in entrapment of parasites in the small capillaries of the lung. Recruitment of eosinophils to the airway and deposition of eosinophil major basic protein results in development of airway disease. Similarly, injection of *O. volvulus* antigens into the corneal stroma induces eosinophil recruitment to the cornea, although it is preceded

by neutrophil migration to this site. IL-5 deficient mice do not produce eosinophils in response to immunization with either parasite. However, whereas IL-5 deficiency is sufficient to ablate airway disease, neutrophils are recruited to the cornea in place of eosinophils and mediate keratitis. Similarly, rIL-12 skews the Th response to antigens from either parasite from a Th2- to a Th1-like response. This modulation results in diminished eosinophils in the lungs and reduced airway disease in response to *B. malayi*. In contrast, rIL-12 modulation exacerbates *O. volvulus*: induced keratitis, largely as a result of elevated chemokine production.

The factors underlying the different modulatory effects of IL-5 and IL-12 in these models have yet to be determined, but as injection of *B. malayi* antigens to the cornea has a similar effect to *O. volvulus* antigens (Pearlman, unpublished observations), the contrasting results in the cornea and lungs are more likely to be due to the specific microenvironment of these tissues rather than to innate antigenic characteristics of the parasites. In the keratitis model, we have determined a role for eotaxin in eosinophil recruitment to the cornea and have recently demonstrated that antibody mediated mechanism are essential for both neutrophil and eosinophil recruitment to the cornea (Hall et al., 1999). A comparison of mechanisms for cell recruitment to the cornea and lungs will identify differences specific for each tissue.

## METHODS

### Identification of murine neutrophils by immunohistochemistry

We have used Vector Red mostly for identification of neutrophils and eosinophils, although it can be used for any cell marker. The bright red is contrasted against a blue background. Vector Red is also fluorescent and can be detected in the same wavelength as FITC.

### Tissue Preparation

Tissue samples were fixed in 10% Formalin for at least 24 hr, and processed in a Tissue-Tek VIP tissue processor (dehydration through increasing concentrations of ethyl alcohols (60%, 70%, 95%, 100%), -xylene + paraffin, -paraffin only).

5 μm thick sections were cut and placed on Colorfrost/Plus microscope slides (Fisher) and then air-dried.

### Immunohistochemistry

Slides are deparaffinized in xylene x 2 for 5 min each, hydrated through a series of ethyl alcohols (100% EtOH x 2, 95% EtOH x 2, 80% EtOH x 1 to 2 min each), distilled water for 2 min, and 0.05 M Tris-buffered saline (TBS) for 2 min.

*The tissue sections are kept at room temperature in a humid chamber throughout the procedure.*

Cover slides with Proteinase K (DAKO) 10 min to expose epitopes (optional step). Rinse slides in TBS.

Dilute primary antibody in 1% FCS in 0.05 M TBS. We used rat monoclonal Ab 7/4 with specificity for neutrophils (Serotec, Oxford, UK) at a dilution of 1:100. Add sufficient volume to cover all sections (about 70 to 120 μl)

*Important negative controls are 1) no primary Ab and 2) no secondary Ab.*

Rinse slides with 0.05 M TBS, then by gentle rocking for 5 min in TBS.

Add about 70 to 120 μl of the secondary antibody to each slide, incubate 30 min. The biotinylated secondary antibody, pre-diluted Goat anti-rat (Rat Link, BioGenex. San Ramon, CA).

Rinse slides with 0.05 M TBS, then by gentle rocking for 5 min in TBS.

Add about 70 to 120 μl of pre-diluted alkaline phosphatase conjugated streptavidin (BioGenex) to each slide and incubate 30 min.

Rinse slides with TBS, then by gentle rocking 5 min in TBS.

While the slides are washing, make up the Vector Red Substrate (Vector Laboratories, (this must be fresh, i.e., < 30 min before use):

For 30 slides or less (for >30 slides in parentheses)

- 5 ml (7.5 ml) Tris Stock Solution (0.1M Tris-HCl, pH 8.2)
- 0.012 g (0.018 g) of Levamisole (Sigma) to block endogenous alkaline phosphatase
- 2 drops (3 drops) of reagent #1, mix
- 2 drops (3 drops) of reagent #2, mix
- 2 drops (3 drops) of reagent #3, mix

Use immediately, solution is good for 30 min. If precipitate appears (cloudy, white), this will not affect color development.

After substrate is applied to the slides, cover with foil. Improved staining is obtained by developing the substrate in the dark.

Add Vector Red Substrate to the slides on the staining tray(s) for approximately 6 min (maximum time limit is 30 min) Color development can be detected by microscopy.

Rinse the slides thoroughly in tap water to terminate the reaction.

Counterstain the slides:

- Harris Hematoxylin (Richard-Allen) for 2 min
- Rinse in tap water (approximately 1 min)
- 1 to 3 dips in 1% Acid Alcohol
- Rinse with tap water (approximately 1 min)
- 5 to 20 dips in lithium carbonate solution
- Rinse in tap water (approximately 2 min)
- Rinse in distilled water

Dehydrate slides in series of alcohols and xylene:

80% EtOH x 1, 95% EtOH x 2, 100% EtOH x 2 (10 dips each)

Xylene x 2 (2 min each)

Mount slides with Permount (Fisher) and add 22 x 40 mm (1.5 sec) coverslips (Corning). Lay slides flat to dry at least 1 to 2 hr to dry before examining. Lay slides flat 1 to 2 days before storing.

## ACKNOWLEDGEMENTS

The authors are indebted to the Dr Thomas Nutman for providing unpublished data on human TPE responses. We are also grateful to Drs Kirsten Larson, Gerry Gleich and Jamie Lee for providing anti-MBP sera, Drs Stanley Wolfe and Joe Sypek for providing recombinant IL-12, and Drs Manfred Kopf and Edward Pearce for providing IL-$5^{-/-}$ mice. We are especially grateful to Eugenia Diaconu and Ellen Strine for their excellent technical skills, and to Alan Higgins for providing the IHC protocol and Figure 1. The work presented in this review was funded by NIH grants EY10320, EY11373, EY06913 (LRH), and by Burroughs Wellcome New Investigator Award #0720. Funding was also provided by the Research to Prevent Blindness Foundation and by the Ohio Lions Eye Research Foundation.

## REFERENCES

World Health Organization Expert Committee on Onchocerciasis Third Report. (1987). World Health Organization Expert Committee on Onchocerciasis. Third Report. In Technical Report Series (Geneva: World Health Organization), pp. 167.

Abiose, A. (1998). Onchocercal eye disease and the impact of Mectizan treatment. Annals of Tropical Medicine and Parasitology *92*, S11-S22.

Chakravarti, B., Herring, T. A., Lass, J. H., Parker, J. S., Bucy, R. P., Diaconu, E., Tseng, J., Whitfield, D. R., Greene, B. M., and Chakravarti, D. N. (1994). Infiltration of CD4+ T cells into cornea during development of Onchocerca volvulus-induced experimental sclerosing keratitis in mice. Cell Immunol *159*, 306-14.

Chakravarti, B., Lagoo-Deenadayalan, S., Parker, J. S., Whitfield, D. R., Lagoo, A., and Chakravarti, D. N. (1996). In vivo molecular analysis of cytokines in a murine model of ocular onchocerciasis. I. Up-regulation of IL-4 and IL-5 mRNAs and not IL-2 and IFN gamma mRNAs in the cornea due to experimental interstitial keratitis. Immunology Letters *54*, 59-64.

Chakravarti, B., Lass, J. H., Bardenstein, D. S., Diaconu, E., Roy, C. E., Herring, T. A., Chakravarti, D. N., and Greene, B. M. (1993). Immune-mediated Onchocerca volvulus sclerosing keratitis in the mouse. Exp Eye Res *57*, 21-7.

Chhabra, S., and Gaur, S. (1988). Airway hyperreactivity in tropical pulmonary eosinophilia [see comments]. Chest *93*, 1105-6.

Egwang, T. G., and Kazura, J. W. (1990). The BALB/c mouse as a model for immunological studies of microfilariae-induced pulmonary eosinophilia. Am J Trop Med Hyg *43*, 61-6.

Foster, P., Hogan, S., Ramsay, A., Matthaei, K., and Young, I. (1996). Interleukin 5 deficiency abolishes eosinophilia, airways hyperreactivity, and lung damage in a mouse asthma model [see comments]. J Exp Med *183*, 195-201.

Gallin, M. Y., Murray, D., Lass, J. H., Grossniklaus, H. E., and Greene, B. M. (1988). Experimental interstitial keratitis induced by *Onchocerca volvulus* antigens. Arch Ophthalmol *106*, 1447-1452.

Gavett, S., O'Hearn, D., Li, X., Huang, S., Finkelman, F., and Wills-Karp, M. (1995). Interleukin 12 inhibits antigen-induced airway hyperresponsiveness, inflammation, and Th2 cytokine expression in mice. J Exp Med *182*, 1527-36.

Hall, L., Lass, J., Diaconu, E., Strine, E., and Pearlman, E. (1999). An essential role for antibody in neutrophil and eosinophil recruitment to the cornea: B cell deficient (μMT) mice fail to develop Th2 – dependent, helminth - mediated keratitis. Manuscript submitted.

Hall, L. R., Mehlotra, R. K., Higgins, A. W., Haxhiu, M. A., and Pearlman, E. (1998). An essential role for interleukin-5 and eosinophils in helminth-induced airway hyperresponsiveness. Infect Immun *66*, 4425-30.

Hall, L. R., and Pearlman, E. (1999). Pathogenesis of onchocercal keratitis (River blindness). Clin Microbiol Rev *12*, 445-53.

Hogan, S. P., Koskinen, A., and Foster, P. S. (1997). Interleukin-5 and eosinophils induce airway damage and bronchial hyperreactivity during allergic airway inflammation in BALB/c mice. Immunol Cell Biol *75*, 284-8.

Kopf, M., Brombacher, F., Hodgkin, P., Ramsay, A., Milbourne, E., Dai, W., Ovington, K., Behm, C., Kohler, G., Young, I., and Matthaei, K. (1996). IL-5 deficient mice have a developmental defect in CD5+ B-1 cells and lack eosinophilia but have normal antibody and cytotoxic T cell responses. Immunity *4*, 1-20.

Lawrence, R. A., Allen, J. E., Osborne, J., and Maizels, R. M. (1994). Adult and microfilarial stages of the filarial parasite Brugia malayi stimulate contrasting cytokine and Ig isotype responses in BALB/c mice. J Immunol *153*, 1216-24.

Mehlotra, R. K., Hall, L. R., Higgins, A. W., Dreshaj, I. A., Haxhiu, M. A., Kazura, J. W., and Pearlman, E. (1998). Interleukin-12 suppresses filaria-induced pulmonary eosinophilia, deposition of major basic protein and airway hyperresponsiveness. Parasite Immunol *20*, 455-62.

Ottesen, E. (1995). Immune responsiveness and the pathogenesis of human onchocerciasis. J Infect Dis *171*, 659-671.

Ottesen, E. A., and Nutman, T. B. (1992). Tropical pulmonary eosinophilia. Annu Rev Med *43*, 417-424.

Pearlman, E. (1996). Experimental onchocercal keratitis. Parasitol Today *12*, 261-267.

Pearlman, E., Hall, L. R., Higgins, A. W., Bardenstein, D. S., Diaconu, E., Hazlett, F. E., Albright, J., Kazura, J. W., and Lass, J. H. (1998). The role of eosinophils and neutrophils in helminth-induced keratitis. Invest Ophthalmol Vis Sci *39*, 1176-82 (color micrograph p1641).

Pearlman, E., Hazlett, F. E., Jr., Boom, W. H., and Kazura, J. W. (1993). Induction of murine T-helper-cell responses to the filarial nematode Brugia malayi. Infect Immun *61*, 1105-12.

Pearlman, E., Heinzel, F. P., Hazlett, F. E., Jr., and Kazura, J. W. (1995). IL-12 modulation of T helper responses to the filarial helminth, Brugia malayi. J Immunol *154*, 4658-64.

Pearlman, E., Lass, J. H., Bardenstein, D. S., Diaconu, E., Hazlett, F. E., Jr., Albright, J., Higgins, A. W., and Kazura, J. W. (1997). IL-12 exacerbates helminth-mediated corneal pathology by augmenting inflammatory cell recruitment and chemokine expression. J Immunol *158*, 827-33.

Pearlman, E., Lass, J. H., Bardenstein, D. S., Diaconu, E., Hazlett, F. E., Jr., Albright, J., Higgins, A. W., and Kazura, J. W. (1996). Onchocerca volvulus-mediated keratitis: cytokine production by IL-4- deficient mice. Exp Parasitol *84*, 274-81.

Pearlman, E., Lass, J. H., Bardenstein, D. S., Kopf, M., Hazlett, F. E., Jr., Diaconu, E., and Kazura, J. W. (1995). Interleukin 4 and T helper type 2 cells are required for development of experimental onchocercal keratitis (river blindness). J Exp Med *182*, 931-40.

Pinkston, P., Vijayan, V. K., Nutman, T. B., Rom, W. N., O'Donnell, K. M., Cornelius, M. J., Kumaraswami, V., Ferrans, V. J., Takemura, T., Yenokida, G., and et, a. l. (1987). Acute tropical pulmonary eosinophilia. Characterization of the lower respiratory tract inflammation and its response to therapy. J Clin Invest *80*, 216-25.

Plier, D., Awadzi, K., and Freedman, D. (1996). Immunoregulation in Onchocerciasis: persons with ocular inflammatory disease produce a Th2-like response to *Onchocerca volvulus* antigen. J Infect Dis *174*, 380-386.

Rothenberg, M. E., MacLean, J. A., Pearlman, E., and Leder, P. (1997). Targeted disruption of the chemokine eotaxin partially reduces peripheral blood and antigen induced tissue eosinophilia. J Exp Med *185*, 785-790.

Sur, S., Lam, J., Bouchard, P., Sigounas, A., Holbert, D., and Metzger, W. J. (1996). Immunomodulatory effects of IL-12 on allergic lung inflammation depend on timing of doses. J Immunol *157*, 4173-4180.

Wynn, T., Eltoum, I., Oswald, I., Cheever, A., and Sher, A. (1994). Endogenous interleukin 12 (IL-12) regulates granuloma formation induced by eggs of *Schistosoma mansoni* and exogenous IL-12 both inhibits and prophylactically immunizes against egg pathology. J Exp Med *179*, 1551-1561.

Wynn, T. A., Jankovic, D., Hieny, S., Zioncheck, K., Jardieu, P., Cheever, A. W., and Sher, A. (1995). IL-12 exacerbates rather than suppresses T helper 2-dependent pathology in the absence of endogenous IFN-gamma. J Immunol *154*, 3999-4009.

# 12

# MOLECULAR MECHANISMS OF GRANULOMA FORMATION IN SCHISTOSOMIASIS

Stephen J. Davies and James H. McKerrow
*Tropical Disease Research Unit, Department of Pathology, University of California San Francisco, CA 94121*

## INTRODUCTION

Trematodes of the genus *Schistosoma* reside in the bloodstream of their definitive vertebrate hosts, where they avoid immune destruction and survive for years or decades. Widespread species such as *S. mansoni* and *S. japonicum* remain important causes of intestinal and hepatic schistosomiasis in humans throughout South America, Africa and Asia, while *S. haematobium*, the causative agent of human urinary schistosomiasis, remains prevalent in many areas of Africa. Other species, such as *S. bovis* and *S. mattheei*, are of considerable veterinary concern. Despite their continued presence at intravascular locations for periods of years, the adult parasites themselves provoke remarkably little tissue damage or inflammation. In contrast, the eggs produced by adult schistosomes can cause considerable tissue damage and stimulate intense inflammatory and immune reactions. Indeed, the clinical classification of schistosomiasis as hepatic, intestinal or urinary depends on which organ or system is most severely affected by parasite eggs. Frequently over 100 μm long and possessing a tough proteinaceous shell, schistosome eggs trapped in host tissues such as the liver are not readily eliminated. Consequently, like other focal stimuli of chronic inflammation, parasite eggs become the focus of granulomatous inflammatory reactions known as granulomas. Persistence of eggs and the granulomas that surround them results in fibroblast activation, synthesis of extracellular matrix proteins, formation of intractable fibrotic scars and disruption of tissue architecture, ultimately leading to organ dysfunction and the clinical manifestations of schistosomiasis. In this chapter, we review the current understanding of the molecular mechanisms that drive granuloma formation in schistosomiasis. Experimental *S. mansoni* infections in the laboratory mouse closely resemble those observed in humans, both pathologically and immunologically, and provide an extremely malleable and useful model of the human disease. We will therefore pay particular attention to the latest insights obtained using this murine model of schistosome infection.

## WHY FORM A GRANULOMA?

As schistosome eggs and the host responses to them are the major cause of pathology in schistosomiasis, the formation of circumova granulomas could be viewed as detrimental to the host. In human hepatic schistosomiasis for example, disruption of liver architecture by granulomas, in conjunction with Symmer's fibrosis of portal areas, leads to portal hypertension and portosystemic shunting, a development, which can have grave clinical consequences. Likewise in intestinal and urinary schistosomiasis, egg-associated granulomatous inflammation in the intestinal and urinary tracts is a considerable cause of pathology in both of these diseases. However, despite the negative effects of egg-stimulated granuloma formation, studies utilising mouse models where granuloma formation is disrupted have identified a critical role for granulomas in prolonging survival and protecting the host from egg-associated toxic effects (Amiri et al., 1992; Doenhoff et al., 1981; Fallon and Dunne, 1999). In these models, hepatotoxicity caused by parasite eggs impacting in the liver is manifested by swelling of hepatocytes and intracellular accumulation of lipid (Amiri et al., 1992), sometimes referred to as microvesicular steatosis (Fallon and Dunne, 1999). Hepatocyte toxicity is believed to be mediated by hepatotoxins elaborated by schistosome eggs, such as those described from the eggs of *S. mansoni* (Dunne et al., 1991, and references therein). The granuloma therefore, in conjunction with antibody responses to egg toxins (Doenhoff et al., 1979), likely serves an essential protective role, sacrificing tissue architecture in the long term, in return for enhanced survival in the short term.

## WHAT IS A GRANULOMA?

Granulomas are organised aggregates of inflammatory cells that form around persistent inflammatory stimuli. Focal sources of chronic inflammation which can stimulate granuloma formation include relatively inert foreign bodies such as suture material, in addition to persistent infectious agents like bacteria (*Mycobacterium* spp., *Corynebacterium* spp., *Treponema pallidum*), fungi (*Cryptococcus neoformans*, *Coccidioides immitis*) and parasites. Whatever the inciting cause of the granuloma, the defining cell type of granulomatous reactions is activated macrophages that frequently adopt a modified epithelium-like, or "epithelioid" appearance. In haematoxylin- and eosin-stained sections, these epithelioid cells display pale, pink-staining cytoplasm and large pale oval or elongate nuclei, and may show folding of the nuclear envelope. Not infrequently in the case of circumova granulomas, epithelioid cells may fuse to form giant cells that can attain diameters in excess of 50 μm. Giant cells are comprised of a large mass of cytoplasm and up to twenty or more nuclei, arranged peripherally (Langhan's type giant cell) or haphazardly (foreign body-type giant cell).

Many other cell types are also frequently present in granulomas. This is particularly true in the case of schistosome egg-induced granulomas where, for reasons outlined later in this chapter, eosinophils can represent almost 50% of the granuloma cell population, while macrophages typically represent only 30% of granuloma cells, and the remaining 20% are mostly $CD4^+$ T lymphocytes (Hernandez et al., 1997). However, it is the macrophage component that defines the lesion as a granuloma and furthermore, it is activated macrophages that probably mediate the protective function of granulomas, as they are by themselves protective when other cell types are largely absent (Amiri et al., 1992).

## MINIMUM REQUIREMENTS FOR GRANULOMA FORMATION

In order for a granuloma to form around a schistosome egg, macrophages must be recruited to the site of egg deposition and subsequently activated. To determine which signals are responsible for driving macrophage recruitment and activation in schistosomiasis, various mouse models that are deficient in granuloma formation have been identified and examined in detail. From these studies, it is clear that while the circumova granuloma is largely composed of cells of the innate immune system (i.e. macrophages and eosinophils), an adaptive immune response is essential for recruiting these cells to the site of granuloma formation. The work of Warren et al. (Warren et al., 1967) first demonstrated that circumova granulomas are dependent on adaptive cell-mediated immunity. Buchanan et al. (Buchanan et al., 1973) subsequently showed that it was thymus-dependent T lymphocytes that were essential for granuloma-inducing adaptive immune responses in murine schistosomiasis, since animals deprived of T cells were unable to form circumova granulomas. Subsequent studies by Byram and von Lichtenberg (Byram and von Lichtenberg, 1977) and by Dunne and Doenhoff (Dunne and Doenhoff, 1983) using T cell-deficient nude mice further underlined the importance of T cells in orchestrating granuloma formation. Further, defective granuloma formation in nude mice was accompanied by egg-associated hepatocyte damage, an observation that subsequently allowed for the characterisation of egg antigens that possess hepatotoxic properties (Dunne et al., 1991).

Not surprisingly, other murine models that are deficient in adaptive immune function, such as mice homozygous for the spontaneously occurring severe combined immunodeficiency (*scid*) allele of the DNA-activated protein kinase gene (*Prkdc*), or mice with targeted mutation of the recombination activating (RAG)-1 or RAG-2 genes, display a similar inability to form granulomas (Amiri et al., 1992; Iacomini et al., 1995; Jankovic et al., 1999). While these animals lack both B and T cells, the deficit in granuloma formation is presumably due to a lack of T cells, since B cell-deficient mice do form granulomas (Hernandez et al., 1997; Jankovic et al., 1998). Other mouse models bearing specific mutations in genes that disrupt T cell development, such as T cell receptor (TCR) $\alpha/\beta$ knock-out mice, are also incapable of forming granulomas when infected with *S. mansoni* (Iacomini et al., 1995), further reinforcing the role for T cells already delineated by earlier work. Specifically, $CD4^+$ T helper (Th) cells are essential, since *in vivo* depletion of $CD4^+$ cells with CD4-specific antibodies or disruption of $CD4^+$ T cell development through targeted mutation of major histocompatibility (MHC) class II genes both prevent granuloma formation (Hernandez et al., 1997; Mathew and Boros, 1986). There appears therefore, to be an absolute requirement for T cell-dependent adaptive responses in orchestrating granuloma formation in schistosomiasis. In this respect, the mechanisms of granuloma formation in schistosomiasis are similar to those operating in some other infectious diseases where granuloma formation occurs, e.g. tuberculosis (Caruso et al., 1999; Chan and Kaufmann, 1994) and cryptococcosis (Hill, 1992).

## DIVERSITY OF ADAPTIVE IMMUNE RESPONSES DURING SCHISTOSOMIASIS

Diversity in the phenotypes of effector $CD4^+$ Th cells accounts for the heterogeneity observed in $CD4^+$ T cell-dependent adaptive immune responses (reviewed in Abbas et al., 1996). Type-1 responses are driven by Th1 cells that produce the proinflammatory cytokines interferon (IFN)-γ and lymphotoxin (LT)-α. On the other hand, type-2 responses are driven by Th2 cells that produce interleukin (IL)-4, IL-5 and IL-13. Intermediate Th cell responses that result in mixed type-1/type-2 cytokine expression patterns are frequently described as Th0. The effector functions of type-1 and type-2 responses differ markedly. IFN-γ production by Th1 cells promotes macrophage activation, B cell class switching to complement-fixing and opsonizing antibodies (IgG2a and IgG3 in the mouse, IgG1 and IgG3 in human), and together with IL-2 promotes differentiation of $CD8^+$ cytotoxic T cells. Th2 cells produce IL-4, which induces B cell class switching to IgE production, and IL-5, which activates eosinophils. Type-2 cytokines also stimulate B cell proliferation and production of neutralising antibodies (IgG1 in the mouse, IgG4 in human). Type-1 responses therefore function primarily to induce phagocyte-mediated defence mechanisms, while type-2 responses primarily support humoral immunity. Another important feature of type-1 and type-2 responses is their ability to counter-regulate each other by producing cytokines that inhibit development of the other subset. For example, IFN-γ produced by Th1 cells inhibits proliferation of Th2 cells, while Th2 cells inhibit proliferation of Th1 cells by producing IL-10, a cytokine that can also be produced by macrophages and Th1 cells.

Schistosome infections are complex, in that both type-1 and type-2 responses are induced during the course of infection. Prior to the onset of oviposition, which normally begins around 5 weeks post infection in *S. mansoni*, a type-1 response to worm antigens predominates, characterised by production of IFN-γ (Pearce et al., 1991). There is a preponderance of both Th1 and type-1 $CD8^+$ T (Tc1) cells during the prepatent period (Fallon et al., 1998), and indeed, infection with male parasites alone is sufficient to induce a type-1 response (Leptak and McKerrow, 1997). However, coincident with the onset of egg laying and exposure of the host to egg antigens, a dominant type-2 response develops, where T cell populations that produce type-2 cytokines in response to parasite and egg antigens emerge (Pearce et al., 1991). Originally described in experimentally infected mice, type-2 responses are also prevalent in schistosome-infected people (Araujo et al., 1996). Significantly, emergence of the type-2 response at the onset of patency is accompanied by downregulation of the prepatent type-1 response (Pearce et al., 1991), which is presumably mediated via production of IL-10 (Wynn et al., 1998). Parasite egg antigens are the stimulus driving type-2 response induction (Grzych et al., 1991). Indeed, schistosome eggs are potent autonomous inducers of type 2 responses (Vella and Pearce, 1992), capable of inducing type-2 promoting IL-4 production within hours of exposure to the immune system (Sabin and Pearce, 1995). The early source of IL-4 production in response to parasite eggs has been identified as eosinophils (Sabin et al., 1996).

Much debate has centred on the roles of type-1 and type-2 responses in circumova granuloma formation, stemming largely from inconsistencies in results and controversy surrounding the roles of specific type-1 and type-2 cytokines in granuloma formation.

## ROLES FOR TYPE-1 RESPONSES IN GRANULOMA FORMATION

Resistance to *Mycobacterium* infection requires effective granuloma formation (Emile et al., 1997), which in turn is dependent on type-1 adaptive immune responses (Chensue et al., 1995; Cooper et al., 1995; North and Izzo, 1993). In addition to type-1 responses, signalling by the proinflammatory cytokine tumor necrosis factor (TNF) is also necessary for granuloma formation and control of the infection (Kindler et al., 1989), as demonstrated by mice with targeted mutation of TNF receptor 1 (TNFR1, p55) (Ehlers et al., 1999) or of TNF itself (Bean et al., 1999; Kaneko et al., 1999). The role of TNF in the mycobacterial model of granuloma formation is believed to be twofold. First, synergy between TNF and type-1 cytokine signalling serves to activate macrophages (Flesch et al., 1994). Second, TNF facilitates recruitment of inflammatory cells to sites of granuloma formation by inducing expression of adhesion molecules on endothelial cells. There is evidence that upregulation of adhesion molecules on endothelium by TNF requires T cell activity (Horie et al., 1997). Type-1 responses and TNF therefore appear to co-operate in recruiting macrophages to sites of inflammation and subsequently activating them for granuloma formation.

Several lines of evidence indicate that TNF is also involved in schistosome egg-induced granuloma formation. As outlined above, T cell-deficient animals are incapable of forming granulomas in response to schistosome eggs. However, administration of exogenous TNF to T and B cell-deficient

BALB/c-*Prkdc*$^{scid}$ mice reconstituted granuloma formation, suggesting that TNF can mediate sufficient macrophage recruitment and activation for granuloma formation to occur in the absence of an adaptive immune response (Amiri et al., 1992). This finding suggests that TNF production is key to the genesis of circumova granulomas and that adaptive immune responses in schistosome-infected immunocompetent mice facilitate granuloma formation by promoting TNF expression. Indeed, granuloma macrophages from the livers of infected wild type mice produce TNF, and administration of neutralising anti-TNF antibody significantly reduces granuloma volume (Joseph and Boros, 1993). Similar results have been obtained with other means of reducing TNF levels in vivo, such as treatment with thalidomide, or with a soluble chimeric protein consisting of the extracellular domain of TNF receptor 2 fused to the Fc region of immunoglobulin (TNFR:Fc; D. Ritter *et al.*, submitted). The role of CD4$^+$ T cell responses in egg-induced granuloma formation might therefore be to activate macrophages for TNF production, presumably through synthesis of macrophage-activating cytokines such as IFN-γ. Alternatively, T cells themselves can be a substantial source of TNF (Feldmann et al., 1996). T cells are also capable of producing lymphotoxin (LT)-α, a closely related cytokine that shares approximately 30% amino acid identity with TNF and signals through the same cellular receptors to induce similar responses. The contribution of LT-α to circumova granuloma formation has not been specifically investigated, although our observations suggest that LT-α is not expressed at high levels during murine schistosome infection (Davies and McKerrow, unpublished). Results from *Mycobacterium*-infected TNF knockout mice indicate that while functional redundancy might exist between TNF and LT-α, there are specific functions of TNF for which LT-α is not able to compensate (Bean et al., 1999).

Production of proinflammatory cytokines such as IFN-γ, TNF and LT-α are properties of type-1 CD4$^+$ T cells, rather than type-2 CD4$^+$ T cells, which generally antagonise proinflammatory mechanisms through production of anti-inflammatory cytokines (Abbas et al., 1996). Through their promotion of mononuclear phagocyte-mediated defence mechanisms, type-1 responses, rather than type-2 responses, would be predicted to play an essential role in granuloma formation. This is clearly the case for granulomas induced by intracellular bacteria like *Mycobacterium* and propionibacteria (Senaldi et al., 1996). There is evidence that type-1 responses induced by the schistosome during prepatency do contribute significantly to granuloma formation once egg production commences (Leptak and McKerrow, 1997). However, the prevalence of egg-induced type-2 responses after the onset of egg laying has prompted much speculation on the potential role of type-2 responses in mediating granuloma formation. The evidence supporting a role for type-2 responses in granuloma formation, together with other potential functions of type-2 responses in schistosomiasis, will be discussed below.

## ROLES FOR TYPE-2 RESPONSES IN GRANULOMA FORMATION

Many lines of evidence support some role for type-2 responses in granuloma formation have been presented. For example, mice lacking the STAT-6 signalling molecule, generally considered essential for development of the Th2 phenotype, display greatly reduced granuloma volumes on infection with *S. mansoni* (Kaplan et al., 1998). Similarly, mice depleted of Th2-promoting cytokines IL-4 (Chensue et al., 1992) and IL-13 (Chiaramonte et al., 1999) develop smaller granulomas than normal mice. Further, inhibition of type-2 response development by B7-2 blockade also inhibited granuloma formation (Subramanian et al., 1997). Recently, IL-4Rα-deficient mice, which are nonresponsive to both IL-4 and IL-13, were also found to have greatly reduced granuloma volume around schistosome eggs in the liver during infection (Jankovic et al., 1999), strongly suggesting that granuloma formation is Th2-mediated. In these models, type-1 responses are largely intact, and therefore these results contrast starkly with those outlined above that support a role for type-1 responses in granuloma formation. Noteworthy however, is the fact that well defined, albeit smaller granulomas are produced in the absence of Th2 signalling in these studies.

Other results suggested that type-1 responses may not be essential for granuloma formation. For example, TNF was not found to be important for granuloma formation in the lung in response to egg antigen-coated beads (Chensue et al., 1994), and mice deficient in IFN-γ (Yap et al., 1997) or IFN-γ receptor (Akhiani et al., 1996) also developed normal granulomas. However, while type-2 responses are indeed a dominant feature of schistosome infections after the onset of oviposition, several lines of evidence argue that type-2 responses are not solely responsible for granuloma formation.

First, several murine models that do not mount dominant type-2 responses during patency have been identified, and each one forms ostensibly normal granulomas around schistosome eggs.

(i) Mice deficient in both B7-1 and B7-2 costimulatory molecules formed normal granulomas but produced little IL-4 or IL-10 (Hernandez et al., 1999).
(ii) Likewise, mice deficient in CD28, the T cell receptor for costimulatory B7 molecules, were found to be impaired in their ability to mount type-2 responses to eggs, but formed normal granulomas (King et al., 1996).
(iii) B cell-deficient mice were found to be dramatically impaired in their expression of egg-induced type-2 responses, and yet formed granulomas that were indistinguishable form those in wild type animals (Hernandez et al., 1997).

Taken together, these results suggest that type-2 responses are dispensable for granuloma formation.

Second, despite the prevalence of type-2 responses after the onset of oviposition, there is ample evidence that type-1 responses persist beyond the prepatent period and that these responses play a role in normal granuloma formation.
(i) Egg antigens do not exclusively stimulate type-2 responses, as demonstrated by the fact that a dominant egg antigen can autonomously stimulate type-1 cytokine expression in previously unsensitized mice (Cai et al., 1996).
(ii) Type-1 T cell clones specific for egg antigens have been isolated and these clones are capable of mediating granuloma formation (Chikunguwo et al., 1993).
(iii) During natural infection, both type-1 and type-2 cytokines are expressed in the granulomas of infected animals (Cook et al., 1993). Indeed, large numbers of T cells capable of producing IFN-$\gamma$ are present in the granulomas of infected mice, although their activity appears to be tightly regulated by anti-inflammatory cytokines such as IL-10 and transforming growth factor (TGF)-$\beta$ (Rakasz et al., 1998).

There are two factors that can account for at least most of the inconsistencies reported in the literature on cytokine signalling in schistosome granulomas. First, many of the studies employed a pulmonary model of granuloma formation, where eggs are injected intravenously and subsequently embolize to the lungs (Chensue et al., 1992; Chensue et al., 1994; Chiaramonte et al., 1999). Although a more rapid, convenient and synchronous method for studying granuloma formation, this system completely ignores any contribution which might normally be made by type-1 responses that develop during the course of a natural infection (Leptak and McKerrow, 1997). The pulmonary model also does not take into account any tissue-specific differences in granuloma formation that might exist between liver and lungs. Indeed, granuloma formation in the lungs appears to be quite dependent on type-2 cytokines (Cheever et al., 1994; Chiaramonte et al., 1999), while the response in the liver of infected animals is not (Pearce et al., 1996). Second, many studies have failed to distinguish non-specific or acute inflammatory responses from true granulomas that contain macrophages and epithelioid cells. In natural infections and the pulmonary model, egg antigens stimulate vigorous type-2 responses, resulting in dramatic peripheral and tissue eosinophilia. Not surprisingly therefore, eosinophils are attracted in large numbers to sites of egg deposition and become incorporated into developing granulomas, where they can represent over half of the granuloma cell population. Inhibition of type-2 responses and prevention of eosinophilia can therefore have a dramatic effect on granuloma volume, but the functional significance of this reduction in size may be minor, since granuloma macrophage content may be less affected. In fact, there is little evidence to suggest that eosinophils play any role in mediating the protective functions of granulomas (Sher et al., 1990). If two redundant of overlapping pathways (type-1/TNF and type-2) are induced in a natural infection, then results already presented from previous studies allow an estimate of the contribution of each to the volume of inflammatory granuloma reactions to be made. In BALB/c-*Prkdc*$^{scid}$ mice, which do not produce TNF, injection of recombinant TNF gives granulomas whose volume is 60% of that seen in wild type BALB/c mice (Amiri et al., 1992). In CB17- or C57BL/6-*Prkdc*$^{scid}$ mice, TNF is still produced, and this "leaky" *scid* background still produces egg granulomas 15% the volume of wild type with *S. mansoni* and 60% of the volume with *S. japonicum* (Cheever et al., 1999). In IL-4Rα knockout mice, where IL-4/IL-13 signalling should be eliminated, granulomas 29% the volume of wild type (BALB/c) is still produced (Jankovic et al., 1999). In summary, somewhere between 15% and 60% of "granuloma" volume is probably attributable to the classic type-1/TNF pathway, depending on species of parasite and genetic background of the mouse. Whether the type-2 pathway merely augments this response or can initiate it is the key question yet to be answered.

## OTHER ROLES FOR TYPE-2 RESPONSES IN GRANULOMA FORMATION

Recent data suggest that type-2 responses play a critical regulatory role in granuloma formation. In IL-4-deficient mice, which are defective in anti-egg type-2 responses, animals succumbed to severe cachexia and death during acute schistosomiasis, an effect attributable to uncontrolled TNF production in the absence of IL-4 (Brunet et al., 1997). Similarly, IL-10-deficient mice show increased mortality during the acute phase of the disease, which correlates with enhanced expression of proinflammatory cytokines (Wynn et al., 1998). The primary function of egg-induced type-2 responses might therefore be to control concomitant proinflammatory processes induced by the parasite and subsequently aggravated by egg deposition in host organs. A negative consequence of such control may be the concomitant induction of fibrosis by the type-2 response (Wynn et al., 1995).

## CONCLUSIONS

As emphasised in this review, much controversy still surrounds the part played by adaptive immune responses in schistosome circumova granuloma formation. Hopefully, future studies will clarify the roles of type-1 and type-2 responses in this process. Perhaps the presence of two life cycle stages of the parasite in the liver or portal venous system leads to parallel and redundant pathways to granuloma formation: one the classic type-1/innate pathway involving TNF and the other a unique type-2 mechanism. From the experimental standpoint, the role of type-1 and type-2 responses in macrophage recruitment and activation should be addressed independently of granuloma volume per se. The idea that type-2 responses can mediate granuloma formation in the absence of type-1 responses is intriguing, and schistosome egg-induced granulomas might be a useful model for testing this hypothesis. It is clear that type-2 responses do far more than simply inhibit macrophage function, but can also induce alternative activation of macrophages (Goerdt and Orfanos, 1999). However, whether such alternative activation, in and of itself, can mediate effective granuloma formation by macrophages remains to be tested.

## REFERENCES

Abbas, A. K., Murphy, K. M., and Sher, A. (1996). Functional diversity of helper T lymphocytes. Nature *383*, 787-93.

Akhiani, A. A., Lycke, N., Nilsson, L. A., Olling, S., and Ouchterlony, O. (1996). Lack of interferon-gamma receptor does not influence the outcome of infection in murine schistosomiasis mansoni. Scand J Immunol *43*, 257-62.

Amiri, P., Locksley, R. M., Parslow, T. G., Sadick, M., Rector, E., Ritter, D., and McKerrow, J. H. (1992). Tumour necrosis factor alpha restores granulomas and induces parasite egg-laying in schistosome-infected SCID mice [see comments]. Nature *356*, 604-7.

Araujo, M. I., de Jesus, A. R., Bacellar, O., Sabin, E., Pearce, E., and Carvalho, E. M. (1996). Evidence of a T helper type 2 activation in human schistosomiasis. Eur J Immunol *26*, 1399-403.

Bean, A. G., Roach, D. R., Briscoe, H., France, M. P., Korner, H., Sedgwick, J. D., and Britton, W. J. (1999). Structural deficiencies in granuloma formation in TNF gene-targeted mice underlie the heightened susceptibility to aerosol Mycobacterium tuberculosis infection, which is not compensated for by lymphotoxin. J Immunol *162*, 3504-11.

Brunet, L. R., Finkelman, F. D., Cheever, A. W., Kopf, M. A., and Pearce, E. J. (1997). IL-4 protects against TNF-alpha-mediated cachexia and death during acute schistosomiasis. J Immunol *159*, 777-85.

Buchanan, R. D., Fine, D. P., and Colley, D. G. (1973). Schistosoma mansoni infection in mice depleted of thymus-dependent lymphocytes. II. Pathology and altered pathogenesis. Am J Pathol *71*, 207-18.

Byram, J. E., and von Lichtenberg, F. (1977). Altered schistosome granuloma formation in nude mice. Am J Trop Med Hyg *26*, 944-56.

Cai, Y., Langley, J. G., Smith, D. I., and Boros, D. L. (1996). A cloned major Schistosoma mansoni egg antigen with homologies to small heat shock proteins elicits Th1 responsiveness. Infect Immun *64*, 1750-5.

Caruso, A. M., Serbina, N., Klein, E., Triebold, K., Bloom, B. R., and Flynn, J. L. (1999). Mice deficient in CD4 T cells have only transiently diminished levels of IFN-gamma, yet succumb to tuberculosis. J Immunol *162*, 5407-16.

Chan, J., and Kaufmann, S. H. E. (1994). Immune mechanisms of protection, B. R. Bloom, ed. (Washington, DC: American Society for Microbiology).

Cheever, A. W., Poindexter, R. W., and Wynn, T. A. (1999). Egg laying is delayed but worm fecundity is normal in SCID mice infected with Schistosoma japonicum and S. mansoni with or without recombinant tumor necrosis factor alpha treatment. Infect Immun *67*, 2201-8.

Cheever, A. W., Williams, M. E., Wynn, T. A., Finkelman, F. D., Seder, R. A., Cox, T. M., Hieny, S., Caspar, P., and Sher, A. (1994). Anti-IL-4 treatment of Schistosoma mansoni-infected mice inhibits development of T cells and non-B, non-T cells expressing Th2 cytokines while decreasing egg-induced hepatic fibrosis. J Immunol *153*, 753-9.

Chensue, S. W., Terebuh, P. D., Warmington, K. S., Hershey, S. D., Evanoff, H. L., Kunkel, S. L., and Higashi, G. I. (1992). Role of IL-4 and IFN-gamma in Schistosoma mansoni egg-induced hypersensitivity granuloma formation. Orchestration, relative contribution, and relationship to macrophage function. J Immunol *148*, 900-6.

Chensue, S. W., Warmington, K., Ruth, J., Lincoln, P., Kuo, M. C., and Kunkel, S. L. (1994). Cytokine responses during mycobacterial and schistosomal antigen- induced pulmonary granuloma formation. Production of Th1 and Th2 cytokines and relative contribution of tumor necrosis factor. Am J Pathol *145*, 1105-13.

Chensue, S. W., Warmington, K. S., Ruth, J. H., Lincoln, P., and Kunkel, S. L. (1995). Cytokine function during mycobacterial and schistosomal antigen-induced pulmonary granuloma formation. Local and regional participation of IFN- gamma, IL-10, and TNF. J Immunol *154*, 5969-76.

Chiaramonte, M. G., Schopf, L. R., Neben, T. Y., Cheever, A. W., Donaldson, D. D., and Wynn, T. A. (1999). IL-13 is a key regulatory cytokine for Th2 cell-mediated pulmonary granuloma formation and IgE responses induced by Schistosoma mansoni eggs. J Immunol *162*, 920-30.

Chikunguwo, S. M., Quinn, J. J., Harn, D. A., and Stadecker, M. J. (1993). The cell-mediated response to schistosomal antigens at the clonal level. III. Identification of soluble egg antigens recognized by cloned specific granulomagenic murine CD4+ Th1-type lymphocytes. J Immunol *150*, 1413-21.

Cook, G. A., Metwali, A., Blum, A., Mathew, R., and Weinstock, J. V. (1993). Lymphokine expression in granulomas of Schistosoma mansoni-infected mice. Cell Immunol *152*, 49-58.

Cooper, A. M., Roberts, A. D., Rhoades, E. R., Callahan, J. E., Getzy, D. M., and Orme, I. M. (1995). The role of interleukin-12 in acquired immunity to Mycobacterium tuberculosis infection. Immunology *84*, 423-32.

Doenhoff, M., Musallam, R., Bain, J., and McGregor, A. (1979). Schistosoma mansoni infections in T-cell deprived mice, and the ameliorating effect of administering homologous chronic infection serum. I. Pathogenesis. Am J Trop Med Hyg *28*, 260-3.

Doenhoff, M. J., Pearson, S., Dunne, D. W., Bickle, Q., Lucas, S., Bain, J., Musallam, R., and Hassounah, O. (1981). Immunological control of hepatotoxicity and parasite egg excretion in Schistosoma mansoni infections: stage specificity of the reactivity of immune serum in T-cell deprived mice. Trans R Soc Trop Med Hyg *75*, 41-53.

Dunne, D. W., and Doenhoff, M. J. (1983). Schistosoma mansoni egg antigens and hepatocyte damage in infected T cell-deprived mice. Contrib Microbiol Immunol *7*, 22-9.

Dunne, D. W., Jones, F. M., and Doenhoff, M. J. (1991). The purification, characterization, serological activity and hepatotoxic properties of two cationic glycoproteins (alpha 1 and omega 1) from Schistosoma mansoni eggs. Parasitology *103 Pt 2*, 225-36.

Ehlers, S., Benini, J., Kutsch, S., Endres, R., Rietschel, E. T., and Pfeffer, K. (1999). Fatal granuloma necrosis without exacerbated mycobacterial growth in tumor necrosis factor receptor p55 gene-deficient mice intravenously infected with Mycobacterium avium. Infect Immun *67*, 3571-9.

Emile, J. F., Patey, N., Altare, F., Lamhamedi, S., Jouanguy, E., Boman, F., Quillard, J., Lecomte-Houcke, M., Verola, O., Mousnier, J. F., Dijoud, F., Blanche, S., Fischer, A., Brousse, N., and Casanova, J. L. (1997). Correlation of granuloma structure with clinical outcome defines two types of idiopathic disseminated BCG infection. J Pathol *181*, 25-30.

Fallon, P. G., and Dunne, D. W. (1999). Tolerization of mice to Schistosoma mansoni egg antigens causes elevated type 1 and diminished type 2 cytokine responses and increased mortality in acute infection. J Immunol *162*, 4122-32.

Fallon, P. G., Smith, P., and Dunne, D. W. (1998). Type 1 and type 2 cytokine-producing mouse CD4+ and CD8+ T cells in acute Schistosoma mansoni infection. Eur J Immunol *28*, 1408-16.

Feldmann, M., Brennan, F. M., and Maini, R. N. (1996). Role of cytokines in rheumatoid arthritis. Annu Rev Immunol *14*, 397-440.

Flesch, I. E., Hess, J. H., Oswald, I. P., and Kaufmann, S. H. (1994). Growth inhibition of Mycobacterium bovis by IFN-gamma stimulated macrophages: regulation by endogenous tumor necrosis factor-alpha and by IL-10. Int Immunol *6*, 693-700.

Goerdt, S., and Orfanos, C. E. (1999). Other functions, other genes: alternative activation of antigen- presenting cells. Immunity *10*, 137-42.

Grzych, J. M., Pearce, E., Cheever, A., Caulada, Z. A., Caspar, P., Heiny, S., Lewis, F., and Sher, A. (1991). Egg deposition is the major stimulus for the production of Th2 cytokines in murine schistosomiasis mansoni. J Immunol *146*, 1322-7.

Hernandez, H. J., Sharpe, A. H., and Stadecker, M. J. (1999). Experimental murine schistosomiasis in the absence of B7 costimulatory molecules: reversal of elicited T cell cytokine profile and partial inhibition of egg granuloma formation. J Immunol *162*, 2884-9.

Hernandez, H. J., Wang, Y., and Stadecker, M. J. (1997). In infection with Schistosoma mansoni, B cells are required for T helper type 2 cell responses but not for granuloma formation. J Immunol *158*, 4832-7.

Hernandez, H. J., Wang, Y., Tzellas, N., and Stadecker, M. J. (1997). Expression of class II, but not class I, major histocompatibility complex molecules is required for granuloma formation in infection with Schistosoma mansoni. Eur J Immunol *27*, 1170-6.

Hill, J. O. (1992). CD4+ T cells cause multinucleated giant cells to form around Cryptococcus neoformans and confine the yeast within the primary site of infection in the respiratory tract. J Exp Med *175*, 1685-95.

Horie, Y., Chervenak, R. P., Wolf, R., Gerritsen, M. E., Anderson, D. C., Komatsu, S., and Granger, D. N. (1997). Lymphocytes mediate TNF-alpha-induced endothelial cell adhesion molecule expression: studies on SCID and RAG-1 mutant mice. J Immunol *159*, 5053-62.

Iacomini, J., Ricklan, D. E., and Stadecker, M. J. (1995). T cells expressing the gamma delta T cell receptor are not required for egg granuloma formation in schistosomiasis. Eur J Immunol *25*, 884-8.

Jankovic, D., Cheever, A. W., Kullberg, M. C., Wynn, T. A., Yap, G., Caspar, P., Lewis, F. A., Clynes, R., Ravetch, J. V., and Sher, A. (1998). CD4+ T cell-mediated granulomatous pathology in schistosomiasis is downregulated by a B cell-dependent mechanism requiring Fc receptor signaling. J Exp Med *187*, 619-29.

Jankovic, D., Kullberg, M. C., Noben-Trauth, N., Caspar, P., Ward, J. M., Cheever, A. W., Paul, W. E., and Sher, A. (1999). Schistosome-infected IL-4 receptor knockout (KO) mice, in contrast to IL-4 KO mice, fail to develop granulomatous pathology while maintaining the same lymphokine expression profile. J Immunol *163*, 337-42.

Joseph, A. L., and Boros, D. L. (1993). Tumor necrosis factor plays a role in Schistosoma mansoni egg-induced granulomatous inflammation. J Immunol *151*, 5461-71.

Kaneko, H., Yamada, H., Mizuno, S., Udagawa, T., Kazumi, Y., Sekikawa, K., and Sugawara, I. (1999). Role of tumor necrosis factor-alpha in Mycobacterium-induced granuloma formation in tumor necrosis factor-alpha-deficient mice. Lab Invest *79*, 379-86.

Kaplan, M. H., Whitfield, J. R., Boros, D. L., and Grusby, M. J. (1998). Th2 cells are required for the Schistosoma mansoni egg-induced granulomatous response. J Immunol *160*, 1850-6.

Kindler, V., Sappino, A. P., Grau, G. E., Piguet, P. F., and Vassalli, P. (1989). The inducing role of tumor necrosis factor in the development of bactericidal granulomas during BCG infection. Cell *56*, 731-40.

King, C. L., Xianli, J., June, C. H., Abe, R., and Lee, K. P. (1996). CD28-deficient mice generate an impaired Th2 response to Schistosoma mansoni infection. Eur J Immunol *26*, 2448-55.

Leptak, C. L., and McKerrow, J. H. (1997). Schistosome egg granulomas and hepatic expression of TNF-alpha are dependent on immune priming during parasite maturation. J Immunol *158*, 301-7.

Mathew, R. C., and Boros, D. L. (1986). Anti-L3T4 antibody treatment suppresses hepatic granuloma formation and abrogates antigen-induced interleukin-2 production in Schistosoma mansoni infection. Infect Immun *54*, 820-6.

North, R. J., and Izzo, A. A. (1993). Granuloma formation in severe combined immunodeficient (SCID) mice in response to progressive BCG infection. Tendency not to form granulomas in the lung is associated with faster bacterial growth in this organ. Am J Pathol *142*, 1959-66.

Pearce, E. J., Caspar, P., Grzych, J. M., Lewis, F. A., and Sher, A. (1991). Downregulation of Th1 cytokine production accompanies induction of Th2 responses by a parasitic helminth, Schistosoma mansoni. J Exp Med *173*, 159-66.

Pearce, E. J., Cheever, A., Leonard, S., Covalesky, M., Fernandez-Botran, R., Kohler, G., and Kopf, M. (1996). Schistosoma mansoni in IL-4-deficient mice. Int Immunol *8*, 435-44.

Rakasz, E., Blum, A. M., Metwali, A., Elliott, D. E., Li, J., Ballas, Z. K., Qadir, K., Lynch, R., and Weinstock, J. V. (1998). Localization and regulation of IFN-gamma production within the granulomas of murine schistosomiasis in IL-4-deficient and control mice. J Immunol *160*, 4994-9.

Sabin, E. A., Kopf, M. A., and Pearce, E. J. (1996). Schistosoma mansoni egg-induced early IL-4 production is dependent upon IL-5 and eosinophils. J Exp Med *184*, 1871-8.

Sabin, E. A., and Pearce, E. J. (1995). Early IL-4 production by non-CD4+ cells at the site of antigen deposition predicts the development of a T helper 2 cell response to Schistosoma mansoni eggs. J Immunol *155*, 4844-53.

Senaldi, G., Yin, S., Shaklee, C. L., Piguet, P. F., Mak, T. W., and Ulich, T. R. (1996). Corynebacterium parvum- and Mycobacterium bovis bacillus Calmette- Guerin-induced granuloma formation is inhibited in TNF receptor I (TNF- RI) knockout mice and by treatment with soluble TNF-RI. J Immunol *157*, 5022-6.

Sher, A., Coffman, R. L., Hieny, S., Scott, P., and Cheever, A. W. (1990). Interleukin 5 is required for the blood and tissue eosinophilia but not granuloma formation induced by infection with Schistosoma mansoni. Proc Natl Acad Sci U S A *87*, 61-5.

Subramanian, G., Kazura, J. W., Pearlman, E., Jia, X., Malhotra, I., and King, C. L. (1997). B7-2 requirement for helminth-induced granuloma formation and CD4 type 2 T helper cell cytokine expression. J Immunol *158*, 5914-20.

Vella, A. T., and Pearce, E. J. (1992). CD4+ Th2 response induced by Schistosoma mansoni eggs develops rapidly, through an early, transient, Th0-like stage. J Immunol *148*, 2283-90.

Warren, K. S., Domingo, E. O., and Cowan, R. B. T. (1967). Granuloma formation around schistosome eggs as a manifestation of delayed hypersensitivity. Am J Pathol *51*, 735-756.

Wynn, T. A., Cheever, A. W., Jankovic, D., Poindexter, R. W., Caspar, P., Lewis, F. A., and Sher, A. (1995). An IL-12-based vaccination method for preventing fibrosis induced by schistosome infection. Nature *376*, 594-6.

Wynn, T. A., Cheever, A. W., Williams, M. E., Hieny, S., Caspar, P., Kuhn, R., Muller, W., and Sher, A. (1998). IL-10 regulates liver pathology in acute murine Schistosomiasis mansoni but is not required for immune down-modulation of chronic disease. J Immunol *160*, 4473-80.

Yap, G., Cheever, A., Caspar, P., Jankovic, D., and Sher, A. (1997). Unimpaired down-modulation of the hepatic granulomatous response in CD8 T-cell- and gamma interferon-deficient mice chronically infected with Schistosoma mansoni. Infect Immun *65*, 2583-6.

# 13

# EXPERIMENTAL APPROACHES TO STUDYING THE IMMUNOLOGY OF PARASITIC DISEASES

Edward J. Pearce and Phillip Scott
*Department of Microbiology and Immunology, College of Veterinary Medicine, Cornell University, Ithaca, NY 14853-6401 and Department of Pathobiology, School of Veterinary Medicine, University of Pennsylvania, Philadelphia PA 19104*

## OVERVIEW AND BACKGROUND INFORMATION

The immune system plays a central role in recognizing and controlling the spread and impact on the host of infectious organisms. In the case of parasitic infections this task is complicated by the fact that many eukaryotic parasites have evolved sophisticated mechanisms for evading the immune response. Consequently, many diseases caused by parasites are chronic in nature and result in prolonged and excessive immune responsiveness which in itself can be deleterious. The study of immune responses to parasites has been important not only because of the potential for improving human and animal health through immunological means such as vaccination, but also because the extreme nature of many parasite-induced responses allow insights into the functioning of the immune system itself.

### The immune system

The immune response is initiated by a series of cells that include dendritic cells, macrophages, neutrophils, and natural killer (NK) cells and, which are capable of detecting and responding to foreign organisms such as parasites. The recognition occurs through various pattern recognition receptors, which either directly or indirectly in collaboration with the complement system recognize molecular structures not normally found in mammals. This early innate response is complex and involves the expression by the responding cells of new surface and soluble molecules that serve two key functions. First, through inflammatory cascades they limit the growth and spread of the organism, and second, they play a central role in initiating the adaptive, highly specific immune response by B and T lymphocytes. These cells proliferate and differentiate and produce the molecules, antibodies (Ab) and cytokines, which act to neutralize or destroy the parasite while at the same time controlling the magnitude, longevity and nature of the immune response itself. Since many of the molecules produced by T cells or by other cells in response to signals from T cells are toxic, they can also directly cause pathologic changes. The successful response combines the production of the appropriate Ab and or cytokines to kill or at least control the parasite, with concomitant regulatory elements that prevent overt host damage.

## EXPERIMENTAL APPROACHES

### Different parasites elicit different types of responses

Being such a varied group of protozoan and metazoan organisms with intra- and extracellular habitats, parasites elicit the entire range of immune responses. Consequently, we have made some attempt to be comprehensive in terms of the techniques that we utilize on the Course. However, due to the nature of our own research programs, we have tended to focus on contemporary assays of innate and T cell mediated immunity and it is on these techniques that we will concentrate here.

### Immune response induction

We have examined innate and early adaptive immune responses to *Leishmania major* and *Schistosoma mansoni,* and the immune response and immunopathology that develops during established infection with these pathogens. For analysis of early responses, we generally inject parasites either subcutaneously (s.c.) into the rear footpad and examine responses in the draining popliteal lymph nodes (LN), or inject parasites intraperitoneally (i.p.), and subsequently recover cells from the site of infection using peritoneal lavage. For infection studies, we often analyze the immune response ongoing in the spleen, or if the site of infection is clearly demarcated (as for example is the case in *L. major* infected mice, where parasites have been injected into the footpad), the draining LN. We have also removed sites of infection or areas of parasite-induced pathology and analyzed immune responses in these sites.

**Immunizations and Infections**

***Leishmania***

We have studied the immune response of mice to infections with *L. major*, *L. amazonensis*, and *L. mexicana*. Each of these parasites can be readily grown as promastigotes in several different liquid culture media. Variations in the media and the passage protocol vary among both species and strains of *Leishmania*. Our standard protocol for *L. major* (MHOM/IL/80/Friedlin) is shown below.

**Maintenance in culture.** *L. major* is grown in Grace's insect culture medium (Life Technologies, Grand Island, NY) with 20% fetal bovine serum (HyClone, Inc., Logan, UT; endotoxin <= 0.125 EU/ml) and 2 mM glutamine. A growth curve to establish the logarithmic and stationary phases of the parasites needs to be performed for each new strain of parasite to develop a passage strategy. As a standard procedure for *L. major* (MHOM/IL/80/Friedlin), we passage the organisms during logarithmic growth phase by splitting the cultures 1:100 on Monday and Wednesday, and 1:200 on Friday. Since *Leishmania* parasites have the potential to lose virulence after long-term in vitro passage, a stock of early passage parasites is maintained in liquid nitrogen. These are derived from amastigotes harvested from lesions of experimentally infected mice.

**Isolation of metacyclic parasites.** When *Leishmania* parasites reach stationary growth phase, a certain percentage of the parasites will transform into the infective form, which is termed a metacyclic promastigote. The percentage of parasites that transform is highly dependent on both the parasite strain and culture conditions, and may be anywhere between 1% to 40%. This transformation is accompanied by a change in the sugars exposed on the parasite surface, which apparently allow the parasite to detach from the sandfly midgut. This change in the surface of the organisms is utilized to separate the metacyclic parasites from the non-metacyclic parasites. Essentially, stationary promastigotes are harvested and metacyclic stage parasites are negatively selected with peanut agglutinin (*Arachis hypogae* agglutinin; Sigma Chemical Co., St. Louis, MO; Sacks *et al.*, 1984)).The procedure outlined below is appropriate for *L. major*, but not other *Leishmania* species, due to species differences in the sugars expressed on the parasite surface.

Harvest parasites when in stationary growth phase (usually between 4 and 6 days of culture). The parasites (usually 10 ml) are put in a 50 ml tube, and 40 ml RPMI is added. A sample of the parasites is taken to determine the total number at the beginning. This is important, since it is used to calculate the percent metacyclic organisms obtained.

Procedure:

1. Parasites are spun at 3,000 rpm for 15 min.
2. The supernatant is discarded and the parasites are resuspended in 5 ml RPMI.
3. Peanut agglutinin (PNA) is added to make a 50 μg/ml final concentration.
4. The parasites are incubated at room temperature for 15 to 20 min.
5. Fill tube with RPMI to 50 ml mark.
6. Centrifuge at low speed, approximately 800 rpm for 5 to 6 min.
7. Collect supernatant. Be sure not to harvest any of the pellet.
8. Spin as in step 2.
9. Resuspend in 5 ml PBS, count. Calculate the percent metacyclics recovered.
10. Add 45 ml PBS and spin as in step 2.
11. Wash an additional 2 times.

*Comment:*
*To maintain a low endotoxin level, all of the media must be screened for endotoxin, and new plasticware should be used.*

***Schistosoma mansoni***
Schistosome eggs are purified from the livers of heavily infected mice using the technique of Dalton *et al.* (1997), and stored at -70°C in sterile PBS at 5 x $10^4$ per ml. Mice are injected with 50 μl of this suspension s.c. per footpad,

or with 100 μl i.p., via a 23 g needle. Soluble egg antigen (SEA) is made by Dounce homogenizing parasite eggs in PBS on ice; multiple strokes are required for this process as the eggs are resilient. After 15 strokes we check an aliquot of the disrupted eggs microscopically to assess the percentage of eggs that are broken. We continue homogenizing until >80% are disrupted. At this point the homogenate is first centrifuged at maximum speed in a refrigerated bench top centrifuge and the supernatant from this step is ultracentrifuged at $10^5$ g for 1 hr at 4°C. The final supernatant is sterilized by passage through a 0.2 μm syringe filter, assessed for protein content, adjusted to 1 mg/ml, aliquoted and stored at -70°C.

For infections with this parasite it is necessary to maintain in the laboratory the intermediate host which is a fresh water snail *Biomphalaria glabrata.* In the USA infected snails are available to NIH-funded investigators working on schistosomes through an NIH-NIAID contract (NO2-AI-55270). Infected snails are maintained in aquaria at 26°C in the dark. To recover the infectious schistosome cercariae from the snails, the snails are moved to 30°C under a bright light for 1 hr. Under these conditions, the cercariae leave the snails and enter the surrounding water where they can be collected and counted. Experimental infections are initiated by anesthetising mice and, using the ring method, exposing them to an aliquot of water containing a known number of parasites.

**Recovery of responding cells**

**Lymph node and spleen cell recovery and preparation of single cell suspensions**

Procedure:

1. Put clean fine scissors and forceps in 70% ethanol.

2. Euthanize mouse. *In the USA, $CO_2$ inhalation is the recommended procedure.* Spray mouse with 70% ethanol.

3. Flick excess alcohol off instruments and use to remove spleen and/or LN as cleanly as possible.

4. Put organs into sterile tubes containing "wash" medium. Put tubes on ice. Wash medium is DMEM plus 100 U/ml penicillin, 100 μg/ml streptomycin, 2 mM L-glutamine, 30 mM HEPES, and 2% FBS.

5. All of the above work can be done on an open bench. From here on the work should be performed in a sterile hood.

6. Pour organs into 70 μm sterile nylon mesh resting in top of open sterile 50 ml tube. Rinse organ with wash medium (squirt from a 12 ml syringe with 19 g needle).

7. Discard the wash medium that drained into the 50 ml tube. Using the plunger of a sterile 3 or 6 ml syringe, squash organ through mesh to create a single cell suspension. Rinse well with wash medium and bring single cell suspension to 20 ml in tube.

8. Spin at 1,500 rpm at 4°C for 5 min (or 1,000 rpm at 4°C for 10 min).

9. Discard supernatant, resuspend cell pellet by thwacking bottom of tube.

   LN: Add 10 ml of wash medium and count viable cells (see PEC and cell counting protocols).

   Spleen: To lyse erythrocytes add 2 to 4 ml per spleen of red blood cell lysing buffer (Sigma). Leave for 2 min at room temp, then bring cells to 25 ml with wash medium. Spin cells (1,500 rpm, 4°C, 5 min), discard supernatant, resuspend pellet, bring to 25 ml with wash medium, and count viable cells.

10. Remove a volume of the cell suspension which contains the number of cells that you will need (plus a little more just in case) for your experiment. Pellet cells and resuspend at desired concentration in complete tissue culture medium (CTCM) (DMEM containing 10% FBS,100 U/ml penicillin and 100 μg/ml streptomycin, 2 mM L-glutamine, 25 mM HEPES and $5 \times 10^{-5}$ M 2-mercaptoethanol). The cells are now ready to use.

**Recovery of peritoneal exudate cells (PEC)**

1. Sacrifice mice, and apply 70% alcohol to fur to limit dispersion of hairs. Make a slit in the skin over the navel and pull back the skin to reveal the body wall over the peritoneal area.

2. Inject 10 ml of ice-cold wash wash media into the peritoneal cavity by inserting needle at the bottom of the cavity. Use a 10 ml syringe with a

18 gauge needle; usually 10 ml of media can be injected. Begin injecting medium as soon as needle has penetrated cavity. Be careful to make one clean entry hole with the needle, as this will help minimize back-leakage of the injected media.

3. Pick up needle to extend peritoneum into a "tent." Make sure bevel of needle is down, and very slowly withdraw fluid. You should be able to recover almost all of the fluid you put into the cavity. Squirt fluid back into cavity and withdraw again: this serves to flush out as many cells as possible.

4. Put cell suspension into a sterile 50 ml tube on ice, and continue with other mice.

5. When cells are collected from all the mice, spin the tube at 1,000 rpm for 10 min and resuspend in a small volume of final media (10 ml) and count. Adjust cells to appropriate concentration for experiments.

*Comments:*

1. *Macrophages will readily adhere to plastic and glass. To limit this adherence keep PEC on ice until they are ready to be plated and always use polypropylene tubes.*
2. *Macrophages can be harvested from normal animals (resident macrophages) or from mice that have received an injection of an irritant, such as thioglycollate (elicited macrophages). The recovery from normal mice is about $10^6$ macrophages per animal. The recovery from a thioglycollate-injected animal can be up to $3 \times 10^7$ cells per animal.*
3. *To maintain a low endotoxin level, all of the media must be screened for endotoxin, and new plasticware should be used. Living cells are counted using a standard haemocytometer. Prior to counting, 20 μl of the cell suspension is mixed with 20 μl of 0.4% trypan blue in PBS. Dead cells take up trypan blue and thus a measure of cell viability can be obtained.*

**Analysis of responding cell phenotype**

**Flow cytometry**

1. Wash single cell suspension and resuspend in FACS buffer (0.1% BSA or 1% FBS in PBS, 0.08% NaN3) at $1 \times 10^7$ cells/ml. Filter cells through

70 μm cell strainer or nylon mesh (optional). Dispense 50 to 100 μl per FACS tube (Falcon #2008), staining 5 x$10^5$ to 1 x $10^6$ cells/test.

2. For mouse cells, incubate cells with Fc-block (1 μg/test; mAb anti-FcR, 2.4G2, from Pharmingen) and/or normal rat IgG (Sigma) for 15 min on ice. Do not wash. For human cells heat inactivated human serum (1 to 2%) may be added to FACS buffer to block non-specific binding of Ab to Fc receptors.

3. Add 20 μl of fluorescent-labeled Ab (e.g., anti-CD4-FITC) at the appropriate working dilution (to be determined empirically). Incubate on ice for 25 to 30 min in the dark. Be sure to include controls labeled with isotype control Ab. Alternatively, for antibodies that are not directly conjugated to a fluorochrome, cells can be stained in 2 steps. For example, cells are incubated with biotinylated Ab specific for the appropriate surface marker. After incubating for 25 min followed by 1 wash in FACS buffer to remove unbound Ab, Ab present on the surface of cells can be detected by incubating the cells for an additional 25 min with a streptavidin-conjugated fluorochrome (e.g., streptavidin-PE).

4. Wash once with FACS buffer (add 2 ml of FACS buffer to each tube; spin at 1,500 rpm for 5 min, 4°C; discard supernatant).

5. Resuspend cells in 0.3 to to 0.5 ml FACS buffer for acquisition.

6. If excluding dead cells during acquisition, add 5 μl of propidium iodide at 100 μg/ml in PBS to each tube.

*Comments:*
*Combinations of antibodies labelled with different fluorochromes (e.g. FITC; PE and Cy5) can allow simultaneous analysis of at least 3 different markers in an individual cell type within any given population.*

*If acquisition cannot be done on the same day as staining, stained cells may be fixed. After final wash, resuspend cells in 0.5 ml of 1% formaldehyde in PBS. Cells can be kept up to 7 days at 4°C in the dark. Wash out formaldehyde with 2 ml of FACS buffer prior to acquisition. Resuspend cell pellet in 0.3 to 0.5 ml of FACS buffer for acquisition. If you have 96 well plate carriers for your centrifuge, you can scale down and perform the entire staining procedure in a U-bottom 96 well plate.*

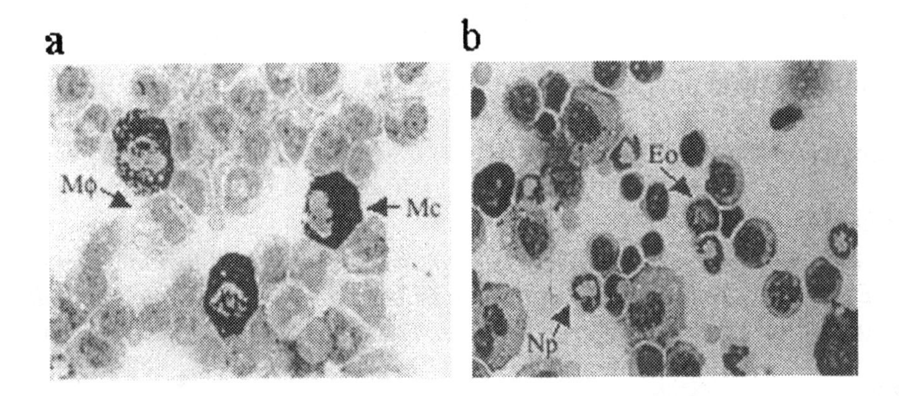


**Figure 1.** A DiffQuick stained cytospin of cells taken from the peritoneal cavity of a mouse 2 hr following PBS injection (panel a), or schistosome egg injection (panel b). Mφ, macrophage; Mc, mast cell; Eo, eosinophil; Np, neutrophil. The small mononculear cells in panel b are probably lymphocytes.

**Cytospins**

1. Label slides clearly with pencil and place in a cytospin carrier along with the cytofunnel.

2. Add 50 μl FCS to each before adding 50 μl of cell suspension containing about $10^5$ cells total. Set up duplicates.

3. Secure top of cytospin and spin for 10 min at 1,250 rpm. (Note: on some machines 1,250 rpm = setting of 125).

4. Carefully remove carrier and funnel.

5. Air dry slides.

6. You can stain cells with a variety of stains. Most commonly used is "Diff Quick" (American Scientific Products) which allows easy visualization of most cell types. Other stains can be more specific for particular cells.

7. Blot excess stain on a paper towel and place slide in Coplin jar containing deionized water for 30 sec. Do not agitate.

8. Rinse briefly in running deionized water, blot excess water, and air dry.

9. Coverslip slides.

*Comments:*
*It is a good idea to always use fresh methanol for the initial fixation step in the Diff Quick procedure. Cytospins can be used for immunostaining as well as simple dye-based cell differentiation studies.*

**Assays of cellular responsiveness**
Induced immunological responsiveness can be measured in a variety of ways. A focus of recent courses has been on cytokine production, which can be measured by ELISA, ELISPOT or bioassay for secreted proteins, flow cytometric analysis of intracellular cytokine or competitive RT-PCR or RNase protection assays (RPA) for cytokine gene transcripts. In addition, cultured cells can be assayed for proliferation using 3H-thymidine incorporation into DNA as an indicator of cell division, or flow cytometrically by monitoring the diminution of fluoresence intensity in dividing cells labelled before division with the fluorescent marker CFSE. Lastly, direct effector mechanisms such as NK cell or CD8 cell mediated cytotoxicity and Ab can be examined. We will focus our discussion on one technique from each category. In addition we will provide details on how to measure the production of reactive oxygen and nitrogen intermediates by cells, since these are important immunological mediators.

**Cytokine analysis**
**Cell culture**
Cells recovered from immunized or infected animals are usually resuspended at $5x10^6$ to $10^7$ per ml in a complete tissue culture medium (DMEM containing 10% FBS, 100 U/ml penicillin and 100µg/ml streptomycin, 2 mM L-glutamine, 25 mM HEPES and $5x10^{-5}$ M 2-mercaptoethanol. Cells are

then plated at 200 μl per well of a 96-well plate, 500 μl per well of a 48 well plate or 1 ml per well of a 24 well plate. The advantages of 96 well plates are the potential for numerous replicates and ease of transfer of supernatants to another 96 well plate (e.g. an ELISA plate). The 48 and 24 well plates allow the generation of greater volumes of supernatant from each experimental condition. Typically we compare unstimulated cells (cells placed into culture without any additional stimuli) with cells stimulated with parasite antigen at 20 to 50 μg/ml, or with soluble or plate bound mAb anti-CD3 at 0.5 μg per well of a 96 well plate (and proportionally more as the volume and surface area increases in the larger well plates). Antigens are typically made by washing parasites into PBS, disrupting the parasites by sonication (protozoa) or Dounce homogenization (metazoa), and ultracentrifugation for 1 hr at $10^5$ g (see above). The resulting supernatants are filter-sterilized and stored frozen in aliquots. Anti-CD3 can be purchased from Pharmingen. Wells are coated with mAb anti-CD3 (0.5 μg per well in 30 μl of sterile PBS) by incubation for 2 hr at 37°C followed immediately prior to addition of cells by removal of the mAb solution and one wash with sterile PBS.

**Cytokine ELISA**

These assays are used to measure cytokine that has accumulated in culture supernatants. For IL-2, which is rapidly utilized, we measure levels in supernatants from 24 hr cultures, while other cytokines are usually measured in 72 hr supernatants. These assays involve the use of a cytokine specific capture antibody, which is bound to the wells of a 96 well ELISA plate, and a second cytokine specific antibody, usually biotinylated, that is used to detect cytokine that is bound to the immobilized antibody. Streptavidin-peroxidase is then used to detect the bound second antibody and the reaction is developed with a peroxidase substrate such as ABTS or TMB. The degree of color development, which is proportional to the concentration of cytokine as indicated by a standard curve which utilizes a recombinant cytokine in known quantities, is typically measured using a microplate reader. These assays generally work best with monoclonal antibodies (mAbs) and a wide range of mAb pairs for different cytokines are available from various commercial sources (e.g. Pharmingen). Clearly, one requirement is that mAbs in the pair recognize different epitopes on the cytokine. An alternative approach is to use a mAb to capture the cytokine, but a polyclonal secondary antibody made in a different species than the mAb followed by a tertiary peroxidase conjugate that is specific for the species donating the polyclonal reagent. This approach is most commonly used for the murine IFN-γ ELISA.

An example of the set up for a typical cytokine ELISA is shown in Table 1. Buffer conditions and concentrations of antibodies are typical.

**Flow cytometric measurement for intracellular cytokines**

1. Cells will have either been freshly recovered from the host or cultured for some time with or without antigen or T cell mitogen. Add 50 ng/ml (final conc.) of PMA, 500 ng/ml (final conc.) of ionomycin and 10 μg/ml (final conc.) Brefeldin A for 4 to 6 hr before harvesting.

2. Harvest and pool cells that have been stimulated under the same conditions. Aliquot an equal volume of cells to each tube to be stained - you don't need to count the cells just mix them gently and aliquot them. You can do the washes directly in the FACS tubes, as outlined below. Alternatively, you can put the cells into either 96 well plates or microfuge tubes. In the latter cases, increase the number of washes at each step to 2 times.

3. Wash 1 time with FACS buffer (see above). Pour off supernatant and resuspend cells by thwacking.

4. Incubate cells with Fc block (Pharmingen) or normal rat Ig (both at 1 μg/test) for 5 to 15 min on ice. Do not wash.

5. Incubate cells with your primary antibody for 25 min on ice in the dark.

6. Wash cells 1 time with FACS buffer.

7. If the primary mAb is biotinylated, you will need to add the streptavidin-CyChrome (or other fluorochrome) for 25 min on ice in the dark.

8. Wash 1 time with FACS buffer.

9. Fix cells in 500 μl of 1% formaldehyde; be sure to vortex gently while adding formaldehyde to prevent cell clumps. Incubate on ice in the dark for 15 min.

10. Wash cells 1 time with FACS buffer.

11. Resuspend cells in 1 ml of fresh saponin buffer (0.1 to 0.2% saponin in FACS buffer) and allow to sit for 10 min. Spin and resuspend cells by thwacking.

12. Add 20 μl mAb anti-cytokine-PE (or other flurochrome) diluted in saponin buffer. Incubate on ice in the dark for 25 min.

13. Wash 1 time with saponin buffer.

14. Wash cells with 2 ml of FACS buffer.

15. Resuspend cells in 300 μl of FACS buffer for acquisition.

**RNA protection assay (RPA) to measure cytokine gene transcripts**

RNA can be recovered from lymphoid organs, diseased and/or infected organs or isolated cells. Tissue for RNA extraction can either be immediately homogenized in a proprietry RNA isolation reagent such as RNA-Stat or placed into such a reagent and immediately frozen in an ethanol/dry ice bath. The LN can then be stored at -70$^0$C until you are ready to isolate the RNA. It is very important that the LNs that are to be used for RPA are collected rapidly and either immediately homogenized in RNA Stat or frozen. The protocol outlined below is adapted from the recommended protocol which comes with the kit from the manufacturer (Pharmingen).

**RNA isolation**

1. Follow RNA STAT method- resuspend pellet in 50 μl of water.
2. Add 5 μl of RNA and 195 μl water to a new tube (40x) dilution.
3. Read A260/A280. The ratio should be about 1.4 to 1.6.
4. Calculate concentration of RNA: $A_{260}$ x Dilution Factor x 40 μg/ml
5. Calculate volume of stock for 10 μg of RNA.
6. Aliquot that volume into an microfuge tube to be dried down in under vacuum.

*Comments:*
*RNA work. Be sure to use gloves to handle all material related to this protocol. Use only new sterile plasticware. Use appropriate safety precautions when handling* $^{32}$P.

**Probe synthesis and hybridization**

1. Bring the [$\alpha$-$^{32}$P] UTP, GACU pool, DTT, 5x transcription buffer and RPA template to room temperature.

2. Add the following in order to a 1.5 ml microfuge tube:
   10 µl [$\alpha$-$^{32}$P] UTP
   1 µl GACU pool
   2 µl DTT
   4 µl 5X Transcription Buffer
   1 µl RPA Template
   1 µl Rnasin
   1 µl T7 polymerase

3. Mix, spin (up and down in microfuge) and incubate at 37°C for 1 hr (water bath or hybridization oven).

4. Terminate the reaction by adding 2 µl of RNase-free DNase. Mix, spin (up and down in microfuge) and incubate at 37°C for 30 min (water bath or hybridization oven).

   *At this point you should start drying-down RNA samples in speed vac so they will be ready for the next step. Use 5 to 10 µg of RNA per sample and remember to set up 2 tubes with tRNA (2 µl of a 2 mg/ml stock).*

5. Add in order the following to each tube:
   50 µl 20 mM EDTA
   100 µl of phenol: chloroform: isoamyl alcohol (25:24:1)
   2 µl yeast tRNA (2 mg/ml)

6. Mix by vortexing into an emulsion and spin at room temperature in microfuge for 5 min (at maximum).

7. Transfer the upper aqueous phase to a new tube and add 100 µl of Phenol: chloroform: isoamyl alcohol (25:24:1).

8. Mix by vortexing into an emulsion and spin at room temperature in microfuge for 5 min (at maximum).

9. Prepare RNA spin column by draining (remove top and bottom and allow to drip into tube). Centrifuge drained column for 2 min at 3,000 rpm. Add 100 μl of TE buffer, pH 8.0, and spin for 2 min at 3,000 rpm. The column is now ready to go.

10. Transfer upper aqueous phase from Step 7 to top of column. Centrifuge for 4 min at 3,000 rpm.

11. Count 1 μl of flow-through in the β counter for one minute.

12. Calculate dilution of probe for $6 \times 10^5$ cpm per sample ($6 \times 10^4$ cpm/μl). Assume 10 μl volume per sample (plus a few extra).

Probe: __________ Volume of probe: __________
Number of samples: ________ x 10 μl = ______________ Total volume
CPM per μl of probe: ______________
($6 \times 10^4$ cpm/μl total volume)/CPM per μl of Probe = ________ (Probe volume)
Total volume needed - probe volume = ____________ (hyb buffer volume)

13. Dilute probe in hybridization buffer.

14. Add 10 μl of diluted probe to RNA samples.

15. Mix by pipetting or vortex. Quick spin in microfuge.

16. Heat samples at 90°C for 3 min (heat block). Vortex and heat again for 1 minute. Incubate for 12 to 16 hr at 56°C (heat block).
    Time in: __________
    Time out: __________

**RNase Treatment**

1. Prepare the RNase cocktail. Add 18 μl of RNase A + T1 mix to 2.5 ml of RNase Buffer. This is good for 20 samples (well, actually 25, but be safe).

2. Add 100 μl of RNase cocktail to each experimental sample and one of the tRNA controls. Leave 1 tRNA control without RNase. Spin for 10 seconds at maximum in a microfuge. Incubate for 45 min at 30°C (hybridization oven).

3. Label new 1.5 ml microfuge tubes and aliquot 120 µl of 4 M ammonium acetate into each tube.

4. Prepare the proteinase K cocktail (18 µl per sample). For 20 reactions (plus a few extra again) add:
   410 µl proteinase K buffer
   15 µl proteinase K (20 mg/ml)
   25 µl yeast tRNA (10 mg/ml)

5. Add 18 µl of proteinase K cocktail to each sample, vortex and quick spin. Incubate for 15 min at 37°C.

6. Add 130 µl of phenol: chloroform: isoamyl alcohol (25:24:1). Vortex into an emulsion and spin in the microfuge at room temperature for 5 min.

7. Carefully extract the upper aqueous phase by setting the pipette at 120 µl. Avoid the organic interface. Transfer to tube containing 4 M ammonium acetate and add 650 µl of ice cold 100% ethanol. Incubate at -70°C for 30 min (or longer- this is a good time to pour gels if you are doing this alone).

8. Spin samples for 15 min at 4°C.

9. Aspirate the supernatant using a vacuum or just pour off the supernatant. It is not critical to get all the supernatant but do your best.

10. Add 100 µl of ice cold 90% ethanol and spin for 5 min at 4°C.

11. Completely remove the supernatant- make sure to get it all this time. Allow the tubes to air-dry for a bit.

12. Resuspend the pellet in 7.5 µl of 1x loading buffer by vortexing.

13. Heat samples for 3 min in a 90°C heating block. Vortex and heat again. Immediately quench on ice.

14. Load on gel. Dilute the undigested tRNA sample 1:100 before loading.

15. Run gel at 50 Watts constant power until it is done (about 2.5 hr). Do not let tracking dye go into buffer.

**Gel resolution**

1. Use a sequencing gel system. Thoroughly wash a set of glass plates, comb and spacers (1 mm thick). Silicanize the plates and comb.

2. Use 50 ml of the Sequenase for one large gel. Combine the following:
   5 ml of 10x Buffer
   10 ml of Concentrate
   35 ml of Diluent
   400 μl of 10% APS
   20 μl of TEMED

3. Pre-run the gel for 1 hr at 40 watts constant power. If you are running 2 gels off one power supply, double the power. Try to time this so you don't pre-run for longer than 1 hr but don't let the gel sit for too long.

4. Separate glass plates and put onto filter paper.

4. Dry gel. Set up cassettes with film. Put in -70°C for 7 days.

**Flow cytometric analysis of cell division**

1. Wash cells in serum-free PBS two times and resuspend at 2 x $10^7$ cells/ml in PBS. You may lose 40 to 60% of your cells during this process.

2. Dilute CFSE (#C-1157, Molecular Probes) 1:2,000 in PBS at room temperature. Mix cells and diluted CFSE 1:1. Ensure that you mix as you are adding the CFSE to the cells.

3. Incubate at room temperature on an agitator (ideally an end-over-end mixer or alternatively a rocker) for 8 min.

4. Quench reaction by adding FBS 1:1 to the cells. Mix well and let sit for 1 min before centrifuging. Subsequently, wash 2 times with 10% FBS in PBS. Let cells sit in the 10% FBS solution for 5 min before each centrifugation.

5. Recount cells and resuspend at 2.5 x $10^6$ cells/ml in CTCM. Add 500 μl per well in a 48 well plate. Stimulate cells as you would for a cytokine

analysis assay with, for example, antigen, anti-CD3, etc. Include control wells with cells unstained with CFSE; these will be important for setting up the flow cytometer.

6. After a period of culture (to be determined empirically for your system try 2 to 4 days in a first attempt), analyze the cells using flow cytometry.

*Comments:*
*The stock solution of CFSE (5-(and-6)-carboxyfluoresceindiacetate succinimidyl ester) is at 5 mM in DMSO, and is stored aliquoted at -20°C. This technique can be combined with staining for intracellular cytokines (see above).*

**NK cell cytotoxicity assay**
The ability of murine NK cells to lyse the YAC-1 cell line is a standard technique used to measure NK cell activity. YAC-1 cells are Moloney leukemia virus induced lymphoma cell line, (ATTC Number TIB-160). These cells grow as a suspension culture in RPMI-1640 plus 10% heat inactivated FBS and are subcultured approximately once per week.

The measurement of cytolysis is dependent upon the ability to specifically label the YAC-1 target cells with $Cr^{51}$. The better the labeling of the target cells the more likely the assay is to succeed. When the cells are lysed the $Cr^{51}$ is released into the medium and can be measured using a γ-counter. The specific cytolysis is then calculated based on the amount of $Cr^{51}$ incorporated and the levels of $Cr^{51}$ released naturally during the course of the experiment.

Because $Cr^{51}$ is bound by FBS it is important that the YAC-1 cells are labeled in the absence of FBS. Cells are washed twice in RPMI 1640 without FBS. YAC-1 cells (preferably still in the log phase of growth) are resuspended at a cell density of $10^7$ per ml in serum-free RPMI and labeled. Radioisotope (100 µl $Cr^{51}$) is added to 200 ml of cells in screw-cap polypropylene tubes and incubated for 60 min at 37°C. The $Cr^{51}$ is in the form of sodium chromate in aqueous solution at a specific activity of 250-500 mCi/mg. Cells are agitated every 15 to 20 min. After 60 min, RPMI with 10% FBS is added to cells and cells are washed three times. This removes excess $Cr^{51}$ both by dilution and binding to FBS, thus reducing the background counts of cells alone.

**$Cr^{51}$ Release Assay**

After the final wash the labeled cells are resuspended at a cell density of 5 x $10^4$ cells per ml and 100 µl of the cells are plated out in 96 well, round bottomed wells. Each Effector:Target ratio is performed in triplicate .

Effector cells are prepared at a cell density of 2 x $10^6$ per ml and serial dilutions of the effector cells are made; 100 µl of the effector cells are added to the target cells. By using different ratios it is easier to compare the activity of different samples.

Several controls are necessary: target cells alone (non-specific Cr51 release) as well as target cells plus 100 µl of 0.1% Triton X-100. The Triton lyses all the cells and allows the maximum $Cr^{51}$ release to be estimated. Also add 100 µl of media to the cells alone to give a final volume of 200 µl.

*Effector:Target Ratio*

| 1 | 2 | 3 | 4 | 5 | 6 | 7 |
|---|---|---|---|---|---|---|
| Cells | Triton 2.5:1 | 5:1 | 10:1 | 20:1 | 40:1 | 80:1 |

After target and effector cells are plated together, the culture plate is spun at 500 rpm, no brakes, for 5 min to rapidly bring the cells together, and then left at 37°C in 5% $CO_2$ for 4 to 6 hr. Incubation times can be increased but lead to high background levels of isotope release and so decrease sensitivity of the assay. After incubation plates are spun again at 1,000 rpm, no brakes, for 5 min and 100 µl of the supernatant is harvested and counted. An alternative to the manual collection of supernatants from individual wells is the use of the Skatron supernatant collection system.

Specific Activity is calculated by: $\frac{\text{Experimental - non specific}}{\text{Maximum - non specific}} \times 100$

**Measures of effector function**

**Macrophage killing**

The effector phase of resistance to *Leishmania* involves macrophage activation by IFN-γ, which leads to the upregulation of inducible nitric oxide (NO) synthetase (iNOS; NOS2), NO production, and parasite killing. Effector function can be examined by establishing macrophage cultures and infecting these cells with *Leishmania* under distinct conditions and then measuring NO in the culture supernatants and counting the number of infected cells and the numbers of parasites per cell.

Procedure:

PEC collection media: RPMI, 2% FBS, 25 mM Hepes, 2x pen-strep, glutamine.

Final Macrophage Media: RPMI, 10% FBS, 25 mM Hepes, pen-strep, glutamine.

1. Sacrifice mice by CO2 inhalation. Harvest cells from the peritoneal cavity (see above). All the cells should be kept on ice. The cells can be harvested on the open bench, but the test tubes should be capped except when putting cells into them. The test tubes used should be polypropylene since macrophages do not stick to this surface. Keeping the cells on ice also helps reduce macrophage adherence, as well as maintaining viability. Cells from multiple mice are combined into one tube.
2. Count the cells using a hemacytometer.
3. Adjust the cells to $10^6$ per ml in final media.

   *The cells can be either plated into a sterile 96 well flat bottom plate, or into 5 ml snap-cap polypropylene tubes as outlined below.*

*Comment:*
*These types of experiment absolutely require sterile techniques.*

**Suspension cultures**

1. First add 0.5 ml of cell suspension per polypropylene tube.

2. Add IFN-γ (final concentration of 100 U/ml) or other cytokines or mediators of interest to macrophages. Leave some tubes without added cytokine as controls.

3. Place all tubes into the 37°C $CO_2$ incubator for 4 hr.

   *For this 4 hr incubation macrophages will be maintained at 37°C*

4. After 4 hr add 50 µl of *Leishmania* parasites to appropriate tubes. (Infection ratio of 2 parasites to 1 macrophage).

5. Incubate tubes at 34°C (skin temperature) or at 37°C in a $CO_2$ incubator for 72 hr.

   *Some cutaneous species of Leishmania will not survive at 37°C, such as L. amazonensis and L. mexicana. L. major will survive at 37°C.*

6. Harvest cells by cytospin. Allow slides to air dry prior to staining.

7. Stain by Diff Quik and examine under the microscope. Diff Quik-Fix the slides in methanol for 5 min, then add into solutions 1, 2 and 3 for 1 min each. Wash slides under running water, and air dry (see above).

8. Quantitate the in vitro infection by microscopy. Determine the percent infected macrophages, the number of parasites per infected macrophage, and the number of parasites per 100 macrophages.

**Macrophage adherence cultures (96 well plates)**

1. Add 250 μl of macrophage cell suspension per well of a flat bottom sterile 96 well plate.

2. Add IFN-γ or other mediators of interest to appropriate wells. Incubate plate for 4 hr in a 37°C $CO_2$ incubator.

3. Remove media, and immediately add 100 μl of final media to every well. You do not want the wells to dry.

4. Add IFN-γ, plus or minus other mediators plus 50 μl of *Leishmania* suspension, the appropriate wells. Add media to wells to bring volume in every well to 250 μl.

5. Incubate plates at 34°C or 37°C in a $CO_2$ incubator for 72 hr. At 72 hr, supernatant will be collected for an NO assay. This supernatant could also be used to measure other macrophage derived mediators such as IL-12 or IL-1. When harvesting supernatants be careful not to disturb cells on bottom of plate.

**Measurement of NO**

Solution A: Sulfanilamide (1%) in 2.5% $H_3PO_4$ (Phosphoric Acid) (Sigma S9251)

Solution B: Napthylethylenediamine dihydrochloride (0.1%) in 2.5% $H_3PO_4$

Solution C: Mix solution A and solution B 1:1 immediately before the assay

> *Do not use if any precipitate, color or cloudiness is seen in either of the solutions. Store solutions in dark glass bottles at 4°C.*

Add 100 μl of supernatants containing $NO_2$ in 96 well plates and prepare standard curve (see below). Add 100 μl of solution C. Allow reaction to proceed for a minimum of 10 minutes before reading 600 nm on an ELISA plate reader.

**Standard Curve**

Stock of $NaNO_2$ is 10 mM (keep frozen and protect from light) Prepare plates as follows: In duplicate, add 100 μl of medium to 12 wells. In the first 2 wells add 90 μl more of medium. To the first wells add 10 μl stock $NaNO_2$. Mix well and serially dilute by removing 10 μl and adding to the next well. Concentration of $NaNO_2$ in the first well is 500 μM. Normally, the limit of detection for this standard curve is 4.0 μM.

**Measurement of superoxide**

1. Add 100 μl media or 100 μl media + PMA (1000 ng/ml) to wells of a 96 well plate. Also add 200 μl media to 2 wells to serve as media blanks.

2. Add 10 μl of superoxide dismutase (SOD; 4,000 U/ml) to half the wells containing PMA.

3. Add 100 μl/well of the cell suspension (at 4 x $10^6$ cells/ml). The final concentration of cells should now be 2 x $10^6$ cells/ml; PMA at 500 ng/ml and SOC at 200 U/ml.

4. Incubate the plate for 30 min at 37°C.

5. Remove 100 μl of supernatant from each well and reserve supernatant in another 96 well plate.

6. Add 100 μl NBT (nitroblue tetrazolium) solution to the appropriate wells. Also add NBT to the 2 media blanks.

7. Incubate plate at 37°C for about 1 hr. NBT reacts with superoxide to form an insoluble blue formazan inside the cells.

8. Make cytospins (see above) of cells from wells set aside for this purpose. These cytospins should allow you to identify the cell type producing superoxide.
9. Add 100 µl 10% SDS/10 mM HCl to remaining wells and to the blank wells. Return to 37°C.

10. After 2 hr the blue formazan should be dissolved in the SDS/HCl. Measure absorbance at 570 nm on a microplate reader. Use the media blanks as blanks for the plate reader and the wells without NBT as blanks for each cell preparation.

**Antibody titres**
1. Coat a 96 well ELISA plate with antigen by adding 50 µl per well of a 10 µg/ml solution of antigen diluted in carbonate buffer (0.1M sodium carbonate; 0.1M sodium bicarbonate). Coat overnight at 4°C.

   *It may be necessary to use more or less antigen depending on the situation. PBS can substitute for carbonate buffer in some situations.*

2. Wash wells 3 times with T-PBS (PBS, 0.05% tween 20).

3. Block unsaturated protein binding sites by adding 200 µl per well of PBS containing either 5% FCS, 3% (w/v) dried milk powder or 1% BSA, and incubating at 37°C for 1 hr.

4. Wash wells 3 times with T-PBS.

5. To all wells add 50 µl of diluent (5% FBS in T-PBS). To the first column add 50 µl of a 1:20 dilution (in PBS) of serum, or 50 µl of undiluted culture supernatant samples (or of other source of antibody). Do two-fold serial dilutions across the plate (1 through 11) by removing 50 µl from wells in column 1 and adding to wells in columns 2. Mix. Remove 50 µl from column 2 and add to wells in column 3. Mix. Leave column 12 as buffer only; this will be the blank. Incubate for 1 to 2 hr at 37°C.

6. Wash wells 6 times with T-PBS.

7. Add 50 μl per well of biotinylated anti-Ig, diluted 1:2000 in diluent (H and L chain specific antibody will detect all isotypes of immunoglobulin, or alternatively isotype-specific reagents can be purchased from a number of suppliers, e.g. Jackson Immunochemicals). Incubate 1 hr at 37°C.

8. Wash wells 6 times with T-PBS.
9. Add 50 μl per well of streptavidin-peroxidase (or -phosphatase) diluted 1:1000 in diluent. Incubate 30 min at 37°C.
10. Wash wells extensively. Add 50 μl per well of peroxidase substrate (we use ABTS) or phosphatase substrate. Let reaction develop and read at 405 nm.

*Comment:*
*Details of the ELISA protocol must be worked out empirically for each antigen and antibody.*

**Assays of parasite infection intensity**
Since a primary function of the immune response is to limit pathogen invasion or replication it is often necessary to measure the success or failure of an immune repsonse by quantitating infection intensity following exposure to a well defined experimental infection. In the Course we have focused on obtaining measures of infection intensity with schistosomes and leishmania.

***Leishmania***
Our method of choice has been dilution analysis of parasites within lesions.
1. Sacrifice animal, cut off foot and place in disinfectant (e.g. Wescodyne) for 10 to 20 min.

2. Place foot into 70% ethanol for 10 to 20 min.

3. Remove skin with forceps, wash in petri dish with PBS containing 2x pen/strep.

4. Remove lesion with scalpel blade and place into tissue grinder.

5. Needed with PBS, 2x pen/strep. Place all washings into test tube; fill to 15 ml.

6. Spin at 500 to 800 rpm for 5 min.

7. Harvest supernatant and transfer to another tube.

8. Spin at 3,000 rpm, discard supernatant and resuspend pellet by thwacking.

9. Add 2 ml of Complete parasite media.

10. Resuspend and remove 220 μl and add in triplicate to the first well of 96 well plates.

11. Add 200 μl of media to other wells; remove 20 μl for first well and add to second well. Carefully mix, and continue dilutions.

12. First row will represent 1:10 dilution; second row 1:100, etc. Leave plate at room temperature and visually assess parasite growth 5 to 7 days later.

***Schistosomes***

We have routinely analyzed adult parasite numbers and parasite egg numbers.

To quantitate worm numbers:

1. Give infected mice a fatal injection of pentobarbitol sodium, i.p.; Include heparin (2000 U/ml) in the anesthetic. Once mouse is deeply anesthetized (with failure to respond to external stimuli), rinse in water to remove loose hair, make an incision in the skin over the abdominal midline. Pull skin back to bare the abdomenal wall and rib cage. Open abdomen.

2. Spray interior with citrated saline (150 mM NaCl, 50 mM sodium citrate) to inhibit clot formation. Open rib cage to reveal heart. Be careful not to damage heart or to cut major blood vessels.

3. Lay animal tail to left over the edge of a Petri dish such that its intestines are laying in the dish and the bulk of the body is outside of the dish.

4. Cut portal vein with scissors and rinse the vein and surrounding areas with citrated saline.

5. Inject 20 ml of citrated saline into the left ventricle using a 20 ml syringe and 20 g needle. Use forceps to steady the needle in the heart. The perfusate, containing the adult worms, will come out of the cut portal vein into the Petri dish.

6. Be careful to rinse the intestines and other organs to ensure there are no worms remaining adherent.

7. Take petri dish and rinse contents through a 70 μm nylon seive (as used for making single cell suspensions from LN and spleen, see above). Invert filter and reverse flush with citrated saline using a syringe and needle, and collect worms into a gridded dish for subsequent counting using a dissecting microscope.

*Comment:*
*This technique takes some practice.*

**To count parasite eggs**

1. Remove and weigh the whole liver. Remove the intestine from the pyloric to where the caecum begins. Place into separate 50 ml tubes. You can freeze these tissues if you need.

2. Add 30 ml of 4% KOH per gram of live tissue. Add a total volume of 20 ml of 4% Potassium Hydroxide to the intestine. Incubate at 37°C for 12 to 18 hr (no longer). The tissue will dissolve leaving only the eggs intact.

3. Resuspend and count the eggs in several 100 μl aliquots; use a dissecting scope and gridded dish to accomplish this.

4. Calculate the number of eggs per organ.

## DISCUSSION AND CONCLUSION

Studying immune responses during infection with a parasite is similar to doing any other type of immunology except that the parasite dictates the nature of the questions being asked. We have provided a description of some of the assays that have been part of the Immunology section in the Biology of Parasitism Course over the last several years. The techniques that we have highlighted are those that we believe are particularly useful for investigating the cellular immune response to parasites, and which are frequently used in our laboratories. These methods have been adopted from a variety of sources, and represent how these assays are currently performed in our laboratories. For more in depth information on any of these techniques, we recommend Current Protocols in Immunology (published by John Wiley & Sons, Inc.) and for detailed information on the use of mice in experimental studies we suggest consulting The Biology and Medicine of Rabbits and Rodents (John E. Harkness and Joseph E. Wagner, published by Williams and Wilkins).

## ACKNOWLEDGEMENTS

Teaching the immunology section of the Biology of Parasitism Course is a team effort and one in which we have been fortunate to have the following outstanding colleagues: Steve Reiner, Eric Denkers, Chris Hunter, Merle Elloso, Laura Rosa Brunet, Brian Hondowicz, Dan Brown, Joao Pedras Vasconcelos, Anne LaFlamme, Colby Zaph and Amy Straw. Several of the techniques outlined above were prepared for the course by these individuals. We have also received generous assistance from several companies without which we would not have been able to perform many of our experiments. In particular we would like to thank Becton Dickinson for loaning us a Facscalibur flow cytometer for 2 weeks each summer.

## REFERENCES

Dalton, J.P., Day, S.R., Drew, A.C., and Brindley, P.J. (1997). A method for the isolation of schistosome eggs and miracidia free of contaminating host tissues. Parasitology, 115, 29-32.

Sacks, D.L., and Perkins, P.V. (1984). Identification of an infective stage of Leishmania promastigotes. Science, 223, 1417-1419.

**IL-4 (murine) ELISA**

| REAGENT | Concentration (μg/ml) | volume (μl) | Temp | Time (min) | Washes |
|---|---|---|---|---|---|
| **Coat:** 11B11 in PBS | 1 | 100 | $4^0$<br>$37^0$ | O/N<br>120 | none |
| **Block:** 5% Newborn calf serum in PBS (NCS/PBS) | | 150 | $37^0$ | $\geq 60$ | |
| **Wash:** TPBS | | | | | 5 |
| **Samples or Standard** diluted in 5% NCS/PBS | | 100 | $37^0$ | $\geq 60$ | |
| **Wash:** TPBS | | | | | 5 |
| **2nd Antibody** (Biotinylated anti-IL-4 (BVD6 or BVD4) | 1:2,000 in 5% NCS/PBS | 100 | $37^0$ | $\geq 60$ | |
| **Wash:** TPBS | | | | | 5 |
| **3rd "Antibody"** (Peroxidase labelled Streptavidin) | 1:2000 in 5% NCS/PBS | 100 | $37^0$ | 30 | |
| **Wash:** TPBS | | | | | 10 |
| **Substrate:** ABTS | | 100 | $37^0$ | $\geq 10$ | |

# APPENDIX

Participants in the Biology of Parasitism Course 1998. **Students**: Fabio Alves, David Artis, Katrin Henze, Bolyn Hubby, Laurence Lambert, Jenny Lovett, Emily Lyons, Kai Matuschewski, Pius Nde, Monika Oli, Kimmie Sue Paul, Lawrence Proulx, Myriam Ramirez, Tina Saxowsky, Kendra Speirs, Ross Waller, Colby Zaph. **Faculty and assistants** (in photo): Deirdre Brekken, Maristela de Camargo, Ann Dell, Ayman Hussein, Wolf Malkusch (Carl Zeiss, Inc.), Edward Pearce, Meg Phillips, Elle Salko, Murray Selkirk, Christian Tschudi, Stepanka Vanacova, C. C. Wang.

Participants in the Biology of Parasitism Course 1999. **Students**: Véronique Angeli, Hernan Aviles, Antonio Barragan, Adrian Batchelor, Rusty Bishop, Abdoulaye Djimde, Caroline Dobbin, Franco Falcone, Cristina Gavrilescu, Stacy Jones, Jacqui Montgomery, Isabel Santori, Cátia Sodré, Leah Stern, Laurent Toe, Zefeng Wang. **Faculty and assistants** (in photo): Taylor Chapple, Caitlin Chipperfield, Gunnar Mair, Edward Pearce, Christian Tschudi, Elisabetta Ullu, Annemarie van der Wel, Andrew Waters.

# INDEX

'nted in the USA
3IA information can be obtained
ww.ICGtesting.com
W011547240923
64LV00007B/883

9 780792 378235